Compendium of Plant Genomes

Series Editor

Chittaranjan Kole, Raja Ramanna Fellow, Government of India, ICAR-National Research Center on Plant Biotechnology, Pusa, New Delhi, India

Whole-genome sequencing is at the cutting edge of life sciences in the new millennium. Since the first genome sequencing of the model plant Arabidopsis thaliana in 2000, whole genomes of about 100 plant species have been sequenced and genome sequences of several other plants are in the pipeline. Research publications on these genome initiatives are scattered on dedicated web sites and in journals with all too brief descriptions. The individual volumes elucidate the background history of the national and international genome initiatives; public and private partners involved; strategies and genomic resources and tools utilized; enumeration on the sequences and their assembly; repetitive sequences; gene annotation and genome duplication. In addition, synteny with other sequences, comparison of gene families and most importantly potential of the genome sequence information for gene pool characterization and genetic improvement of crop plants are described.

Interested in editing a volume on a crop or model plant?

Please contact Prof. C. Kole, Series Editor, at ckoleorg@gmail.com

More information about this series at https://link.springer.com/bookseries/11805

Jian Gao
Editor

The Moso Bamboo Genome

Springer

Editor
Jian Gao
Key Laboratory of Bamboo and Rattan
Science and Technology
State Forestry and Grassland
Administration
International Centre for Bamboo
and Rattan (ICBR)
Beijing, China

ISSN 2199-4781 ISSN 2199-479X (electronic)
Compendium of Plant Genomes
ISBN 978-3-030-80835-8 ISBN 978-3-030-80836-5 (eBook)
https://doi.org/10.1007/978-3-030-80836-5

This Springer imprint is published by the registered company Springer Nature Switzerland AG
The registered company address is: Gewerbestrasse 11, 6330 Cham, Switzerland

This book series is dedicated to my wife Phullara and our children Sourav and Devleena

Chittaranjan Kole

Preface to the Series

Genome sequencing has emerged as the leading discipline in the plant sciences coinciding with the start of the new century. For much of the twentieth century, plant geneticists were only successful in delineating putative chromosomal location, function, and changes in genes indirectly through the use of a number of "markers" physically linked to them. These included visible or morphological, cytological, protein, and molecular or DNA markers. Among them, the first DNA marker, the RFLPs, introduced a revolutionary change in plant genetics and breeding in the mid-1980s, mainly because of their infinite number and thus potential to cover maximum chromosomal regions, phenotypic neutrality, absence of epistasis, and codominant nature. An array of other hybridization-based markers, PCR-based markers, and markers based on both facilitated construction of genetic linkage maps, mapping of genes controlling simply inherited traits, and even gene clusters (QTLs) controlling polygenic traits in a large number of model and crop plants. During this period, a number of new mapping populations beyond F2 were utilized and a number of computer programs were developed for map construction, mapping of genes, and mapping of polygenic clusters or QTLs. Molecular markers were also used in the studies of evolution and phylogenetic relationship, genetic diversity, DNA fingerprinting, and map-based cloning. Markers tightly linked to the genes were used in crop improvement employing the so-called marker-assisted selection. These strategies of molecular genetic mapping and molecular breeding made a spectacular impact during the last one and a half decades of the twentieth century. But still they remained "indirect" approaches for elucidation and utilization of plant genomes since much of the chromosomes remained unknown and the complete chemical depiction of them was yet to be unraveled.

Physical mapping of genomes was the obvious consequence that facilitated the development of the "genomic resources" including BAC and YAC libraries to develop physical maps in some plant genomes. Subsequently, integrated genetic–physical maps were also developed in many plants. This led to the concept of structural genomics. Later on, emphasis was laid on EST and transcriptome analysis to decipher the function of the active gene sequences leading to another concept defined as functional genomics. The advent of techniques of bacteriophage gene and DNA sequencing in the 1970s was extended to facilitate sequencing of these genomic resources in the last decade of the twentieth century.

As expected, sequencing of chromosomal regions would have led to too much data to store, characterize, and utilize with the-then available computer software could handle. But the development of information technology made the life of biologists easier by leading to a swift and sweet marriage of biology and informatics, and a new subject was born—bioinformatics.

Thus, the evolution of the concepts, strategies, and tools of sequencing and bioinformatics reinforced the subject of genomics—structural and functional. Today, genome sequencing has traveled much beyond biology and involves biophysics, biochemistry, and bioinformatics!

Thanks to the efforts of both public and private agencies, genome sequencing strategies are evolving very fast, leading to cheaper, quicker, and automated techniques right from clone-by-clone and whole-genome shotgun approaches to a succession of second-generation sequencing methods. The development of software of different generations facilitated this genome sequencing. At the same time, newer concepts and strategies were emerging to handle sequencing of the complex genomes, particularly the polyploids.

It became a reality to chemically—and so directly—define plant genomes, popularly called whole-genome sequencing or simply genome sequencing.

The history of plant genome sequencing will always cite the sequencing of the genome of the model plant Arabidopsis thaliana in 2000 that was followed by sequencing the genome of the crop and model plant rice in 2002. Since then, the number of sequenced genomes of higher plants has been increasing exponentially, mainly due to the development of cheaper and quicker genomic techniques and, most importantly, the development of collaborative platforms such as national and international consortia involving partners from public and/or private agencies.

As I write this preface for the first volume of the new series "Compendium of Plant Genomes," a net search tells me that complete or nearly complete whole-genome sequencing of 45 crop plants, eight crop and model plants, eight model plants, 15 crop progenitors and relatives, and three basal plants is accomplished, the majority of which are in the public domain. This means that we nowadays know many of our model and crop plants chemically, i.e., directly, and we may depict them and utilize them precisely better than ever. Genome sequencing has covered all groups of crop plants. Hence, information on the precise depiction of plant genomes and the scope of their utilization are growing rapidly every day. However, the information is scattered in research articles and review papers in journals and dedicated Web pages of the consortia and databases. There is no compilation of plant genomes and the opportunity of using the information in sequence-assisted breeding or further genomic studies. This is the underlying rationale for starting this book series, with each volume dedicated to a particular plant.

Plant genome science has emerged as an important subject in academia, and the present compendium of plant genomes will be highly useful to both students and teaching faculties. Most importantly, research scientists involved in genomics research will have access to systematic deliberations on the plant genomes of their interest. Elucidation of plant genomes is of interest not only for the geneticists and breeders, but also for practitioners of an array of plant science disciplines, such as taxonomy, evolution, cytology, physiology, pathology, entomology, nematology, crop production, biochemistry, and obviously bioinformatics. It must be mentioned that information regarding each plant genome is ever-growing. The contents of the volumes of this compendium are, therefore, focusing on the basic aspects of the genomes and their utility. They include information on the academic and/or economic importance of the plants, description of their genomes from a molecular genetic and cytogenetic point of view, and the genomic resources developed. Detailed deliberations focus on the background history of the national and international genome initiatives, public and private partners involved, strategies and genomic resources and tools utilized, enumeration on the sequences and their assembly, repetitive sequences, gene annotation, and genome duplication. In addition, synteny with other sequences, comparison of gene families, and, most importantly, the potential of the genome sequence information for gene pool characterization through genotyping by sequencing (GBS) and genetic improvement of crop plants have been described. As expected, there is a lot of variation of these topics in the volumes based on the information available on the crop, model, or reference plants.

I must confess that as the series editor, it has been a daunting task for me to work on such a huge and broad knowledge base that spans so many diverse plant species. However, pioneering scientists with lifetime experience and expertise on the particular crops did excellent jobs editing the respective volumes. I myself have been a small science worker on plant genomes since the mid-1980s and that provided me the opportunity to personally know several stalwarts of plant genomics from all over the globe. Most, if not all, of the volume editors are my longtime friends and colleagues. It has been highly comfortable and enriching for me to work with them on this book series. To be honest, while working on this series, I have been and will remain a student first, a science worker second, and a series editor last. And, I must express my gratitude to the volume editors and the chapter authors for providing me the opportunity to work with them on this compendium.

I also wish to mention here my thanks and gratitude to the Springer staff, particularly Dr. Christina Eckey and Dr. Jutta Lindenborn for the earlier set of volumes and presently Ing. Zuzana Bernhart for all their timely help and support.

I always had to set aside additional hours to edit books beside my professional and personal commitments—hours I could and should have given to my wife, Phullara, and our kids, Sourav and Devleena. I must mention that they not only allowed me the freedom to take away those hours from them but also offered their support in the editing job itself. I am really not sure whether my dedication of this compendium to them will suffice to do justice to their sacrifices for the interest of science and the science community.

New Delhi, India Chittaranjan Kole

Preface

The human genome has now been sequenced for 20 years. Genomic information has become a common tool for human beings to study the genetic regulation of organisms. It is obvious that more in-depth research is needed on the growth and development mechanism, wood property formation, and carbon sequestration mechanism of moso bamboo, which has the largest cultivated area and the strongest economic utilization in the world.

With the continuous progress of genome analysis technology, the genome of moso bamboo (*Phyllostachys edulis*) was deciphered in 2013. Since the publication of the genome sequence of moso bamboo, scientists in China and other countries have done a lot of in-depth exploration on the genetic regulation of moso bamboo. This is where this book comes in. The purpose of this book is to provide an intuitive, comprehensive, and nontechnical introduction to the topic of moso bamboo genome.

I have several goals in mind with this book. First, it is designed to provide an introduction to the economic utilization in China. The Chinese people have created a large number of products by planting and using moso bamboo, which has brought a lot of wealth to the farmers. Second, as a Chinese, I would like to share the germplasm resources of moso bamboo in China to the world. Thus, this book is intended to be a companion to bamboo lovers all over the world.

A third goal of this book is to cover different aspects of current research on moso bamboo, including genetics, biotechnology, comparative genomics, and more specific topics such as the shoot fast growing and flower development in transcriptional and posttranscriptional levels which lays a theoretical foundation for the breeding of the bamboo.

The book will be a guide for those who are interested in gene discovery, comparative genomics, functional genomics, and molecular breeding. It will be particularly useful for the scientists, breeders, and students of the universities and public sector institutes that are involved in research for the development of bamboo industry for updating the amount of knowledge generated in recent years.

Acknowledgements

I wish to express my deep and sincere appreciation to all the many authors who have contributed to the chapters of this book. Eighteen authors from 11 affiliations on seven provinces and cities in China have contributed. The work of the lead chapter authors, who in many cases recruited other authors, shaped the individual chapters and collated, edited, and revised the contributed texts, is particularly appreciated. Special thanks to the predecessors and colleagues in the research on the resources, genetics, and development of moso bamboo, without whose contributions this book would not be possible. Without these dedicated volunteers, all of them experienced moso bamboo scientists and experts, it would not have been possible to create such a massive collection of knowledge and information. To these writers, we must add the many willing contributors of photographs, tables, and figures that enhance this book so greatly.

The original proposal for creating *The Moso Bamboo Genome* was presented in 2017 by invitation from Prof. Dr. Chittaranjan Chole. Thanks to his enthusiasm and persistent invitation, it is his patiently persuading and ambition of promoting mankind to share the existing scientific research achievements that makes this book be compiled and published.

Behind the formal endorsement of the International Centre for Bamboo and Rattan (ICBR), Ministry of Science and Technology (MOST) of China and National Natural Science Foundation of China (NSFC) have provided the financial support to the projects. We are deeply appreciative of the 13th five-year plan of National Key R&D Project of China (No. 2018YFD0600100), National High Technology Research and Development Program of China (No. 2013AA102607-4), and National Natural Science Foundation of China (No.31570673).

I am most grateful for Prof. Peng Zhenhua and Prof. Jiang Zehui, who are my respect teachers, and they are also advocates and practitioners of cracking the mystery of moso bamboo genome.

I wish to thank all the helpful comments and supports received from scientists on bamboo and model plant research.

My heartfelt thanks go to my husband and my daughter, who give me continuous practical and logistical help and spiritual consolation.

Contents

Contributors

Qingsong Bai Guangdong Academy of Forestry, Guangzhou, Guangdong, China

Zhanchao Cheng Key Laboratory of Bamboo and Rattan Science and Technology, International Center for Bamboo and Rattan, State Forestry and Grassland Administration, Beijing, China

Benhua Fei Key Laboratory of Bamboo and Rattan Science and Technology, International Center for Bamboo and Rattan, State Forestry and Grassland Administration, Beijing, China

Jian Gao Key Laboratory of Bamboo and Rattan Science and Technology, International Centre for Bamboo and Rattan, State Forestry and Grassland Administration, Beijing, China

Wei Ge Beijing University of Agriculture, Beijing, China

Lianfeng Gu College of Forestry, Fujian Agriculture and Forestry University, Fuzhou, Fujian, China

Cheng-Lin Hou College of Life Science, Capital Normal University, Beijing, China

Dan Hou State Key Laboratory of Subtropical Silviculture, Zhejiang Agricultural and Forestry University, Hangzhou, Zhejiang, China

Guanghui Lai Guangde Forestry Bureau of Anhui Province, Guangde, China

Juan Li Key Laboratory of Bamboo and Rattan Science and Technology, International Center for Bamboo and Rattan, State Forestry and Grassland Administration, Beijing, China

Long Li Northwest Agriculture and Forestry University, Yangling, Shanxi, China

Xiangyu Li Key Laboratory of Bamboo and Rattan Science and Technology, International Center for Bamboo and Rattan, State Forestry and Grassland Administration, Beijing, China

Jun Liu Academy of Chinese Medicine, Henan University of Chinese Medicine, Zhengzhou, Henan, China

Shaohua Mu Key Laboratory of Bamboo and Rattan Science and Technology, International Center for Bamboo and Rattan, State Forestry and Grassland Administration, Beijing, China

Xiao-Ye Shen College of Life Science, Capital Normal University, Beijing, China

Yan Xiang Anhui Agricultural University, Hefei, Anhui, China

Lihua Xie School of Chemical and Environmental Engineering, Pingdingshan University, Pingdingshan, Henan, China

Hangxiao Zhang College of Forestry, Fujian Agriculture and Forestry University, Fuzhou, Fujian, China

Ying Zhang Beijing Academy of Agricultural and Forestry Sciences, Beijing, China

Abbreviations

2,4-D	2,4-dichlorophenoxyacetic acid
ACC	1-aminocyclopropane-1-carboxylic acid
AltA	Alternative acceptor
AltD	Alternative donor
AP2/EREBP	AP2/EREBP transcription factor gene
APA	Alternative polyadenylation
AS	Alternative splicing
ATI	Alternative transcription initiation
BAC	Bacterial artificial chromosomes
BambooGDB	The Genome Database of Bamboos
BHLH	Basic helix-loop-helix protein
BPs	Biological processes
BR	Brassinosteroid
bZIP	Basic leucine zipper
CCA1	Circadian clock-associated 1
CCs	Cellular components
CDS	Coding sequence
CesA	*Cellulose synthase*
circRNAs	Circular RNAs
CLIP-seq/HITS-CLIP	Cross-linking immunoprecipitation
CO	*CONSTANS*
CRISPR/Cas9	Clustered regularly interspaced short palindromic repeats-associated endonuclease Cas9 system
CRY1	*Crytochrome-1*
Csl	*Cellulose synthase like*
DBH	Diameter at breast height
DEGs	Differentially expression genes
Dof	The DNA binding with one finger
DPAC	Differential poly(A)-clustering
ELF3	*Early flowering 3*
EMP	Glycolytic pathway
ER	Endoplasmic reticulum
ES	Exon skipping
FMI	Floral meristem identity
FPI	Floral pathway integrator
FT	*Flowering locus T*

G	Germination rate
GA	Gibberellic acid
GABR	Genome Atlas of Bamboo and Rattan
GAMYB	Gibberellin response modulator
Gb	Gigabase
GFP	Green fluorescent protein
GI	*GIGANTEA*
GID	Gibberellin receptors
GO	Gene ontology
GRFs	Growth-regulating factors
HA	Hypocrellin A
HB	Hypocrellin B
HD-Zip	Homeodomain associated with leucine zipper
HMG	High-mobility group box
HSF	Heat shock transcription factor
HSP20	Heat shock transcription factor 20
HSP70	Heat shock transcription factor 70
ICBR	International Centre for Bamboo and Rattan
ICS	Isochorismate synthase
IGS	Intergenic spacer
IR	Intron retention
ISH	In situ hybridization
ITRs	Inverted terminal repeats
JA	Jasmonic acid
Kb	Kilobase
KNOX	Knotted-related homeobox
LHY	Late elongated hypocotyl
LINEs	Long interspersed nuclear elements
LTR	Long terminal repeat
LTREs	Long terminal repeat elements
LTR-retroelements	Long terminal repeat retroelements
m^6A	N^6-methyladenosine
m^6A-LAIC-seq	m^6A-level and isoform-characterization sequencing
MADS	Minichromosome maintenance 1, agamous, deficiens, and serum response factor
MFs	Molecular functions
mGFP	Mutated GFP
miRNAs	Micrornas
MITEs	Miniature inverted-repeat transposable elements
MLE	Mariner-like elements
MYA	Million years ago
MYB	v-myb avian myeloblastosis viral oncogene homolog
NAA	1-naphthaleneacetic acid
NAC	NAM ATAF and CUC2 family
NAT	Natural antisense transcripts
NCRs	Noncoding regions

NES	Nuclear export signal
NGS	Next-generation sequencing
NLS	Nuclear localization signal
NMD	Nonsense-mediated mRNA decay
nt	Nucleotide
ONT	Oxford Nanopore Technologies
ORF	Open reading frame
PacBio	Pacific BioSciences
PAS-Seq	PacBio Iso-Seq and strand-specific polyadenylation site sequencing
PAT-Seq	Poly(A) tag sequencing
PBS	Prime binding site
PCC	Pearson's correlation coefficient
PDT	Photodynamic therapy
PGRs	Plant growth regulators
PHD	Plant homeodomain
PHY	Phtyochromes
PIF	Phytochrome interacting factor
PIF3	Phytochrome-interacting factor 3
PP	Percentage of polymorphism
PPT	Polypurine tract
PRAPI	Post-transcriptional regulation Analysis Pipeline for Iso-Seq
pre-miRNAs	miRNA precursors
pri-miRNA	Primary miRNA
PRR	Pseudoresponse regulators
PTB	Polypyrimidine tract binding protein
PTCs	Premature termination codons
RA	Rheumatic arthritis
ROS	Reactive oxygen species
RPKM	Reads Per Kilobase per Million mapped reads
SA	Salicylic acid
SAM	Shoot apical meristem
SAUR	Small auxin-up RNA
SEM	Scanning electron microscopy
SINEs	Short interspersed nuclear elements
SMRT	Single-molecule real time
SNPs	Single-nucleotide polymorphisms
SPL	SQUAMOSA promoter-binding protein-like
SSM	Step-by-step mutation model
TAPIS	Transcriptome analysis pipeline for isoform sequencing
TCPs	Teosinte Branched 1, Cycloidea, and Proliferating Cell Factors
TDDS	Targeting drug delivery system
TDZ	Thidiazuron
TEs	Transposable elements

TFs	Transcription factors
TOC1	Timing of CAB1
TTF	Trihelix transcription factor
VI	Vigor index
WOX	WUSCHEL-associated homeobox
ZF-HD	Zinc finger related to homeodomain

1 Economic Value and Research Significance of Moso Bamboo

Benhua Fei

Abstract

Bamboo is one of the most important non-timber forest products in the world. Moso bamboo (*Phyllostachys edulis*) is a large woody bamboo with ecological, economic and cultural value in Asia, accounting for ~70% of the total bamboo growth area. Here, we report three major parts, such as resource distribution, uses and economic value, research of ecological management, materials properties and genome in moso bamboo. Moso bamboo research and bamboo industry will become a hot field of scientific research and economic development in the world. Bamboo will make new contributions to the prosperity and development of society in the future.

B. Fei (✉)
Key Laboratory of Bamboo and Rattan Science and Technology, International Center for Bamboo and Rattan, State Forestry and Grassland Administration, Beijing 100102, China
e-mail: feibenhua@icbr.ac.cn

1.1 Distribution of Moso Bamboo

1.1.1 Global Distribution of Moso Bamboo

The *World Checklist of Bamboos and Rattans* (2016) recorded 1642 bamboo species in 23 genera globally, including 1535 woody species and 130 herbaceous species. Among woody bamboos, there are 388 monopodial species, 982 sympodial species and 165 hybrid species. China is home to 647 bamboo species or 39% of the world's total. In total, 549 of the species included in the *Illustrated Record of Bamboos in China* were described in the *World Checklist of Bamboos and Rattans* (2016), including 538 native species.

Bamboos are mostly found growing in tropical and subtropical zones, with a minority of species in temperate and cold zones. Geographically, there are three main bamboo-growing regions, i.e. the Asian-Pacific region, American region and African region. A European-North American region is listed by some scholars as a region for introduced bamboo cultivation.

The Asian-Pacific region is the largest of the three, ranging from New Zealand in the south to mid-Sakhalin in the north and from Pacific Islands in the east to south-western Indian Ocean in the west. Main bamboo-producing countries include China, India, Myanmar, Thailand, Bangladesh, Cambodia, Vietnam, Japan, Indonesia,

J. Gao (ed.), *The Moso Bamboo Genome*, Compendium of Plant Genomes,
https://doi.org/10.1007/978-3-030-80836-5_1

Malaysia, the Philippines, the Republic of Korea, Sri Lanka, etc.

The American region mainly covers the east side of North and South Americas, from southern Argentina in the south to eastern USA in the north. Of 340 million ha forests in the Amazon basin, 10.2 million ha or 3% is covered by bamboos.

In the African region, bamboos are found only in a limited range, starting from southern Mozambique in the south to eastern Sudan in the north. Within this range, the highest concentration of bamboos lies in a strip of evergreen-deciduous mixed forests and tropical rainforests, covering the area from southern Senegal in Africa's west coast to Madagascar in its east coast. Patches of bamboo stands are also found in the Upper Nile valley in Sudan of northern Africa and in temperate montane forests in Ethiopia. Yet, the bamboo resource in Africa is lacking in diversity: a few indigenous species like *Oxytenanthera abyssinica* and *Yushania alpina* are coupled with several introduced species to constitute the dozen of bamboo species found in this continent.

There are no bamboo species native to Europe and only a few to North America. However, in the recent century, European countries (including Britain, France, Germany, Italy, Belgium, Portugal and the Netherlands), the USA and Canada have introduced to their countries a multitude of species from bamboo-producing countries. For instance, the USA has introduced from China 35 species in the genus *Phyllostachys* alone.

Moso Bamboo is a subtropical bamboo species primarily distributed in China and Japan of Asia. Italy, Germany, France, the Netherlands and the UK have introduced moso bamboo from the main bamboo-producing areas in Asia Pacific.

1.1.2 Distribution in China

Bamboo constitutes an important pillar in China's forest resources and acts as a key instrument in developing forestry for ecological conservation and livelihood improvement. China's 9th National Survey of Forest Resources, completed in 2018, showed a total bamboo forest area of 6,411,600 ha which took up 1.98% of China's woodland and 2.94% of the total forest area. 72.96% or 4,677,800 ha of bamboo forests were found to be moso bamboo forests. There were 14,125 million moso bamboo plants in China, with 80.42% or 11.36 billion of them in moso bamboo forests and 19.58% or 2765 million of them growing scattered around.

Moso bamboo is widely distributed in subtropical and north subtropical regions of China. It originated from the Qinling Mountains, Dabie Mountains, Hanshui Basin to the south of the Yangtze River Basin. The southernmost boundary of moso bamboo distribution is Luofu Mountain in Guangdong Province and Dayaoshan Mountain in Guangxi Province, which essentially coincides with 24° north latitude between the two broad tropical regions of South and Central Asia. The northern boundary is along the Northern Slope of Dabie Mountain and the Northern Slope of Tongbai Mountain, passing the Zhaochuan River, an upstream tributary of Danjiang river in the southeast corner of Shaanxi Province, along Dabashan Mountain to the west, and reaching Hui County, Wen County and Wudu in the upstream valley of Jialing River in Gansu Province. To the east is Zhejiang, Fujian, and Taiwan, which is also close to the dividing line between south and central Tropical Asia. The western boundary starts from the eastern edge of the Yunnan-Guizhou Plateau and the upper river valley of the Yangtze River. The horizontal distribution area of moso bamboo is about 23°30′ to 32°20′ north latitude and 104°30′ to 122° east longitude. Main provinces producing moso bamboo include Zhejiang, Jiangxi, Fujian, Hunan, Hubei, Sichuan, Anhui, Jiangsu, Guizhou, Guangdong, Guangxi, Yunnan, Shaanxi, etc. Moreover, the production and distribution area of moso bamboo in Zhejiang, Jiangxi, Fujian, Hunan, etc., are large and wide, respectively. These provinces which have high accumulation and good growth of bamboo forests are the central area of moso bamboo natural distribution (25°30° N, 110°–122° E), accounting for about 81.2% of the total area in China.

At present, wild or scattered moso bamboo forests can be found in the deep mountainous areas of Hunan, Jiangxi, Fujian, northern Guangxi and northern Guangdong. Most of them are mixed in evergreen broad-leaved forests and occasionally form small pure forests. Moso bamboo and *Cunninghamia lanceolata* roughly originate in the same area and require the similar climatic conditions, such as higher temperature, humidity and shelter from winds. This special environment can cause many varieties or variants of moso bamboo.

Moso bamboo forests include natural distribution and artificial management. Its vertical distribution is very complicated and large. Moso bamboo is distributed from more than ten meters above sea level to about 1700 m above sea level. With the change of latitude, longitude, mountain size and pattern, the vertical distribution of moso bamboo also changes, and the maximum and minimum boundaries of the vertical distribution are not uniform. Its general characteristics are high in the south and low in the north, high in the west and low in the east. For example, moso bamboo in Yinping Mountain, Chao County, Anhui Province, is distributed below 700 m above sea level, while the vertical distribution of moso bamboo forests in Guangxi is below 1300 m above sea level. In Dafang County of Guizhou Province, there is still moso bamboo distributed at an altitude of about 1700 m. The upper limit of vertical distribution of Wuyi Mountain in Fujian is about 1350 m above sea level.

In the natural range of moso bamboo, its vertical distribution is very large, but it grows well. Moreover, moso bamboo forests which are distributed in concentrated areas can survive in different environments. For example, moso bamboo forests in Zhejiang Province are mostly distributed in low mountains and hills below 800 m above sea level, accounting for about 93% of the total area in the province, reaching 42 million hectares. In Guangxi, moso bamboo is distributed in low mountains at an altitude of 400–800 m. In the mid-subtropical regions of northern Guangxi, the growth of moso bamboo below 200 m above sea level is also normal, but it is possible to grow normally in low mountainous areas above 700 m above sea level in the subtropical regions of central and southern Guangxi, especially in the subtropical regions of southern and northern Guangxi. In Guizhou, moso bamboo forests are usually concentrated in the middle, middle and lower parts of the mountain at an altitude of 600–900 m and the gentle slopes near the river valley. The distribution of moso bamboo is restricted by temperature and precipitation at altitudes above 900 m and below 600 m (Kuihong and Bohui 1996).

1.1.3 Quality of Moso Bamboo Forests

The quality of a bamboo forest is demonstrated by its biomass amount per unit area, canopy density, average height, average diameter at breast height (DBH) and plant number per unit area, among other indicators. Bamboo forests in China have an average biomass amount of 65.81 t/ha, specifically 70.53 t/ha for moso bamboo forests and 53.10 t/ha for the other bamboo forests, 69.86 t/ha for natural bamboo forests and 59.52 t/ha for bamboo plantations. As to canopy density, the national average is 0.69, specifically 0.70 for moso bamboo forests, 0.67 for the other bamboo forests and 0.69 for both natural bamboo forests and bamboo plantations. China's moso bamboo forests have an average height of 10.1 m, in specific 10.6 m for natural forests and 8.8 m for plantations. The average DBH for moso bamboo forests in China is 9.3 cm, specifically 9.3 cm for natural forests and 9.1 cm for plantations. In average, there are 2429 moso bamboo plants per hectare in China's moso bamboo forests, specifically 2460 in natural forests and 2347 in plantations.

1.1.4 Biomass and Carbon Stock of Moso Bamboo Forests

China's bamboo forests store a total of 422 million tons of biomass, accounting for 2.3% of all that stored by China's forest vegetation (18,364 million tons) and 2.54% of all that stored in forests. In particular, moso bamboo forests store 330 million tons of biomass, taking up 78.20% of that stored by all bamboo forests. In terms of carbon stock, bamboo forests in China store 211 million tons, accounting for 2.35% of all that stored by China's forest vegetation (8.98 billion tons) and 2.60% of all that stored in forests (8125 million tons). In particular, moso bamboo forests store a total of 165 million tons of carbon, taking up another 78.20% of that by all bamboo forests.

1.1.5 Resource Distribution and Changes

China's 1st National Survey of Forest Resources was conducted between 1973 and 1976, and its 9th and latest survey was completed in 2018. These past four decades have seen an increase of bamboo forest area by 3,413,900 ha, registering an annual growth of 81,300 ha at a rate of 1.83%. The 9th survey found moso bamboo forests in 13 provinces. Among them, 4 provinces had an area of moso bamboo forests over 700,000 ha (1,139,600 ha in Fujian, 1,056,500 ha in Jiangxi, 823,100 ha in Hu'nan and 900,600 ha in Zhejiang) and they registered a combined moso bamboo forest area of 3,706,200 ha, accounting for 79.23% of the national total.

1.2 Uses and Economic Value

1.2.1 Uses

Of all bamboo species, moso bamboo boasts the most versatile uses. They have found their way into people's daily life, playing a role in the fields of clothing, food, housing, transport, landscape, decoration, etc. A shining jewel among non-polluting, green resources, moso bamboo is valued as an ideal species in serving as an alternative to timber and as food, medicine, ornamentals and fodder.

1.2.1.1 Used as an Alternative to Timber

Moso bamboo culms are long, straight, with a high strength-to-weight ratio. Many of their properties, including their easiness to split, small shrinkage, good performance in elasticity, rigidity and hardness, long fibres, regular patterns and smooth, shining surface, make them a good material for manifold engineering fields and daily applications. Use as scaffolding and for shed building in modern construction original bamboo culms can not only serve as beams, columns, rafters and walls, but can also. In coastal areas, moso bamboos are used in fishing and in northern China, they are used in well digging. Materials based on moso bamboo could also be used to make a variety of farm implements, tools, furniture, stationery, toys, tableware, sports equipment, musical materials, everyday utensils and decorations, such as mats, covers, dustpans, tile canopies, fans, bamboo weaving products, bamboo filaments, curtains, rulers. Some of these products are China's conventional export commodities. In particular, modern technologies and advanced equipment have been employed in developing moso bamboo-based composite construction materials, which revolutionises the utilization of bamboo by bringing the plant into a new chapter of industrial uses.

1.2.1.2 Used as Food

Moso bamboo shoots were also known as bamboo buds, embryos, sprouts and flesh in ancient times. They can be divided into winter shoots, spring shoots and rhizome shoots, with winter shoots hailed as the best of all three kinds. Moso bamboo shoots are rich in protein, edible cellulose and low in fat, and contain large amounts of vitamins, mineral elements (iron, calcium, phosphorus, etc.) and sugar. China has a tradition of taking bamboo shoots as a kind of vegetable and as health food; hence, in ancient times there was a saying that "there is no feast without bamboo shoots on the table," hinting at how bamboo shoots were treasured by each dynasty as a delicacy. Now, moso bamboo shoots can be enjoyed in a variety of forms, as they are processed into dried shoots, tinned shoots, convenience shoots, shredded shoots, peeled shoot skin, smoked shoots, sour shoots, light-flavoured shoots, pickled shoots, etc.

1.2.1.3 Medicinal Uses

Moso bamboo's shoots, juice, Succus Bambusae, leaves, extractives from various organs and *Phallus indusiatus* all have medicinal values and a high therapeutic effect on many human diseases.

According to Chinese medicine bamboo shoots are slightly cold in nature, sweet in taste, non-toxic, with the functions of clearing heat and removing phlegm, detoxifying and promoting eruption, invigorating the spleen and benefiting *qi*, improving digestion and appetite, lowering blood pressure, preventing the hardening of blood vessels, improving one's looks and preventing cancer. *Phallus indusiatus* is cold in nature, sweet in taste, non-toxic, beneficial to the intestines and stomach, and could help to remove phlegm, lower blood pressure, improve immunity, adjust the level of hematic acid and fatty acids in the elderly and prevent cancer. It can be used to treat sepsis, bacterial enteritis, coughs, obesity, hypertension and other diseases caused by high cholesterol. Bamboo leaves are cool in nature, sweet and pungent in taste, with the functions of cooling the heart and soothing the spleen, removing phlegm and quenching thirst, clearing heat and dispersing depression, detoxifying and clearing the stomach. It is used to kill insects, treat sores and stop vomiting and irritation and can treat wind evil and heat in the upper body, coughing, wheezing, vomiting, hematemesis and all kinds of strokes and pains.

Succus Bambusae and Caulis Bambusae in Taenia are both traditional Chinese medicinal herbs. They are cold in nature, sweet in taste and non-toxic. Caulis Bambusae in Taenia clears the stomach, calms the internal organs and relieves vomiting and chest congestion. Distilled from bamboo juice, Succus Bambusae clears stasis and detoxifies, smoothes channels and collaterals and cools the body. Moso bamboo roots are used to quench thirst, detoxify, replenish deficiency and benefit *qi*. Bamboo caryopsis has a nourishing value and the effect of balancing *yin-yang*, light the body and benefit *qi*.

Extractives from bamboo leaves and culms are rich in flavonoids, phenolic acids and other compounds. They are good at removing reactive oxygen species and blocking nitrosation and have an effect on human body's endogenous system of antioxidant enzyme. They help inhibit lipid peroxidation and enhance immunity and have anti-fatigue, anti-aging effects.

1.2.1.4 Ornamental Uses

Moso bamboo is an evergreen plant with elegant foliage and tall and graceful culms. Its ornamental value is also augmented by its symbolism of modesty, uprightness and noble character. As bamboo culms are strong, upright, unyielding and fearless in face of cold and snow, bamboos are admired for their perseverance under harsh conditions along with pines and plum blossoms and the three are together known as the "Three Friends of Winter" in Chinese culture. The characters that bamboo embodies have been an everlasting source of spiritual strength.

Since ancient times, a great number of poets and scholars have praised bamboo. Su Shi, a greatest poet and artist of the Song dynasty, once said: "I would rather have no meat to eat than no bamboo around my house." Zheng Banqiao, a renowned painter of the Qing dynasty, dedicated a poem to bamboo: "With unswerving faith, bamboos on the rock strike roots deep into clefts. They stand with great tenacity against all the elements and remain strong and firm." Chen Zhi, a master of landscaping, garden design and forestry, praised bamboo's beauty and character with a poem: "I am surrounded by a beautiful, jade-colour bamboo forest. The bamboos look clear under the sun and the moon, make a clear sound as the wind passes through, form clear rhyme as raindrops land on their leaves, give out a clear light with dewdrops, and make for clear scenery after the snow. I am deeply impressed with their beauty, depth and unyielding will. Bamboo is truly essential for my spiritual strength."

Moso bamboo also plays an important part in landscaping, garden design and tourism, which is evidenced by writings from ancient times. Chen Jiru, a Chinese landscape painter and essayist of the Ming dynasty, said in the *Records beside a Small Window* (Xiaochuang Youji): "Bamboos planted after a pavilion will grow sparsely; houses at the end of a patch of bamboos will be quiet and secluded." Wen Zhenheng, a painter and landscape designer of the Ming dynasty, wrote in his best known work *Treatise on Superfluous Things* (Zhang Wu Zhi): "It is best to plant bamboos on a ridge with a small stream running by and a bridge over the stream. Nearby, steps should be built to lead to a higher platform for people to sit or lounge…They will feel like being surrounded by thousands of bamboos."

1.2.1.5 Used in Papermaking

Moso bamboo culms aged 1 or 2 years old can be used for pulp making. Their average chemical composition is as follows: 48.0% of cellulose, 27.2% of lignin, 19.9% of pentose sugars, 1.3% of ash, 0.15% of silicon dioxide, 3.0% of extractives by cold water solvent and 3.2% of extractives by benzene-ethanol solvent. Such culms have fibre length ranging from 1.13 mm to 3.76 mm and averaging 2.03 mm, fibre width ranging from 8.9 mm to 42.6 mm and averaging 23.22 mm, wall/cavity ratio at 6.98 and length/width ratio at 87.5. Moso bamboo's long, thin fibre, thick culm wall, high length/width ratio and strong fibre interlacing capacity make it a good source of material for manufacturing fine, high-strength, thin paper and one of the best monopodial bamboo species to be used in pulping. As the growth age of bamboo culms increases, the content of cellulose, polysaccharides and ash decreases, while the content of lignin, silicon dioxide, extractives by cold water solvent and by benzene-ethanol solvent rises and the wall/cavity ratio goes up. There are few changes to the morphology of cellulose, but the culms no longer perform as well in pulping. In terms of chemical properties, moso bamboos growing at a low latitude are better than those at a high latitude, dense forests are better than open forests, and bamboos growing on good site conditions are better than those on poor site conditions. Therefore, to cultivate moso bamboo stands for pulping purposes, it is essential to choose good site conditions at suitable latitudes and make sure that the stand density is within reasonable range.

1.2.1.6 Used as Fodder

Residues from the processing of young shoot sheaths and shoots are high in crude protein and crude fat and low in crude fibre. They have a feeding value comparable to that of grass and a higher water content, meaning that they can be used directly as daily fodder for livestock and when added with preservatives, they can be processed into silage. Sawdust and filings from the processing of bamboo culms have a low content of crude protein, crude fat and crude ash and a high content of crude fibre, rendering them difficult to directly digest by livestock. However, after an ammoniated treatment, they become softer and their content of crude protein is greatly increased, thus making them more nutritious and suitable as fodder for ruminant animals.

1.2.1.7 Used for Soil Conservation and Wind Blocking

In a moso bamboo stand, underground rhizomes and roots intertwined and interconnected to form a whole, while the aboveground culms stand in great numbers, take care of each other and together form a dense canopy. As an ideal bamboo species for soil and water conservation, moso bamboos can greatly reduce the runoff when being planted in mountainous areas and protect dams, riverbanks and roads against the rushes of flooding when being planted in large patches by the riverside. In addition, moso bamboos planted around houses can both serve a landscaping purpose and work as a windbreak.

1.2.1.8 Other Uses

Moso bamboo's rhizomes, roots, stump and branches can all processed to tap into different uses. Branches can directly made into brooms and other similar utensils. Sheaths can be used as sofa stuffing and as raw materials for ropes, carpets, synthetic fabrics and handicrafts. Exquisite bamboo root carvings, made from discarded roots, are a remarkable example of turning waste into treasure. Sawdust and filings from processing bamboo culms could also be further used in making moulded products.

1.2.2 Economic Value

Native to China, moso bamboo tops all the other existing bamboos in China in terms of distribution area and resource abundance. It is a superior species with strong adaptability, wide-range application and high economic value. With its increasingly extensive role in people's daily life and the national economy and its growing importance in material civilization and cultural life, the economic benefits, social benefits and ecological benefits brought by moso bamboo are more and more prominent. Moreover, in a time when green food gains popularity all over the world, food based on moso bamboo is highly valued. As a more and more popular and favoured natural resource, moso bamboo has attracted attention from a multitude of industries and fields.

The wide-range application of moso bamboo makes it essential to manifold industries. Its unique economic value is important to economic development. The management of high-yield moso bamboo stands and the industry of deep processing can create huge profits. Besides, with the excellent ecological environment of moso bamboo forests, a three-dimensional cultivation of edible fungi, Chinese medicinal herbs, flowers and other economic plants can be developed to help increase the income per mu of the bamboo forest up to 1000–5000 Yuan, sometimes even over 10,000 Yuan. It has indeed a huge potential for increasing production and efficiency. If calculated on the basis of 60% ecological environment availability of bamboo forests, there will be an area of 3.85 million ha of available area added upon the 6.41 million ha of bamboo forests across the country, which is equivalent to an increase of 3.85 million hm^2 of woodland. At the same time, 40–50 bamboos are equal to 1 m^3 wood. About 3 tons of air-dried bamboo can produce 1 ton of paper pulp, and 4 tons of air-dried bamboo can produce 1 ton of rayon pulp. The idea of "substituting wood and plastic with bamboo" opens up great possibilities for resolving the issue of resource and energy shortage. Therefore, bamboo development is not only an effective way and fundamental solution for farmers in the vast forest areas to lift out of poverty and achieve prosperity, but also has far-reaching strategic significance for the revitalization of forestry economy and sustained development of national economy.

1.3 Status of Moso Bamboo Research

1.3.1 Ecological Management of Moso Bamboo Forests

Moso bamboo is an important forest resource in southern China. By the beginning of this century, most of the bamboo forests in China have been managed under a labour-intensive model, which has greatly increased bamboo shoot output and bamboo forest biomass. Therefore, in-depth research on the management techniques will help increase the productivity of moso bamboo forests and improve the quality and quantity of moso bamboo forests. China has carried out a lot of research work on the management of moso bamboo forests, which has given birth to some good management models and measures.

1.3.1.1 Studies on Management Model of Moso Bamboo Forest

For moso bamboo forests, exploring a management model that combines economic and ecological benefits will not only help increase the economic gains of bamboo forests, but will also improve the local ecological environment. Liu (2013) proposed a model that the economic gains per unit area of bamboo stands and various ecological indicators are better than those of the traditional management models, realizing the sustainable management of bamboo stands. However, many studies have shown that the ecological effect of moso bamboo forests is lower than that of mixed forests. Huang (2006) compared the physical and chemical properties of soil in different forests in Jian'ou and found that moso bamboo forests have the highest performance in soil bulk density yet lowest in indicators such as natural water content, maximum water content and soil nutrient content. Based on the studies of the soil physical and chemical properties of different types of moso bamboo forests, researchers found that the soil nutrient content of moso bamboo forests is lower than that of the mixed forests, which could more probably lead to land capability loss (Genqing et al. 2012; Lou 1997). Also, the soil microbial biomass of moso bamboo forests (Lin et al. 2017), litter falls of forest stand (Liao et al. 2010) and soil organic carbon content are also lower than that of the mixed forests of bamboos and broadleaf trees (Qi et al. 2012, 2013). In terms of the growth of bamboo forests, moso bamboo forests have a short shooting duration and a high rate of shoot degradation (Chen et al. 2000). Bamboo forests also perform worse than mixed forests in bamboo culm shape and quality indicators, such as diameter at breast height (DBH), height and clear height (Wu et al. 2007). However, the shoot—young culm height and daily increment of bamboo forests are better than those of mixed forests in the period when height growth speeds up and in the early stage of the vigorous growth period (Hong et al. 2013).

1.3.1.2 Management Model of Moso Bamboo Mixed Forests

For moso bamboo forests, there are problems such as species monotony, structural simplicity, low biodiversity and low productivity. In order to overcome these problems, the mixed management modes among moso bamboo and other plants have been developed. Niu (2016) explored a management model that mixes moso bamboo forests and edible fungi and found that this combination can create more economic returns than conventional monoculture. He then proceeded to propose the best model of cultivating edible fungi under the canopy of moso bamboos. Compared to the moso bamboo forest management model, a combination of broad-leaf plants on the upper slope and moso bamboos on the lower slope has achieved significantly better results in such indicators as DBH, culm height, wall thickness and leaf quality (Cheng et al. 2015; Meng et al. 2016), and a higher capacity for water conservation. Bamboo and wood combination (moso bamboos mixed with broadleaf trees or moso bamboos mixed with Chinese fir) can improve the porosity and air permeability of the soil surface, which is beneficial to maintain soil fertility and prevent soil degradation (Fang et al. 2008). Different mixing proportions of bamboo and wood have different effects on various bamboo growth indicators (bamboo shoot yield, number of new culms, number of mature culms, quality of new culms, etc.). Nevertheless, the indicators of all these different mixed forests are better than those of moso bamboo forests. Yushan (2000) investigated and analysed the forest productivity, understory biomass, soil fertility and economic income of moso bamboo and Chinese fir mixed forests of different densities and put forward a bamboo-fir mixed model with best economic and ecological benefits.

1.3.2 Study on Moso Bamboo Materials Properties

1.3.2.1 Study on Moso bamboo's Properties

Moso bamboo is a high-quality running bamboo species suitable to be used as an alternative to timber and as food. It takes up the largest proportion of the entire bamboo forest area in China and stores a large amount of biomass. The study on its material properties mainly includes anatomical structure, physical, mechanical and chemical properties. The anatomical structure is closely related to its mechanical properties, chemical properties, property improvement as well as processing and utilization.

The quality of bamboo materials is an important indicator in evaluating the quality of bamboo culms. The measurement and analysis of bamboo culms and the exploration into its growth rhythm are helpful to the science-based use of bamboo culms. Based on different usages of bamboo materials, bamboo material property indicators can be divided into two groups. One covers the indicators to evaluate the applicability of structural bamboo, including density, dry shrinkage, drying characteristics and various mechanical performance indicators. The other group covers anatomical indicators and chemical composition to evaluate its quality as paper-making raw materials, including fibrous tissue proportion, basic tissue proportion, transfusion tissue proportion, holocellulose, lignin, cellulose, hemicellulose, extracts and ash. Domestic and foreign scholars have measured the quality indicators of bamboo materials and initiated a large number of studies on the bamboo growth rhythm, as well as how the bamboo material properties respond to external environmental factors and fertilizers.

Since 1971, domestic and foreign scholars have conducted related studies on bamboo vascular bundle and other microstructures, density, dry shrinkage and mechanical properties (Dirtger 1971; Zehui 2002; Mansur 2000). Wang et al. (2004) applied direct scanning with monochromatic X-ray to obtain the successive profile on radial variation in moso bamboo density, providing a theoretical basis for the utilization of bamboo materials based on radial classification. Peng et al. (2009) used CT technology to test the axial and radial density characteristics of moso bamboo and established a mathematical model between the CT measurement value and the moso bamboo density. Cui (2010) took moso bamboos of different growth ages as the research object and studied their major physical indicators such as wall thickness, basic density and dry shrinkage. The results showed that the wall thickness and basic density increase with the growth age, while the dry shrinkage tends to decrease at first and then maintains at a stable level after culms are 2 years old or older. Along the longitudinal height, the wall thickness decreases from the base to the tip, while the basic density as well as oven dry and air-dry radial shrinkage rates increase.

The anatomical structure of bamboo materials is related closely to its processing and utilization. Studying the anatomical structure is an important way to understand fully the basic structure of bamboo materials (Liese 1998; Liese and Weiner 1996; Alvin and Murphy 1988; Murphy and Alvin 1997; Walter 2001). In the 1930s, Japanese scholars initiated the studies on the anatomical structure of bamboo materials (Yoshio 1957). Later, scholars in China and Germany also conducted a large number of studies on bamboo structure, which served as a model for the initial studies on bamboo anatomical structure. The results of these studies made a significant contribution to bamboo classification later on. Regarding the study on bamboo microstructure, in addition to the electron microscopic observation of the micro-morphology of the bamboo culm surface, Bieke (2005) used scanning electron microscopy and cellular ultraviolet micro-spectrophotometry to have a systematic study on the distribution of lignin in the cell walls of moso bamboo and water bamboo. On the basis of the studies on the basic structure of bamboo, Chinese researchers have analysed and tested the anatomical properties of most bamboo species in China's major bamboo production areas. For example, in the early days, a study of Jin (1962) did a preliminary measurement on the fibre length and width of 12 types of bamboos grown in northeast China, providing a reference for the industrialized utilization of bamboo fibre. Ma (1997) measured the fibre morphology, tissue proportion, cellulose content and basic density of moso bamboo as well as the variation patterns of these indicators and pointed out that cultivating and managing small-diameter moso bamboo forests is more beneficial for pulping and papermaking. Zhang (2003) conducted a more detailed study on the changes of fibre wall thickness of uneven-aged bamboos in the growth period. This study found that the fibre wall thickness increases gradually from the outer to the inner of bamboo to a certain point before narrowing down a little. The fibre wall thickness would witness a logarithmic increase as the bamboo age goes up with growth rate climbs up fast for young bamboo but slows down after the age of 7. Ding (2005) discussed the formation mechanism of bamboo fibre cells and the influence of growing season on the growth and development of fibre cells from the perspective of developmental biology, providing more ideas for the study of bamboo anatomical structure. It also used the threshold segmentation method to obtain the change rhythm of the size of the crystallization areas in the microtubule bundle.

1.3.3 Processing and Utilization of Moso Bamboo Materials

1.3.3.1 Bamboo-Based Panels

In order to alleviate wood supply shortage, the idea of "substituting wood with bamboo" was put forward in the mid-1970s. In 1975, the first woven bamboo plywood production line in China was set up within Sichuan Shuangliu Agricultural Material Company. Since then, Chinese Academy of Forestry has carried out research on the production technology, supporting equipment, application technology and basic theories of bamboo curtain composite panels of different grades, moulded products made from bamboo shavings, reconstituted bamboo reconstituted bamboo for structural use and bamboo LVL. It then successfully developed bamboo MDF and, in 2004, built the first building made of bamboo LVL and bamboo plywood in Pingbian Primary School in Yunnan Province. The Institute of Mechanics under Chinese Academy of Sciences developed bamboo-plastic composite laminate using bamboo materials as a reinforcing material. Nanjing Forestry University used bamboo as raw material to successfully develop a high-strength bamboo LVL and used small-diameter bamboo to develop products for structural uses, including reconstituted bamboo, bamboo-wood composite container soleplate, bamboo woven LVL and bamboo-sheet curtain composite panels. In addition, it also carried out development of bamboo OSB and bamboo composite panels. Central South Forestry College successfully developed bamboo-based concrete moulded panels, curtain bamboo plywood, bamboo particle composite panels and high-strength curtain-mat bamboo plywood; in addition, it also developed the radial bamboo strip and curtain bamboo plywood and radial bamboo curtain laminated concrete moulded panels in cooperation with the Fujian Academy of Forestry Sciences. These two products not only have great physical and mechanical performance, but also significantly improve the utilization rate of bamboo materials. The performance indicators of the bamboo shaving panels developed by Zhejiang Forestry College exceed those of the wood shaving panels of first-class standard. With bamboo as raw materials, Southwest Forestry College developed a series of bamboo flooring with great decorative effect and high wear resistance and obtained a national patent for its development.

1.3.3.2 Bamboo Charcoal

Bamboo charcoal is a functional material that has just emerged in recent years. It has good absorbability, humidity control ability, antibacterial properties, anion function and far-infrared emissivity. It has shown great development potential in such fields as clothing and home decoration, textile products and cosmetics.

The porosity of bamboo charcoal gives it great absorbability. Research shows that apart from being an ideal material for removing lead and for removing fluoride in drinking water, bamboo charcoal also has strong performance in the adsorption of phenol and formaldehyde. For example, the maximum absorption value of 250 mesh bamboo charcoal for phenol is 73.4 mg/g. Bamboo charcoal also has many biological uses. According to research, bamboo charcoal particles can improve the physical and chemical properties of soil. Increase the pH, porosity and water holding rate of soil; increase N, P, K, Ca, Mg and Fe content of plants. Decompose and adsorb heavy metal elements such as Cu, Zn and Mn. Promote the growth of plant roots, stems and leaves as well as the formation of flower buds (Weilong et al. 2005). Bamboo charcoal textile is made by adding bamboo charcoal into polyester to produce bamboo charcoal polyester chips, which are then spun into fibres and woven into various textiles. Bamboo charcoal textile products that have appeared on market include bamboo charcoal mattresses, pillows and bedding.

1.3.3.3 Bamboo Vinegar

Bamboo vinegar is a mildly acidic liquid of a light yellow colour. It is produced by collecting the smoke generated from bamboo charcoal production before going through the procedures of condensation, storage and refining. Bamboo vinegar contains more than 200 organic ingredients. According to research, bamboo vinegar has strong bacteriostatic effects and has synergistic effects on certain insecticides; it is oxidation resistant and effective in eliminating phlegm and relieving cough; it can promote growth and delay senescence for plants, such as prolonging the ornamental period of flowers. The bamboo vinegar with a concentration of 0.1–0.01% can promote the growth of *Flammulina velutipes*.

1.3.3.4 Bamboo Fibre

Bamboo fibre textile is a new product developed in recent years. It has great air permeability, unique resilience, UV resistance, instant water absorbability, excellent dyeability, fabric drape, abrasion resistance, antibacterial properties and strong longitudinal and transverse strength. With its top performance in moisture absorption and moisture release among all kinds of fibres, bamboo fibre textile is regarded as the most promising healthy fabrics in the twenty-first century and has attracted great attention from the international textile industry since its introduction. The natural bamboo fibre currently on the market is of the high-strength and low-extension type in terms of mechanical properties. It has significantly better antibacterial and ultraviolet-resistant performance than ramie in terms of healthcare functions. It can rival ramie in terms of heat and moisture comfort. It is better than cotton and flax fibre and comparable to ramie in terms of heat resistance but has a slightly lower thermal stability than ramie; its acid resistance is slightly different from that of ramie (Wang 2005).

1.3.4 Moso Bamboo Genome

Moso bamboo genome size is similar to the maize genome but larger than rice, sorghum, purple false brome and Arabidopsis genome. It is 4.8 times larger than the rice genome which is the model species for grass genomics and 8.4 times larger than the smallest known grass genome of *Oropetium thomaeum* (VanBuren et al. 2018). In 2007, Prof. Jiang Zehui and Prof. Peng Zhenhua from the International Centre for Bamboo and Rattan initiated the genome sequencings of moso bamboo and related studies for the first time. As of 2013, the draft of moso bamboo genome was published in Nature Genetics using whole genome and transcriptome sequencing (Peng et al. 2013). The publication presents a total of 2.05 Gb draft genome with 328 Kb of scaffold N50, covering 95% of the whole genome. In total, 31,987 were predicted genes with an average length of 3350 bp per gene. In 2018, the chromosome-level reference genome of bamboo has been released five years after the draft genome (Zhao et al. 2018a, b), using single-molecule real-time sequencing method for significantly improved genome coverage. The size of the genome is 1908 Mb with a N59 scaffold length of 79.90 Mb, a 243 times improvement of the previous draft genome. The reference reads contained a total of 51,074 high-quality protein-coding genes. In addition, an alternative splicing profile of 25,225 splicing genes was identified by large-scale transcriptome analysis of 26 representative bamboo tissues.

References

Ahmad M (2000) Analysis of calcutta bamboo for structural composite materials. Dissertation for the degree of doctor of philosophy. Wood Science and Forest Products. Blacksburg, Virginia

Alvin KL, Murphy RJ (1988) Variation in fiber and parenchyma wall thickness in culms of the bamboo Sinobambusa tootsik. IAWA Bull n.s 9(4):353–361

Friar E, Kocher G (1991) Bamboo germplasm screening with nuclear restriction fragment length polymorphisms. Theor Appl Genet 82:697–703

Grosser D, Liese W (1971) On the anatomy of asian bamboos, with special reference to their vascular bundles. Wood Sci Technol 5(4):290–312

Isagi Y, Shimada K, Kushima H et al (2004) Clonal structure and flowering traits of a bamboo [Phyllostachys pubescens (Mazel) Ohwi] stand grown from a simultaneous flowering as revealed by AFLP analysis. Mol Ecol 13:2017–2021

Wang K, Huang B (1996) Moso bamboo in China. Zhejiang Science and Technology Press, Hangzhou
Lai CC, Hsiao JY (1997) Genetic variation of Phyllostachys pubescens (Bambusoideae, Poaceae) in Taiwan based on DNA polymorphisms. Bot Bull Acad Sin 38:145–152
Liese W (1998) The anatomy of bamboo culms. Int Netw Bamboo Rattan 7–99
Liese W, Weiner G (1996) Ageing of bamboo culms. Rev Wood Sci Technol 30:77–89
Murphy RJ, Alvin KL (1997) Fibre maturation in the Gigantochloa sortechinii. IAWA J 18(2):14–156
Peng ZH, Lu Y, Li LB, Zhao Q, Feng Q, Gao ZM et al (2013) The draft genome of the fast-growing non-timber forest species moso bamboo (*Phyllostachys heterocycla*)[J]. Nat Genet 45(4):456–461
Vanburen R, Wai CM, Keilwagen J, Pardo JA (2018) Chromosome-scale assembly of the model desiccation tolerant grass Oropetium thomaeum. Plant Dir 2: e00096
Walter S (2001) Non wood forest products in Africa: a regional and national overview. Food & Agriculture Org, Québec, Canada
Jiang Z (2002) China forestry engineering. Jinan Publishing House, Shandong
Jiang Z (2007) Bamboo and Rattan in the world. China Forestry Publishing House, Beijing
Jiang Z, Jiang X (2008) Correlation between wood structure and quality characteristics. China Forestry Publishing House, Beijing
Zhao H, Zhao S, Fei B et al (2017) Announcing the genome atlas of Bamboo and Rattan (GABR) project: promoting research in evolution and in economically and ecologically beneficial plants. GigaScience 6 (7):1–7
Zhao H, Wang S, Wang J et al (2018a) The chromosome-level genome assemblies of two rattans (*Calamus simplicifolius* and *Daemonorops jenkinsiana*). GigaScience 7(9):1–11
Zhao H, Gao Z, Wang L et al (2018b) Chromosome-level reference genome and alternative splicing atlas of moso bamboo (*Phyllostachys edulis*) 7(10):1–10
Peng Z (1999) Chinese *cunninghamia lanceolata*. China Forestry Publishing House, Beijing
Peng Z, Jiang Z (1999) Origin of five needle pine of dabie mountain. China Forestry Publishing House, Beijing

2 Intraspecific Variation of Moso Bamboo

Guanghui Lai, Shaohua Mu,
and Jian Gao

Abstract

Moso bamboo is the most economically and ecologically important bamboo species in East Asia, especially in China, with extensive distribution and long use history. In the process of adapting to different habitats for a long time, moso bamboo has performed a series of morphological and color variations, which have high landscape and handicraft values. There are different intraspecific origins, geographic, ecological and physiological types in moso germplasm. Only intraspecific variation is included in this chapter. Up to now, about 32 forms and/or cultivars of moso bamboo, mainly found from China and Japan, rarely from North America and Europe, are recognized.

G. Lai (✉)
Guangde Forestry Bureau of Anhui Province, Guangde 242200, China
e-mail: lybamb@21cn.com

S. Mu · J. Gao
Key Laboratory of Bamboo and Rattan Science and Technology, International Center for Bamboo and Rattan, State Forestry and Grassland Administration, Beijing 100102, China
e-mail: gaojian@icbr.ac.cn

2.1 Introduction

As the most economically and ecologically important bamboo species native to South China, moso bamboo has a wide distribution range and a long history of cultivation and utilization in East Asia, especially in China. It was first described by Carrière (1866) from a bamboo grove cultivated in France and was named as *Bambusa edulis*. This Latin name was later transferred into the genus *Phyllostachys* by Lehaie (1906) as *Phyllostachys edulis* and became a legal scientific name still recognized so far (Wang and Stapleton 2006; Vorontsova et al. 2016). *Ph. edulis* f. *heterocycla* is the first recognized infraspecific rank of moso bamboo cultivated in Europe for its unique morphology and easy identification (Lehaie 1906). As early as 1737, moso bamboo was introduced into Japan from China (Satow 1899), a close neighbor of the latter with frequent exchange history between the two nations, which can be proved to be the earliest introduction nation. Due to its strong adaptability, high economic value, and wide use, it was quickly spread and was widely cultivated throughout Japan. In the extensive and long-term cultivation in this country, moso bamboo produced some mutations which attracted the attention of Japanese botanists, Most of them have been taxonomically treated (Uchida 1936; Muroi 1941; Nakai 1942; Muroi and Kasahara 1972; Muroi 1974, 1989; Okaurma et al. 1986),

J. Gao (ed.), *The Moso Bamboo Genome*, Compendium of Plant Genomes,
https://doi.org/10.1007/978-3-030-80836-5_2

and were recognized by later researchers and horticulturists (Okamura and Tanaka 1986; Okamura et al. 1986), but some invalid names were revised (Chao and Renvoize 1988; Ohrnberger 1996, 1999). Since the 1970s, due to the need of compiling flora and production, large-scale investigations of bamboo species and their germplasm resources have been carried out in China. Many mutations, about 29 forms and/or cultivars, have been continuously found and been taxonomically named from the main moso distribution and planting areas in this country (Wen 1978, 1982; Wang et al. 1980, 1983; Wang 1984; Lai and Hong 1995; Yang et al. 1997; Yi and Yang 2002; Zhang 2008; Zhang et al. 2012, 2013; Lai 2013a, b; Chen 2013; Wang et al. 2018; Zhang et al. 2020), of which 11 cultivars were recorded in *Flora Reipublicae Popularis Sincae Toms 9 (1)* (Wang et al. 1996). Because many mutations of moso bamboo are produced under natural conditions rather than artificial breeding, all these cultivars were restored into the form rank by the present first author after 15 years (Lai 2012), which was later recognized by some scholars (Ma et al. 2014; Shi et al. 2021). For better protection and utilization, the germplasm resource of these mutations has been collected and preserved in several famous bamboo gardens, such as Taiping Experimental Centre of International Centre for Bamboo and Rattan, Anji Bamboo Expo Park in Zhejiang, Guangde National Forest Germplasm Resource Bank of Bamboos in Anhui, and Changzhou Special Bamboo Breeding Centre in Jiangsu, of which rare germplasm was preserved in private bamboo gardens of Changjiang delta. Since the 1960s, moso bamboo forests in the northern part of Guangxi, China sporadically blossomed and fruited almost every year. The seeds were collected and were then distributed to many places around the world for seedling production. In the process of seedling breeding, two new cultivars with variation in the color of their leaves or frost-resistant selection were discovered from bamboo groves cultivated in North America, and have been applied in landscaping (Haubrich 1983; Ohrnberger 1996, 1999). On the basis of preceding works and a recent study of the present authors, about 32 forms and/or cultivars of moso bamboo, including variations in the color of culms, branches, and leaves, also in the shape of culms or branches, or a combination of both, have been recognized. As excellent plants for landscaping and handicraft making and good materials for scientific research and breeding, they should be better protected and expanded for reproduction and utilization.

2.2 Intraspecific Variation of Moso Bamboo

2.2.1 *Phyllostachys edulis* f. *edulis*

Maozhu (Chinese), Môsô (Japanese) Fig. 2.1.

Phyllostachys edulis (Carr.) H. de Lehaie, Le Bambou 1: 39. 1906.—*Bambusa edulis* Carr., Rev. Hort. 37: 380. 1866; *Phyllostachys pubescens* Mazel ex H. de Lehaie, Le Bambou 1: 7. 1906; *Phyllostachys pubescens* Mazel ex H. de Lehaie f. *purpurescens* H. D. L., Zhulei Yanjiu 10: 3. 1977; *Phyllostachys pubescens* Mazel ex H. de Lehaie f. *rigida* H. D. L., Zhulei Yanjiu 10: 3. 1977; *Phyllostachys pubescens* Mazel ex H. de Lehaie f. *lutea* Wen, Bull. Bot. Res. 2 (1): 76. 1982.

Rhizomes monopodial. Culms diffuse, to 20 m or taller, to 20 cm in diameter; internodes to 40 cm or longer in the middle of culm, gradually shortened and thickened toward the base of culm, initially pruinous, densely puberulent, soon glabrescent, with total length flattened or grooved along the branch-bearing side; nodes with 2 (rarely 1) branches; nodal ridge occult at nodes without branches, prominent at branch-bearing nodes; sheath scar setose. Culm leaves deciduous; sheaths thickly coriaceous, tawny or purple-brown when fresh, with dense dark brown spots, densely brown hairy; auricles small, semiovate or irregularly prominent; oral setae extremely developed, curving; ligule strongly prominent, convex, margin with long and thick cilia; blade narrowly triangular, lanceolate or narrowly lanceolate, tinted with green, initially erect, soon reflexed. Foliage leaves 2–4 per ultimate branchlet; sheaths glabrous; auricles

Fig. 2.1 *Phyllostachys edulis*. (*Photo* courtesy of Guanghui Lai)

absent; oral setae initially developed, straightly extended, later deciduous; ligule comparatively developed, 1–3 mm long, apically arcuate; blade small, thin, lanceolate or linear-lanceolate, 3–11 cm long, 0.4–1.2 cm wide, abaxially basally pubescent, with 3–6-paired secondary veins. Inflorescence interactant; flowering branchlets lateral on branch nodes, spicate, subtended by some bracts that are imbricate and upward gradually enlarged, the superiors spathaceous; spathes narrowly oblong, 16–30 mm long, chartaceous, multiveined, dorsally puberulous, marginally ciliate, apex with a linear reduced leaf 7–18 mm long; bracklets ca. 10 mm long, narrowly lanceolate, membranaceous, abaxially 2-keeled and puberulous. Pseudospikelets 1–3 in each fertile spathe, sessile; glume 1, oblong-lanceolate, 1.5–2.8 cm long, pubescent, apex with a subulate reduced leaf; rachilla puberulous, apically awnlike. Florets 1 per pseudospikelet; lemma 2.2–2.4 cm long, apically acute, distally and marginally pubescent; palea slightly shorter than lemma, abaxially 2-keeled and puberulous, apically bifid; lodicules 3, membranaceous, oblong, lanceolate or semiovate, ca. 5 mm long, 1 mm wide; stamens 3; anthers ca. 1.2 cm long, yellow; filaments separate, ca. 4 cm long, white, glabrous; ovary oblong-ovate, ca. 3 mm long, yellowish, glabrous, with stipe 1.5–5.0 mm long; style 1, ca. 7 mm long, glabrous; stigmas 3, plumose. Caryopsis fusisorm, yellow-brown, brown or dark brown, 5–15 mm long, 1.2–2.5 mm in diameter, ventrally flattened and grooved, with persistent base of style, apically beaked. New shoots in Mar-Apr. Flowing in May-Aug. Fruit in Sept. $2n = 48$*.

China: native to this country, distributed in the east of Qinghai-Tibet Plateau, the south of Qinling Mountains to the south of Huaihe River, and Taiwan, mountain slopes, below alt. 1600 m or more; also cultivated in some places of the Yellow River Basin. Other regions: introduced into Japan, Korea, Vietnam, Cambodia, Philippines, the USA, Brazil, France, and England. This is the most economically important bamboo in China, widely cultivated for its versatile culm, delicious shoot, and useful forest environment.

2.2.2 *Phyllostachys edulis* f. *abbreviata*

Diemaozhu (Chinese) Fig. 2.2.

Phyllostachys edulis (Carr.) H. de Lehaie f. ***abbreviata*** G. H. Lai, Subtrop. Pl. Sci. 42 (1): 56. 2013.

Fig. 2.2 *Phyllostachys edulis* f. *abbreviata*. (*Photo* courtesy of Guanghui Lai)

The middle and lower internodes of culm irregularly shortened and locally depressed; nodes asymmetric, adjacent ones usually not connected, sometimes connected on one side more or less like bowknot, the other side not swollen.

China: native to Guangde, Anhui, rare in moso forest; introduced into Yangzhou, Jiangsu.

2.2.3 *Phyllostachys edulis* f. *anjiensis*

Anji Jinmaozhu (Chinese) Fig. 2.3.

Phyllostachys edulis (Carr.) H. de Lehaie f. ***anjiensis*** (P. X. Zhang) G. H. Lai, Subtrop. Pl. Sci. 42 (1): 59. 2013.—*Ph. heterocycla* (Carr.) Mitford f. *anjiensis* P. X. Zhang, World Bamb. Ratt. 6 (2): 28. 2008.

Culm leaf sheaths light tawny, gray-tawny, or golden, with light purple-brown longitudinal stripes (particularly near the margins) and dense purple-brown spots toward the upper part, but sparse on the middle and lower part.

China: native to Anji, Zhejiang, introduced into Guangde, Anhui, rather rare.

2.2.4 *Phyllostachys edulis* f. *aureovariegata*

Shima-môsô (Japanese).

Phyllostachys edulis (Carr.) H. de Lehaie f. ***aureovariegata*** (Uchida) D. Ohrnberger, Bambus-Brief 1990 (2): 17. 1990.—*Ph. pubescens* Mazel ex H. de Lehaie f. *aureovariegata* Uchida, Bull. Sci. Res. Alumni Assoc. Morioka Coll. Agr. 12: 84. 1936; *Phyllostachys edulis* f. *aureostriata* K. Ueda, Bull. Kyoto Univ. For. 30: 4. 1960, nom. nud.

Culms and branches green. Foliage leaves blades with yellow longitudinal stripes.

Japan: native to Echigo, Honshu, in cultivation, very rare.

Fig. 2.3 *Phyllostachys edulis* f. *anjiensis*. (*Photo* courtesy of Guanghui Lai)

Fig. 2.4 *Phyllostachys edulis* f. *bicolor*. (*Photo* courtesy of Guanghui Lai)

2.2.5 *Phyllostachys edulis* f. *bicolor*

Lücao Maozhu (Chinese), Kimmei-môsô (Japanese) Fig. 2.4.

Phyllostachys edulis (Carr.) H. de Lehaie f. ***bicolor*** (Nakai) G. H. Lai, Journ. Anhui Agr. Sci. 40 (8): 4624. 2012.—*Sinoarundinaria pubescens* var. *bicolor* Nakai, Tennen Kinenbutsu Chosa-Hokoku 19: 36. 1942; *Phyllostachys pubescens* Mazel ex H. de Lehaie f. *viridisulcata* Wen, Bull. Bot. Res. 2 (1): 76. 1982.

Culms and branches yellow, internode with a green groove along the branch-bearing side and a few green narrow longitudinal stripes outside groove (one branchless with a green broad longitudinal stripe on one side). Some of foliage leaf blades with light yellow narrow longitudinal stripes. Culm leaf sheath lighter in color, the spots so.

China: distributed in the main moso producing areas, small clumps in bamboo forest. Japan: originally found in Kusaka, Kochi, cultivated in some gardens for ornament, rare. Other regions: introduced into France and the USA, rather rare.

2.2.6 *Phyllostachys edulis* f. *curviculmis*

Qinglongzhu (Chinese) Fig. 2.5.

Phyllostachys edulis (Carr.) H. de Lehaie f. ***curviculmis*** H. X. Wang et J. S. Peng, Journ. Bamb. Res. 31 (1): 73. 2018.

Culms to 15 m tall, to 13 cm in diameter, sigmoid-curved.

China: native to Fengxin and Zixi, Jiangxi, growing in mixed forest, rare; introduced into Huangshan, Anhui, rather rare.

2.2.7 *Phyllostachys edulis* f. *epruinosa*

Youmaozhu (Chinese) Fig. 2.6.

Phyllostachys edulis (Carr.) H. de Lehaie f. ***epruinosa*** G. H. Lai, Journ. Bamb. Res. 14 (2): 7 0. 1995.

Internodes of young culm glabrous, not pruinous or with an inconspicuous white powdery ring below nodes, oily-glossy, sometimes tinted

Fig. 2.5 *Phyllostachys edulis* f. *curviculmis*. (*Photo* courtesy of Haixia Wang)

Fig. 2.6 *Phyllostachys edulis* f. *epruinosa*. (*Photo* courtesy of Guanghui Lai)

with purplish patches. Culm leaf sheaths purple-brown with red.

China: native to Guangde, Anhui, small clumps in moso forest, rather rare.

2.2.8 *Phyllostachys edulis* f. *exaurita*

Mayizhu (Chinese) Fig. 2.7.

Phyllostachys edulis (Carr.) H. de Lehaie f. ***exaurita*** T. G. Chen, World Bamb. Ratt. 11 (6): 25. 2013.

Culm leaves without auricles and oral setae. Culms slender and flexuose, on some upper internodes irregularly curved and one-sided, or on a certain internode deflexed.

China: native to Mt. Daiyun, Fujian, introduced into Changzhou, Jiangsu, Anji, Zhejiang and Guangde, Anhui, rare.

2.2.9 *Phyllostachys edulis* f. *gracilis*

Jinsi Maozhu (Chinese) Fig. 2.8.

Phyllostachys edulis (Carr.) H. de Lehaie f. ***gracilis*** (Hsiung) Chao et Renv., Kew Bull. 43 (3): 420. 1988.—*Ph. pubescens* Mazel ex H. de Lehaie f. *gracilis* W. Y. Hsiung, Act. Phytotax. Sin. 18 (2): 178. 1980.

All the culms in a clump smaller, 7–8 m tall, 3–4 cm in diameter, with thick wall; basal internodes relatively longer.

China: Yixing, Jiangsu, Guangde and Jingxian, Anhui, distributed in flake on the upper edge of moso forest.

2.2.10 *Phyllostachys edulis* f. *heterocycla*

Guijiazhu (Chinese), Kitsukou-chiku (Japanese) Fig. 2.9.

Phyllostachys edulis (Carr.) H. de Lehaie f. ***heterocycla*** (Carr.) V. N. Vassil., Trans. Sukhumi Bot. Gard. 9: 23. 1956.—*Bambusa heterocycla* Carr., Rev. Hort. 49: 354. 1878; *Phyllostachys edulis* (Carr.) A. et C. Riv. var. *heterocycla* (Carr.) Makino f. *subconvexa* Makino ex Tsuboi, Ill. Jap. Sp. Bamb. (2nd ed.) 21. 1916; *Phyllostachys pubescens* Mazel ex H. de Lehaie var. *biconvexa* Nakai, Journ.

Fig. 2.7 *Phyllostachys edulis* f. *exaurita*. (*Photo* courtesy of Guanghui Lai and Tianguo Chen)

Fig. 2.8 *Phyllostachys edulis* f. *gracilis*. (*Photo* courtesy of Guanghui Lai)

Fig. 2.9 *Phyllostachys edulis* f. *heterocycla*. (*Photo* courtesy of Shaohua Mu and Guanghui Lai)

Jap. Bot. 9 (1): 29. 1933; *Phyllostachys heterocycla* (Carr.) Mitford cv. Kikko-chiku, Mitford ex Ohwi in Fl. Jap. 136. 1965; *Phyllostachys heterocycla* (Carr.) Mitford cv. Kiko, Crouzet in Les Bamb. 75. 1981.

Culms in the lower part or below the middle part with alternately oblique nodes and asymmetrically shortened internodes with a bulge like tortoise-shell on one side and connected adjacent nodes on the other. The main branch on normal nodes is obviously elongated.

China: occasionally found in the main moso producing areas, single culm, several culms, or a small clump in bamboo forest. Japan: first record from Musashi and Ishikawa, Honshu, cultivated in some gardens. Other regions: introduced into France, England and the USA, rather rare. A famous horticultural bamboo, cultivated as ornamental and making handicrafts with culms.

2.2.11 *Phyllostachys edulis* f. *holochrysa*

Huangpi Maozhu (Chinese), Ôgon-môsô (Japanese) Fig. 2.10.

Phyllostachys edulis (Carr.) H. de Lehaie f. ***holochrysa*** (Muroi et K. Kasahara) D. Ohrnberger ex G. H. Lai, Subtrop. Pl. Sci. 42 (1): 64, 2013.—*Ph. heterocycla* (Carr.) Matsum. f. *holochrysa* Muroi et K. Kasahara, Journ. Himeji Gakuin Wom. Coll. Jap. 1974 (1): 4. 1974, nom. inval.

Culms and branches initially brightly yellow, glossy, sometimes tinted with purple-red patches, later golden in the following year and dark yellow in the third or more years; a very few internodes with 1 or 2 green longitudinal stripes; some of foliage leaf blades with white longitudinal stripes; culm leaf sheaths lighter in color than type form, and so in spot.

Fig. 2.10 *Phyllostachys edulis* f. *holochrysa*. (*Photo* courtesy of Shaohua Mu and Guanghui Lai)

China: originated from Gunagde, Anhui and Shangyu, Zhengjinag, introduced into Changzhou, Jiangsu and Lin'an, Zhejiang, rare. Japan: cultivated in Fukuoka, Kyushu, and Kyoto, Honshu, rare. This bamboo, with beautiful culms, has high value for ornament.

2.2.12 *Phyllostachys edulis* f. *huamozhu*

Huangpi Huamaozhu (Chinese) Fig. 2.11.

Phyllostachys edulis (Carr.) H. de Lehaie f. ***huamozhu*** (Wen) Chao et Renv., Kew Bull. 43 (3): 420. 1988.—*Ph. pubescens* Mazel ex H. de Lehaie f. *huamozhu* Wen, Act. Phytotax. Sin. 16 (4): 99. 1978; *Phyllostachys pubescens* Mazel ex H. de Lehaie cv. Tao Kiang, W. C. Lin in Bull. Taiwan For. Res. Inst. 98: 21, f. 13, 14. 1964; *Phyllostachys pubescens* Mazel ex H. de Lehaie f. *grammica* W. Y. Hsiung, Nanlin Keji 1975 (4): 25. 1975, nom. inval.; *Phyllostachys heterocycla* (Carr.) Mitford cv. Tao, B. M. Yang, et al. in Hunan Zhulei 123. 1993.

Culms yellow with green longitudinal stripes varying in width. Some of foliage leaf blades with few yellowish longitudinal stripes.

China: Occasionally found in the main moso producing areas, several culms or a small clump in bamboo forest; also cultivated in Chiayi and Kaohsiung, Taiwan.

2.2.13 *Phyllostachys edulis* f. *luteosulcata*

Huangcao Maozhu (Chinese), Gimmei-môsô (Japanese) Fig. 2.12.

Phyllostachys edulis (Carr.) H. de Lehaie f. ***luteosulcata*** (Wen) Chao et Renv., Kew Bull. 43 (3): 420. 1988.—*Ph. pubescens* Mazel ex H. de Lehaie f. *luteosulcata* Wen, Bull. Bot. Res., 2

Fig. 2.11 *Phyllostachys edulis* f. *huamozhu*. (*Photo* courtesy of Guanghui Lai)

Fig. 2.12 *Phyllostachys edulis* f. *luteosulcata*. (*Photo* courtesy of Shaohua Mu and Guanghui Lai)

(1): 76. 1982; *Phyllostachys heterocycla* (Carr.) Matsum. f. *gimmei* Muroi et Kasahara, Rep. Fuji Bamb. Gard. No.17: 8. 1972, nom. inval.

Culms and branches green, internode with a yellow or yellowish groove along the branch-bearing side (one branchless with a yellow or yellowish broad longitudinal stripe on one side). Foliage leaf blades green without stripes.

China: distributed in the main moso producing areas, small clumps in bamboo forest. Japan: rarely cultivated in some places. Europe: introduced into France, very rare.

2.2.14 *Phyllostachys edulis* f. *mira*

Huaguizhu (Chinese) Fig. 2.13.

Phyllostachys edulis (Carr.) H. de Lehaie f. ***mira*** (P. X. Zhang, G. H. Lai et X. Q. Hua) T. P. Yi, Journ. Sichuan For. Sci. Techn. 36 (2): 25. 2015.—*Phyllostachys edulis* (Carr.) H. de Lehaie 'Mira', P. X. Zhang, G. H. Lai et X. Q. Hua in Journ. Bamb. Res. 32 (2): 4. 2013; *Phyllostachys edulis* (Carr.) H. de Lehaie 'Lücaoguijiazhu', T. G. Chen in World Bamb. Ratt. 12 (2): 23. 2014.

Culms in the lower part or below the middle part with alternately oblique nodes and asymmetrically shortened internodes with a bulge like tortoise-shell on one side and connected adjacent nodes on the other; over the whole length yellow, with green narrow longitudinal stripes varying in width. Some of foliage leaf blades with few yellowish narrow longitudinal stripes.

China: native to Anji, Zhejiang, rather rare, occasionally found in *Phyllostachys edulis* f. *heterocycle* forest; also cultivated in Changzhou, Jiangsu, extremely rare.

2.2.15 *Phyllostachys edulis* f. *nabeshimana*

Lüpi Huamaozhu (Chinese), Tatejima-môsô (Japanese) Fig. 2.14.

Phyllostachys edulis (Carr.) H. de Lehaie f. ***nabeshimana*** (Muroi) Chao et Renv., Kew Bu11.

Fig. 2.13 *Phyllostachys edulis* f. *mira*. (*Photo* courtesy of Guanghui Lai)

Fig. 2.14 *Phyllostachys edulis* f. *nabeshimana*. (*Photo* courtesy of Guanghui Lai)

43 (3): 420. 1988.—*Sinoarundinaria pubescens* (Mazel) Ohwi f. *nabeshimana* Muroi, Hyogo. Chuto. Hakubutsu. Zasshi No. 7: 362. 1941.

Culms green, internodes with yellowish or yellowish-green narrow longitudinal strips varying in width. Foliage leaf blades green without stripes.

China: Anhui, Zhejiang, occasionally found in moso forest of other provinces. Japan: initially founded in Yamegun, Fukuoka, rarely cultivated in some places. Other regions: introduced into the USA and France, rather rare.

2.2.16 *Phyllostachys edulis* f. *obliquinoda*

Qiangzhu (Chinese) Fig. 2.15.

Phyllostachys edulis (Carr.) H. de Lehaie f. ***obliquinosa*** (Z. P. Wang et N. X. Ma) D. Ohrnberger ex G. H. Lai, Journ. Anhui Agr. Sci. 40 (8): 4624. 2012.—*Ph. heterocycla* (Carr.) Mitford var. *pubescens* (Mazel) Ohwi f. *obliquinoda* Z. P. Wang et N. X. Ma, Journ. Nanjing Univ. (Nat. Sci. Ed.) 1983 (3): 493. 1983, nom. inval.; *Phyllostachys heterocycla* (Carr.) Mitford cv. Flexuosa, B. M. Yang et Zeng in Hunan Zhulei 123. 1993; *Phyllostachys edulis* (Carr.) H. de Lehaie 'Dance Frock', D. Ohrnberger in Bamb. World, *Phyllostachys* (3rd ed.) 60. 1996.

Culms usually smaller, at most to 7 m tall and to 6 cm in diameter, with alternately oblique and not connected adjacent nodes, or Z-shaped flexuose below the middle part, internodes not inflated.

China: native to Yixing Jiangsu and Guangde, Anhui, a small clump in moso forest or in cultivation, rather rare.

2.2.17 *Phyllostachys edulis* f. *obtusangula*

Meihuazhu (Chinese) Fig. 2.16.

Phyllostachys edulis (Carr.) H. de Lehaie f. ***obtusangula*** (S. Y. Wang) D. Ohrnberger,

Fig. 2.15 *Phyllostachys edulis* f. *obliquinoda*. (*Photo* courtesy of Guanghui Lai)

Fig. 2.16 *Phyllostachys edulis* f. *obtusangula*. (*Photo* courtesy of Peixin Zhang)

Bambus-Brief 1990 (2): 18. 1990.—*Ph. pubescens* Mazel ex H. de Lehaie f. *obtusangula* S. Y. Wang, Guihaia 4 (4): 319. 1984.

Culms longitudinally 5–7-obtusangular and slightly quincuncial in cross section.

China: native to Yueyang, Hunan, extremely rare.

2.2.18 *Phyllostachys edulis* f. *pachyloen*

Houpi Maozhu (Chinses) Fig. 2.17.

Phyllostachys edulis (Carr.) H. de Lehaie f. ***pachyloen*** (G. Y. Yang et al.) Y. L. Ding ex G. H. Lai, Subtrop. Pl. Sci. 42 (1): 60. 2013.—*Ph. heterocycla* (Carr.) Mitford cv. Pachyloen, G. Y. Yang et al. in Act. Agr. Univ. Jiangxi. 19 (4): 97. 1997; *Phyllostachys pubescens* Mazel ex H. de Lehaie f. *pachyphloea* Cai, Jiangxi For. Sci. Techn. 10 (1): 58, 1982, nom. nud.

Culms slightly quadrangular or elliptical in cross section, with incrassated wall 3–4 cm thick at the base, 2.5 cm thick at the 1.3 m height, 1.4–1.8 cm thick at the middle, nearly solid at the upper; sheath scar in branchless nodes conspicuously prominent; internodes sometimes obviously longitudinally ribbed; specific gravity of fresh timber > 1.

China: native to western Jiangxi, distributed in flake on the edge of moso forest; introduced into some bamboo gardens in Anhui, Jiangsu, and Zhejiang, in cultivation, rare.

2.2.19 *Phyllostachys edulis* f. *porphyrosticta*

Banmaozhu (Chinese) Fig. 2.18.

Phyllostachys edulis (Carr.) H. de Lehaie f. ***porphyrosticta*** G. H. Lai, World Bamb. Ratt. 11 (4): 21. 2013; *Phyllostachys heterocycla* f. *mixta* X. X. Kang et al., Shanghai Zhuzhong Tuzhi 155. 2011, nom. nud.; *Phyllostachys heterocycla* cv. Pubescens (Carr.) H. de Lehaie f. *porphyrosticta* G. H. Lai, X. Q. Hua et W. W. Zhou, Journ. Zhejiang For. Sci. Techn. 32 (5): 73. 2012, nom. inval.

Young culm internodes initially green, gradually appearing purple spots from the autumn,

Fig. 2.17 *Phyllostachys edulis* f. *pachyloen*. (*Photo* courtesy of Haixia Wang and Guanghui Lai)

Fig. 2.18 *Phyllostachys edulis* f. *porphyrosticta*. (*Photo* courtesy of Guanghui Lai)

later spots continuously densified into purple or brownish patches, not occupied entire internodes.

China: first found in Hangzhou, Zhejiang, originated from seedlings, rather rare, introduced into Zhouzhi, Shaanxi; also distributed in Liangping, Chongqing and Beichuan, Sichuan, rare.

2.2.20 *Phyllostachys edulis* f. *purpureoculmis*

Anji Zimaozhu (Chinese) Fig. 2.19.

Phyllostachys edulis (Carr.) H. de Lehaie f. ***purpureoculmis*** P. X. Zhang, G. H. Lai et H. F. Zhang, World Bamb. Ratt. 10 (3): 32. 2012.

Young culm initially green, later gradually appearing light purple-brown tiny spots from basal internodes upward, spots continuously darkened and densified, so that, old culms almost completely purpurascent at last.

China: native to Anji, Zhejiang, extremely rare, endangered.

2.2.21 *Phyllostachys edulis* f. *purpureosulcata*

Xiaofeng Zijin Maozhu (Chinese) Fig. 2.20.

Phyllostachys edulis (Carr.) H. de Lehaie f. ***purpureosulcata*** P. X. Zhang, G. H. Lai et H. F. Zhang, World Bamb. Ratt. 10 (3): 32. 2012.

Young culms and young branches green, internodes gradually purpurascent-blackened along longitudinal grooves the on branch-bearing side, purple and yellowish striped outside groove.

China: native to Anji, Zhejiang, extremely rare, endangered.

2.2.22 *Phyllostachys edulis* f. *quadrangulata*

Fangmaozhu (Chinese) Fig. 2.21.

Phyllostachys edulis (Carr.) H. de Lehaie f. ***quadrangulata*** (S. Y. Wang) Ohrnberger, Bambus-Brief 1990 (2): 18. 1990.—*Ph.*

Fig. 2.19 *Phyllostachys edulis* f. *purpureoculmis*. (*Photo* courtesy of Peixin Zhang)

Fig. 2.20 *Phyllostachys edulis* f. *purpureosulcata*. (*Photo* courtesy of Peixin Zhang and Yingchu Yu)

Fig. 2.21 *Phyllostachys edulis* f. *quadrangulata*. (*Photo* courtesy of Peixin Zhang and Guanghui Lai)

pubescens Mazel ex H. de Lehaie f. *quadrangulata* S. Y. Wang, Guihaia 4 (4): 319. 1984; *Phyllostachys heterocycla* (Carr.) Mitford var. *pubescens* (Mazel) Ohwi f. *quadrangularis* S. C. Li et S. Q. Chen, Bamb. Res. 8 (1): 57. 1989, nom. nud.; *Phyllostachys heterocycla* (Carr.) Mitford cv. Tetrangulata, B. M. Yang et al. in Hunan Zhulei 124. 1993; Z. P. Wang et al. in Fl. Reip. Pop. Sin. 9 (1): 277. 1996.

Culms longitudinally 4-obtusangular and slightly quadrangular in cross section.

China: native to Yueyang, Hunan, rare; also found in Ningguo, Anhui, a small clump in moso forest.

2.2.23 *Phyllostachys edulis* f. *tubaeformis*

Shengyinzhu (Chinese) Fig. 2.22.

Phyllostachys edulis (Carr.) H. de Lehaie f. ***tubaeformis*** (S. Y. Wang) D. Ohrnberger, Bambus-Brief 1990 (2): 18. 1990.—*Ph. pubescens* Mazel ex H. de Lehaie f. *tubaeformis* S. Y. Wang, Guihaia 4 (4): 319. 1984.

Internodes gradually shortened and thickened toward base of culm, thicker in upper part than in lower part, slightly oblate in cross section, sometimes irregularly sunken on one side above node, slightly cymbiform from lateral view; adjacent nodes wavily alternately oblique, not connected; culms sometimes trumpet-shaped near base.

China: native to Yueyang, Hunan, extremely rare; initially introduced into Yiyang, Hunan, later into Huangshan and Guangde, Anhui, Anji and Hangzhou, Zhejiang, rarely in cultivation.

2.2.24 *Phyllostachys edulis* f. *tumescens*

Liuzhi Maozhu (Chinese).

Phyllostachys edulis (Carrière) H. de Lehaie f. ***tumescens*** (Yi et L. Yang) T. P. Yi, Journ. Sichuan For. Sci. Techn. 36 (2): 26. 27. 2015.—

Fig. 2.22 *Phyllostachys edulis* f. *tubaeformis*. (*Photo* courtesy of Shaohua Mu)

Ph. heterocycla (Carr.) Mitford f. *tumescens* Yi et L. Yang, Journ. Bamb. Res. 21 (1): 9. 2002.

Culms smaller, 6–7 m tall, 7–8 cm in diameter. Nodes of lower branches on lower culm extremely swollen, tuberculiform.

China: Changning, Sichuan, sparsely scattered in moso forest, extremely rare, endangered.

2.2.25 *Phyllostachys edulis* f. *ventricosa*

Fodu maozhu (Chinese) Fig. 2.23.

Phyllostachys edulis (Carr.) H. de Lehaie f. ***ventricosa*** (Z. P. Wang et N. X. Ma) D. Ohrnberger, Bambus-Brief 1990 (2): 19 (1990).—*Ph. heterocycla* (Carr.) Mitford f. *ventricosa* Z. P. Wang et N. X. Ma, Journ. Nanjing Univ., Nat. Sci. Ed. 1983 (3): 493. 1983; *Phyllostachys edulis* (Carr.) H. de Lehaie 'Buddha's Belly', D. Ohrnberger in Bamb. World, *Phyllostachys* (3rd ed.) 59. 1996.

Internodes below the middle culm ventricose at the middle, slightly tympaniform; nodes not oblique.

China: native to Anji, Zhejiang, mixed in moso forest, rather rare.

2.2.26 *Phyllostachys edulis* f. *venusta*

Huagan Jinsi Maozhu (Chinese) Fig. 2.24.

Phyllostachys edulis (Carr.) H. de Lehaie f. ***venusta*** G. H. Lai, Journ. Bamb. Res. 14 (2): 8. 1995.

All the culms in a clump smaller, 7–8 m tall, 3–4 cm in diameter, with thick wall, green, internodes with yellowish or yellowish-green narrow longitudinal strips varying in width, basal ones relatively longer.

China: native to Guangde, Anhui and Yixing, Jiangsu, small clumps mixed in *Phyllostachys edulis* f. *gracilis* forest, rare.

Fig. 2.23 *Phyllostachys edulis* f. *ventricosa*. (*Photo* courtesy of Dunyi Dong)

Fig. 2.24 *Phyllostachys edulis* f. *venusta*. (*Photo* courtesy of Guanghui Lai)

2.2.27 *Phyllostachys edulis* 'Anderson'

Phyllostachys edulis (Carr.) H. de Lehaie **'Anderson'**, D. Ohrnberger in Bamb. World, *Phyllostachys* (3rd ed.) 58. 1996.—*Ph. pubescens* Mazel ex H. de Lehaie 'Anderson', Amer. Bamb. Soc. Newsl. 16 (4): 10. 1995.

Culms to 22.5 m tall, to 18 cm in diameter. A very frost-resistant form, tolerating −6 °F (≈21.1 °C) in the USA.

USA: in cultivation, rare.

2.2.28 *Phyllostachys edulis* 'Bicanna'

Bazizhu (Chinese) Fig. 2.25.

Phyllostachys edulis (Carr.) H. de Lehaie **'Bicanna'**, W. G. Zhang et al. in Journ. Bamb. Res. 39 (1): 65. 2020.

Culms nearly tetragonal, on one side and the opposite side simultaneously sunken, with two interlinked lacunae, as '8' shape from the cross section view; wall not thickened.

China: native to Nanchang, Jiangxi, also found in Guixi and Tacheng, Jiangxi, Guangde, Anhui, rare, origined from *Phyllostachys edulis* f. *pachyloen* forest.

2.2.29 *Phyllostachys edulis* 'Jinsilonglinzhu'

Jinsi Longlinzhu (Chinese) Fig. 2.26.

Culms in the lower part or below the middle part with alternately oblique nodes and asymmetrically shortened internodes with a bulge like tortoise-shell on one side and connected adjacent nodes on the other; over the whole length green, with yellow narrow longitudinal stripes varying in width.

China: Yangzhou, Jiangsu, in cultivation, also found in Guangde, Anhui, extremely rare.

Fig. 2.25 *Phyllostachys edulis* 'Bicanna'. (*Photo* courtesy of Wengen Zhang and Guanghui Lai)

Fig. 2.26 *Phyllostachys edulis* 'Jinsilonglinzhu'. (*Photo* courtesy of Yingchun Yu)

2.2.30 *Phyllostachys edulis* 'Moonbeam'

Phyllostachys edulis (Carr.) H. de Lehaie **'Moonbeam'**, D. Ohrnberger in Bamb. World, 66. 1996; *Phyllostachys pubescens* Mazel ex H. de Lehaie 'Albovariegata', Haubrich in Amer. Bamb. Soc. Newsl. 4 (3): 2. 1983, inval.

Foliage leaves blades green, with white stripes.

USA: in cultivation, rare, originated from seed obtained from China.

2.2.31 *Phyllostachys Edulis* 'Oboro'

Oboro-mousou (Japanese) Fig. 2.27.

Phyllostachys edulis (Carr.) H. de Lehaie **'Oboro',** D. Ohrnberger in Bamb. World. Annot. Nom. Liter. Sp. Hi. Low. Tax. 213. 1999.—*Ph. heterocycla* (Carr.) Mitford f. *oboro* Muroi and Y. Tanaka, Journ. Himeji Gakuin Wom. Coll. no. 17: 2. 1989, inval.; *Phyllostachys heterocycla* (Carr.) Mitford f. *pubescens* (Mazel) Muroi 'Gold stripe', G. Cooper in Amer. Bambo Soc. Newsl. 16 (4): 17. 1995, nom. nud.

Culm internodes green, with extremely narrow numerous yellow stripes; nodes almost green. Foliage leaves blades green, without stripes.

China: Anhui, Zhejiang, occasionally found in moso forest of other provinces. Japan: Nagoya, Aichi, also cultivated in other places of Honshu, rare.

2.2.32 *Phyllostachys edulis* 'Okina'

Okina-môsô (Japanese) Fig. 2.28.

Phyllostachys edulis (Carr.) H. de Lehaie **'Okina',** D. Ohrnberger in Bambus-Brief 1990 (2): 17. 1990.—*Ph. heterocycla* (Carr.) Mitford f. *pubescens* (Mazel) Muroi 'Okina', Muroi et H. Okamura in H. Okamura et Y. Tanaka, Hort. Bamb. Sp. Jap. 20, 113. 1986.

Culm internodes yellow-green; nodes green. Foliage leaves blades with numerous green

Fig. 2.27 *Phyllostachys edulis* 'Oboro'. (*Photo* courtesy of Jian Gao and Guanghui Lai)

Fig. 2.28 *Phyllostachys edulis* 'Okino'. (*Photo* from internet)

stripes on white ground, occasionally with white stripes on green ground.

Japan: first found in Hyogo, rather rarely cultivated. Europe: introduced into Germany, in cultivation, very rare.

References

Carrière ÉA (1866) Plantes nouvelles, rares ou peu connues. Rev Hort 37:379–380

Chao CS, Renvoize SA (1988) Notes on some speceis of *Phyllostachys* (Gramineae–Bambusoideae). Kew Bull 43(3):420–422

Chen TG (2013) A new form of *Phyllostachys edulis* (Poaceae: Bambusoideae). World Bamb Ratt 11 (6):25–26

Haubrich R (1983) Variegated bamboos. Amer Bamb Soc Newsl 4(3):2

Lai GH (2012) A reconsideration of taxonomic position of some infraspecific mutants in the genus *Phyllostachys* (Bambusoideae). J Anhui Agr Sci 40 (8):4623–4625

Lai GH (2013a) Contributions to the taxonomy of the genus *Phyllostachys* (Gramineae: Bambusoideae). Subtrop Pl Sci 42(1):59–60

Lai GH (2013b) New scientific name of two forma of *Phyllostachys* spp. and New Synonym of Sinobambusa tootsik f. albostriata. World Bamb Ratt 11(4):21–22

Lai GH, Hong Y (1995) Documents for Bambusoideae in Anhui Province. J Bamb Res 14(2):7–8

Lehaie JH (1906) Le Bambou son Étude, sa Culture, son Emploi (Nos.1). Mons: Typo-Litho A. Libert 7–39

Ma NX, Lai GH, Zhang PX, Zhang HL (2014) The genus *Phyllostachys* in China. Zhejiang Science & Technology Press, Hangzhou, pp 92–107

Muroi H (1941) Bambusaceae novae nipponicae (1). Hyogo-ken Chutokyoiku Hakubutsugaku Zasshi (7):362

Muroi H (1974) The varieties of Bambusaceae with stripe in Japan. J Himeji Gakuin Wom Jun Coll (1):4

Muroi H (1989) Forma and cultivariety in Bambusaceae. J Himeji Gakuin Wom Jun Coll (17):2

Murol H, Kasahara K (1972) *Bambusa*, *Chimonobambusa*, *Phyllostachys*. Rep Fuji Bamb Gard (17):8

Nakai T (1942) *Sinoarundinaria pubescens* vat. *bicolor*. Tennen kinenbutsu chosahokoku 19:36

Ohrnberger D (1996) The bamboos of the world—*Phyllostachys* (3rd ed). Bambus Buch, Berlin, pp 58–60

Ohrnberger D (1999) The bamboos of the world, annotated nomenclature and literature of the species and the higher and lower Taxa. Elservier Science B.V, Amsterdam, pp 207–213

Okamura H, Tanaka Y (1986) The horticultural bamboo species in Japan: the characteristic and utilization of ornamental bamboo species with illustration. Hata Okamura, Kobe, pp 15–21, 113

Okamura H, Tanaka Y, Konishi M, Kashiwagi H (1986) Illustrated horticultural bamboo species in Japan. Haato Limited, Shinzaiki Wakaiama, pp 343–345

Satow E (1899) The cultivation of bamboos in Japan. Trans Asiat Soc Jpn 27(3):35

Shi JY, Zhang YX, Zhou DQ, Ma LS, Yao J (2021) Illustrated flora of bambusoideae in China, vol 1. Science Press & Springer Nature Singapore Pte Ltd., Bejing & Singapore, pp 471–481

Uchida S (1936) New species of Bambuseae in the Tohoku districts of Japan. Bull Sci Res Alumni Ass Morioka Imp Coll Agr For 12:84–85

Vorontsova MS, Clark LG, Dransfield J, Gorvaerts R, William JB (2016) World checklist of bamboos and rattans. International network for Bamboo and Rattan & the Board of Trustees of the Royal Botanic Garden, Kew, Beijing & London, p 151

Wang SY (1984) Three new forms of *Phyllostachys pubescens* Mazel. Guihaia 4(4):319–320

Wang ZP, Ye GH, Yu ZH, Zhu ZD, Zhao QS, Chen SY, Yao CY, Zhao HR, Lu JL (1996) *Phyllostachys* Sieb. et Zucc. In: Geng BJ, Wang ZP (eds) Flora reipublicae popularis sinicae tomus 9 (1). Science Press, Beijing, pp 274–279

Wang HX, Cheng P, Zeng QN, Yu L, Peng JS (2018) A new form of moso bamboo. J Bamb Res 37(1):73–74

Wang ZP, Stapleton CMA (2006) *Phyllostachys* Sieb. et Zucc. In: Wu ZY, Raven PH (eds) Flora of China, vol 22. Science Press & Missouri Botanical Garden Press, Beijing & St. Louis, pp 163–180

Wang ZP, Yu ZH, Ye GH, Zhu ZD, Zhao QS, Chen SY, Yao CY, Zhao HR (1980) A taxonomical study of *Phyllostachys*, China (count.). Act Phytotax Sin 18 (2):177–178

Wang ZP, Ye GH, Ma NX (1983) Some new taxa of Bambusoideae from China. J Nanjing Univ Nat Sci (3):493

Wen TH (1978) New taxa of *Phyllostachys* from western Chekiang. Act Phytotax Sin 16(4):98–99

Wen TH (1982) New taxa of *Phyllostachys* from Zhejiang. Bull Bot Res 2(1):76

Yang GY, Li ZY, Du TZ, Hu YB, Leng DJ, Zou WS (1997) A new cultivated variety of moso bamboo. Act Agr Univ Jiangxi 19(4):97–98

Yi TP, Yang L (2002) A new forma of *Phyllostachys heterocycla* (Carr.) Mitford var. *pubescens* (Mazel) Ohwi from Sichuan, China. J Bamb Res 21(1):9

Zhang PX (2008) New type of Moso in Anji County, Zhejiang Province. World Bamb Ratt 6(2):27

Zhang WG, Li XM, Guo CC, Yang GY, Yu F (2020) A new cultivar of *Phyllostachys edulis*—'Bicanna'. J Bamb Res 39(1):65–67

Zhang PX, Lai GH, Zhang HF (2012) Two new forms of *Phyllostachys edulis* (Gramineae: Bambusoideae) from Anji, Zhejiang. World Bamb Ratt 10(3):32–33

Zhang PX, Hua XQ, Lai GH (2013) A rare new form of *Phyllostachys edulis* (Gramineae: Bambusoideae) from Anji, Zhejiang. J Bamb Res 32(2):4–5

Biological Traits of Moso Bamboo

3

Jian Gao

Abstract

The morphology and forming rules of bamboo plants are the basis of cultivation and utilization. Bamboo vegetative organs are divided into aboveground and underground parts. There are bamboo clums, branches and leaves on the ground. The underground part is rhizome and roots. Bamboo plants have the habit and function of flowering, fruiting and reproducing, but flowering is rare, and they have many characteristics different from other plants. The main vegetative organs, such as shoot, leaf, rhizome and sexual organs, such as flora structure, seed nutrient contents were introduced. The shoot and culm high growth rules and the anatomical structure of shoot apical meristem at the stage of explosive growth were emphasized on.

3.1 Introduction

The morphology and growth of moso bamboo have its unique biological characteristics. The processing and utilization for different products are strongly influenced by their morphological, anatomical, physical and chemical properties. The whole body of moso bamboo is treasure, and all tissues and organs can be used by human beings. The main objective of this chapter is to describe morphology and anatomy of various organs, focusing on the general discussion of the shoot height growth and characteristics of development. Moso bamboo shoots can be used as food and the culm organs will develop into timber.

3.2 Rhizome Characteristics of Moso Bamboo

Evergreen, woody and monopodial moso bamboo (*Phyllostachys edulis* (Carrière) J. Houz.), as a species from genus *Phyllostachys*, comes under the sub-family Bambusoideae in the family Poaceae. It has aerial and underground stems. The stems growing above the ground are known as bamboo culms, the part commonly used in daily life and as an industrial raw material; the part of stems that grows under the ground is called rhizomes, which spread widely underground and send out fibrous roots and shoots from their nodes to form a bamboo grove. Bamboo's rhizomes, roots and shoots interact with each other and together carry out a series of life activities including the absorption of water and minerals and the synthesis, distribution and consumption of organic matter. The most remarkable feature of moso bamboo should be its

J. Gao (✉)
Key Laboratory of Bamboo and Rattan Science and Technology, International Center for Bamboo and Rattan, State Forestry and Grassland Administration, Beijing 100102, China
e-mail: gaojian@icbr.ac.cn

J. Gao (ed.), *The Moso Bamboo Genome*, Compendium of Plant Genomes,
https://doi.org/10.1007/978-3-030-80836-5_3

Fig. 3.1 Rhizome and root of moso bamboo. Photograph courtesy of Feihu Xi, Fujian Agricultural and Forestry University

root system comprising one group of rhizomes. Rhizomes mean stems migrating from the central plant to exploit novel territory. When rhizomes diffuse under the ground, their upward spread will be discontinued to form fresh culms. Such phenomenon occurs every year, which can be seen when new shoots sprout from soil. Influenced by bamboo species and growing conditions, it is often seen in spring. Such behavior in moso bamboo rhizomes belongs to running.

Running Bamboos (Monopodial)—Running bamboos have self-propagating rhizomes capable of traveling underground, and finally breaking through the surface to form new culms. Rhizomes move horizontally and can move through 20 ft of soil in the single season. Rhizome growth direction and distance cannot be predicted. It mostly occurs in temperate zone, among which the most famous genera are *Phyllostachys* and *Pleioblastus*. Most varieties can endure cold and survive below freezing temperature. Running bamboos quickly spread if left unchecked. It may cause trouble if you try to plant running bamboos in a separate part in the garden. The most frequently used solution is to establish the rhizome barrier surrounding the central plant to cease its sprawl.

Nodes with internodes in between segment the rhizomes of moso bamboo. Every node has a bud that can develop into either shoots or new rhizomes; yet, until the bud begins differentiation, there is little difference between a shoot bud and a rhizome bud. The aboveground and underground parts of moso bamboo grow alternately. Rhizome sheaths usually do not start to grow until the young culm branches and foliates. Their growth reaches a peak in August and September and then stops in November. During a period of 5–6 months, the growth of rhizome sheaths is active in alternation with shoot emergence and shoot growth. As bamboo shoots start to grow, rhizome sheaths wither and drop off. In the next year, when moso bamboos have a leaf replacement, side buds sprout out from where the former sheaths fall off and develop into new rhizomes. These new rhizomes come into their strongest growing period in June and July and then cease to grow due to the large amount of shooting in the bamboo forest. Fibrous roots form from rhizome nodes and play an essential role in collecting water, fertilizer and minerals from the soil. The depth of bamboo rhizome systems varies between 5 and 25 cm, depending on site conditions: in rich, loose and thick soil, rhizomes tend to run deeper, while in thin soil with a high gravel content, rhizomes grow closer to the surface. Typically, rhizomes can live for more than 10 years, yet those aged from 3 to 7 years old with a depth of around 20 cm exhibit stoutness and a yellow color and are proven to have the strongest capacity for shooting and bamboo formation (Fig. 3.1).

3.3 Shoot Growth Characteristics

On bamboo rhizomes, some of the side buds will differentiate to become shoot buds and later grow into a shoot. Differentiation usually starts around the Beginning of Autumn (Liqiu) in August and reaches its full speed around the White Dew (Bailu) in September. By early winter, the newly developed shoots, named winter shoots, will look stout with yellow, puberulent sheaths. Due to the low temperature in winter, shoots fall into a dormant state and grow very slowly. By spring, as the temperature rises, the growth rate of bamboo shoots speeds up. They come up out of the ground around the Vernal Equinox (Chunfen) in mid-March, known as spring shoots.

New shoots usually occur in spring, depending on variety. Fresh shoots are susceptible to environmental damages during the first few weeks upon shooting. Actually, most shoots turn hardened culms after almost three growing seasons. A good solution is to protect new shoots from potential hazards, like wind and animal. The bud of new shoots may be easily trodden in the first few days. People should be especially careful while getting close to new shoots. New shoots are wrapped by overlapping sheaths often in brown, with a layer of fuzz. The sheaths shield culm's soft outer tissue and supply hormones necessarily required by fast growth. Shoot growth will be restricted upon the removal of sheaths.

Spring shoots grow and come up at the highest vitality from the Clear and Bright (Qingming) to the Grain Rain (Guyu) in the first half of April and after the Beginning of Summer (Lixia), there are no more shoots emerging from the soil. Shoot degradation, a phenomenon where some of the emerged bamboo shoots cannot grow into shoots, can occur for a variety of reasons, including nutrient deficiencies in the bamboo stand, low temperature, drought, pests, poor health conditions of the mother plant, rhizome or culm bud and inappropriate stand structure or felling practices. For shoots that will later degrade, initially they exhibit slow growth, weakened physiological activities and reduced nutritional content; subsequently, they stop growing and lose their physiological vitality; and eventually, the shoots dry up and decay. Therefore, to minimize the occurrence of such a phenomenon, it is necessary to strengthen the management of bamboo stands, optimize stand structure, improve site conditions, identify and dig out shoots that are degrading, undersized or weak in a timely fashion. The easiest way to identify degrading shoots is to check the tip of shoot sheaths early in the morning: degrading shoots have no dew at the tip.

3.3.1 Daily Growth of Moso Bamboo Shoot

Height of five single moso bamboo shoots was surveyed at 8 A.M. Height increment in entire growth process was recorded every morning in April 2–June 20 (Fig. 3.2). Corresponding growth rate demonstrated the bell curve (Fig. 3.3a) and early growth rate was relatively low (April 9–April 24). When approaching the average height of 102.8 cm till April 24, daily shoot height progressively entered the first growth peak (77.8 cm/day), with average height of 307.0 cm (April 28). Daily growth rate remained high at 61.3–77.8 cm (April 28–May 5; Fig. 3.3b). Daily shoot height increment in later stage of growth was larger compared to early stage of growth. Eventually, shoot tips burst out of sheaths, branched and came into leaf on May 12 (1375 cm), suggesting the finalization of shoot growth. According to the constant measures for height of bamboo shoots, eight representative growth stages of winter bamboo shoots (50, 100, 300, 600, 900 and 1200 cm) were chosen for anatomical variation analysis.

3.3.2 Dynamic Anatomical Variation in Shoot Tip Growth Stages

In shoot growth, the apical meristem possesses forceful division capability, which comprises vegetative cone and subapical meristem (Fig. 3.4a). At S1, shoot tips followed the pyramid-like pattern with no lateral buds (Fig. 3.4a). With the growth of bamboo shoots, shoot tips progressively bifurcated, and lateral buds emerged at S2–S7 on the top of bamboo joints (Fig. 3.4b–g). In early stage of growth (S1–S5), many cell nuclei were seen in ground tissue and vascular tissue, proving the existence of meristematic tissue (Fig. 3.5a–e). As the shoots grew, the amount of nuclei was reduced to a minority that was observable at S7 (Fig. 3.5f–g). In contrast to cells at S1, those at S4 and S6 had a larger size and more fat droplets (Fig. 3.6). The endoplasmic reticulum and mitochondria

Fig. 3.2 Examples for moso bamboo shoots at seven growth stages (**a–g** denoted S1–S7). Bars, 1 cm (**a**); 15 cm (**b**); 30 cm (**c**); 1 m (**d**); 1.5 m (**e**); 3 m (**f**); and 4 m (**g**) (Li et al. 2018)

could be seen on the verge of cells at S4 and S6, because of the existence of one big central vacuole (Fig. 3.6b, c). To sum up, different from constantly active apical meristem, cell division rate decelerated in intercalary meristems in later stage of growth.

3.4 Culm Growth Characteristics

Moso bamboos have relatively tall culms that can reach a height of 20 m or more and a diameter as large as 20 cm. Culms are characterized by nodes with internodes in between, with the highest internode length approximating 40 cm or over. Toward the base of the culm, internodes shorten gradually while culm wall thickens. Nodes comprise flat culm-rings and protruding sheath-rings. Young culms are puberulent and covered with a thick layer of white powder, with pubescent sheath-rings and inconspicuous culm-rings. As moso bamboo grows rapidly, culms shed their white powder and fine soft hairs and turn from green to yellowish green and sheath-rings become more protruding, with prominent auricles and mostly green blades. Moso bamboo's culm wall is around 1 cm thick. For nodes without branches, the culm-ring remains inconspicuous, sits below the sheath-ring or ridges from the slender culm. Culm sheaths have a thick leathery texture, with the back of the sheath proper presenting a yellow-brown or purple-brown color with dark brown spots or patches of dense, brown bristles.

3.5 Characteristics of Leaves

Bamboo leaf organs usually have two kinds, one is the stem leaves, which are attached to the bamboo culm, usually called shoots sheath; The second is the nutrition leaf that is born on the branch. The bamboo leaves mentioned in this book only refer to vegetative leaves. Moso bamboo leaves are mostly lanceolate, with 2–4 pieces per ultimate branch. They have inconspicuous auricles. Leaf length varies from 4 to 11 cm while length width ranges from 0.5 to 1.2 cm. Logarithm 1, leaf blade abaxially glabrous, leaf tip shape gradually acuminate, leaf base shape cuneate, petiole length short, leaf margin serrate sparse on one side, dense on the other. Abaxial bamboo leaves are proximally pubescent along the midrib (Fig. 3.6).

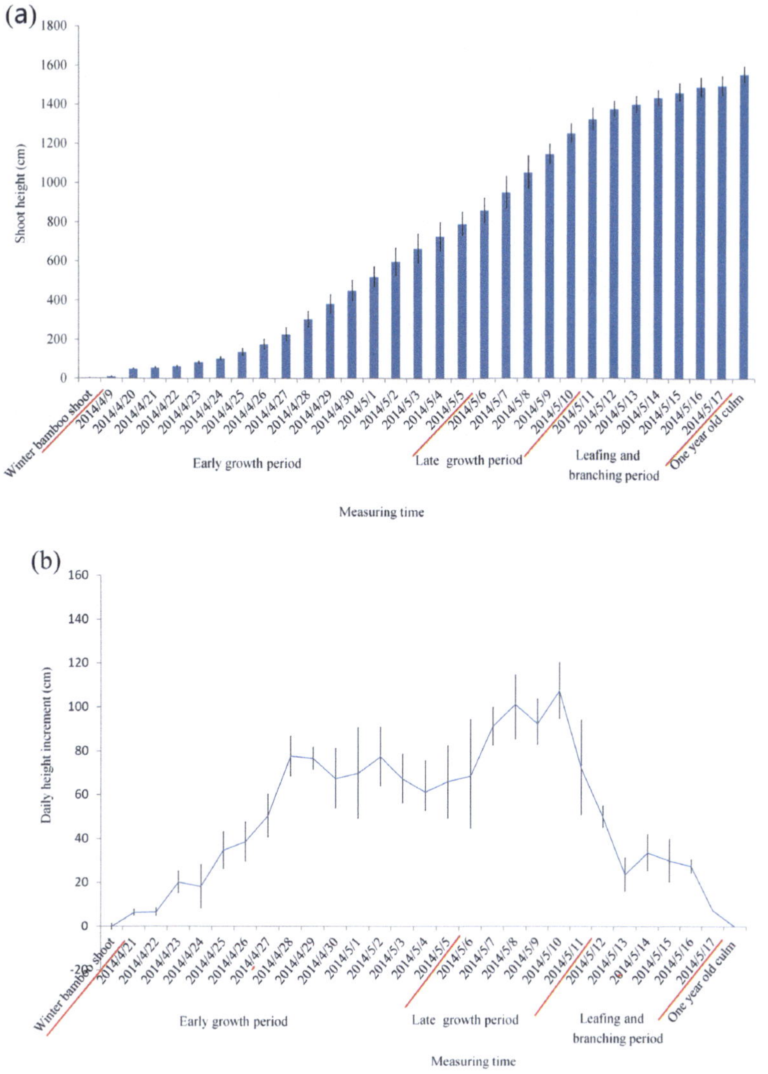

Fig. 3.3 Daily height variation (**a**) and daily height increment of bamboo shoots (**b**). The horizontal axis suggests date of survey, while vertical axis denotes shoot height (**a**) and daily height increment (**b**), separately (Li et al. 2018)

3.6 Characteristics of Flowers

Different from most plants and trees known by people, a large number of bamboo species rarely bloom. The typical flowering season may last for few decades. *Phyllostachys bambusoides*, also called Japanese Timber, possesses the longest flowering season of 130 years. Though some bamboo species indeed bloom per year, it is not a norm. Such long flowering season remains mysterious to even botanists and researchers. The topic has been discussed for centuries, which may be dated back to ancient scientific documents. Though lots of theories have been raised by 2000, no persuasive interpretation for such special flowering mechanism is given. Recently, studies on the flowering and flower development of moso bamboo have been greatly improved by prof. Gao's Lab (Qi et al. 2012; Gao et al. 2014; Ge et al. 2017; Cheng et al. 2018, 2019).

Moso bamboo features rare sexual reproduction and unstable flowering habit; whereas as moso bamboo blooms synchronously and experiences large-scale death upon flowering (Fig. 3.7), the expansion of Chinese moso bamboo industry is greatly restricted. The inflorescence of moso bamboo is complex spike, about

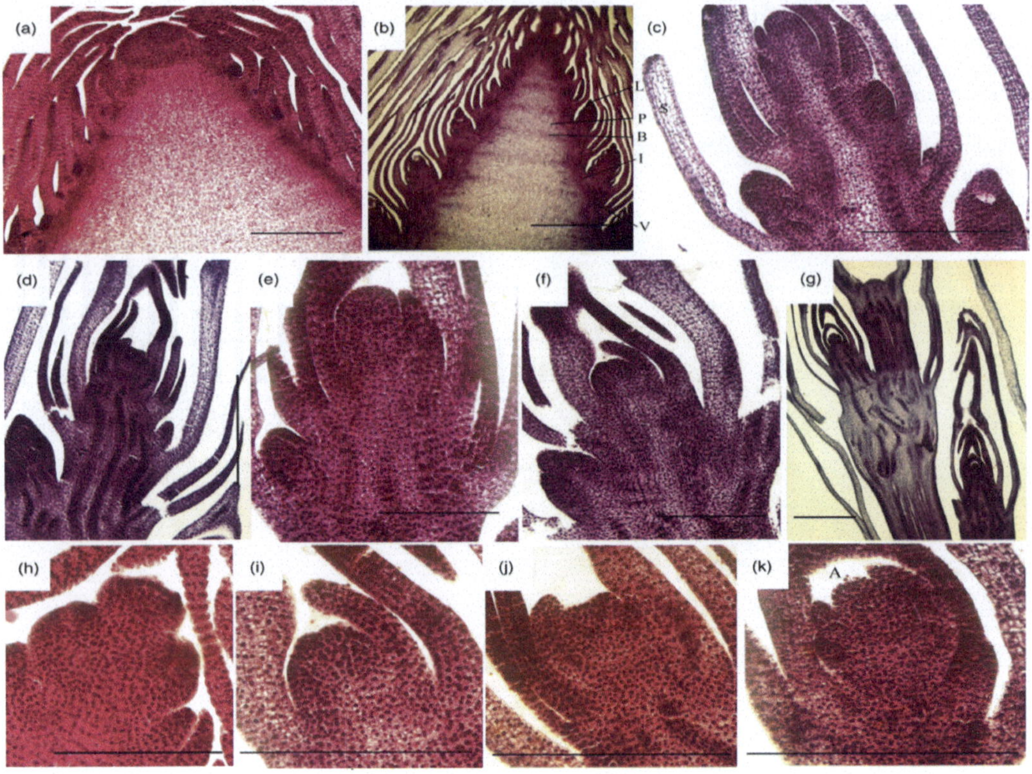

Fig. 3.4 Transverse section for moso bamboo shoot tips at seven growth stages (S1–S7). **h–k** Apical meristems at S1, S3, S4 and S6. Bars, 2 mm (**a–g**); 500 lm (**h–k**). L represents lateral bud, I represents internode, B represents bamboo joint, A represents apical bud, V represents vascular bundle, S represents bamboo sheath and P represents pith (Li et al. 2018)

7–9 cm long. Each inflorescence contains several pseudo-spikelets (Fig. 3.8), and each false spikelet contains 1–6 florets. The floret is composed of lemma, palea, scales, pistil and stamen. The flowering branchlets of moso bamboo are spicate, 5–6 cm long, with 4–6 scaly bracts that are gradually larger at the base. There are usually more than 10 spathes, with 1–3 pseudo-spikelets per fertile spathe. Each spikelet has 1 floret. There is 1 piece of glume which is 15–28 mm long, pubescent, with a small subulate blade at the apex. The lodicules are lanceolate, ca. 5 mm in length and 1 mm in width. The filaments are around 4 cm long with 12 mm-long anthers. There are 3 plumous stigmas (Fig. 3.9). The caryopsis is narrowly elliptic in shape, 4.5–6 mm in length and 1.5–1.8 mm in diameter, with a persistent style base at the apex. For moso bamboos, flowering takes place between May and August.

3.7 Characteristics of Moso Bamboo Seeds

Despite the long flowering cycle of moso bamboo, its seed yield following flowering can be substantial (Fig. 3.10). According to folk records in China, *P. edulis* seeds could assuage hunger. Eating *P. edulis* seeds may nourish body and make people feel light and handy (Li 1590). The mature period of seeds is from August to October. The caryopsis is oblong and reddish brown (Fig. 3.11). Plump *P. edulis* seeds in red have a strong fragrance (Li 1978). The thousand-grain weight of seeds stored for 2 years totaled 21.56 ± 0.62 g, and that of fresh seeds totaled 23.45 ± 0.56 g, whereas *P. edulis* seeds are not easily harvested or stored, thus limiting the culture, development and exploitation of its seedlings. The seeds easily mildew and reduce

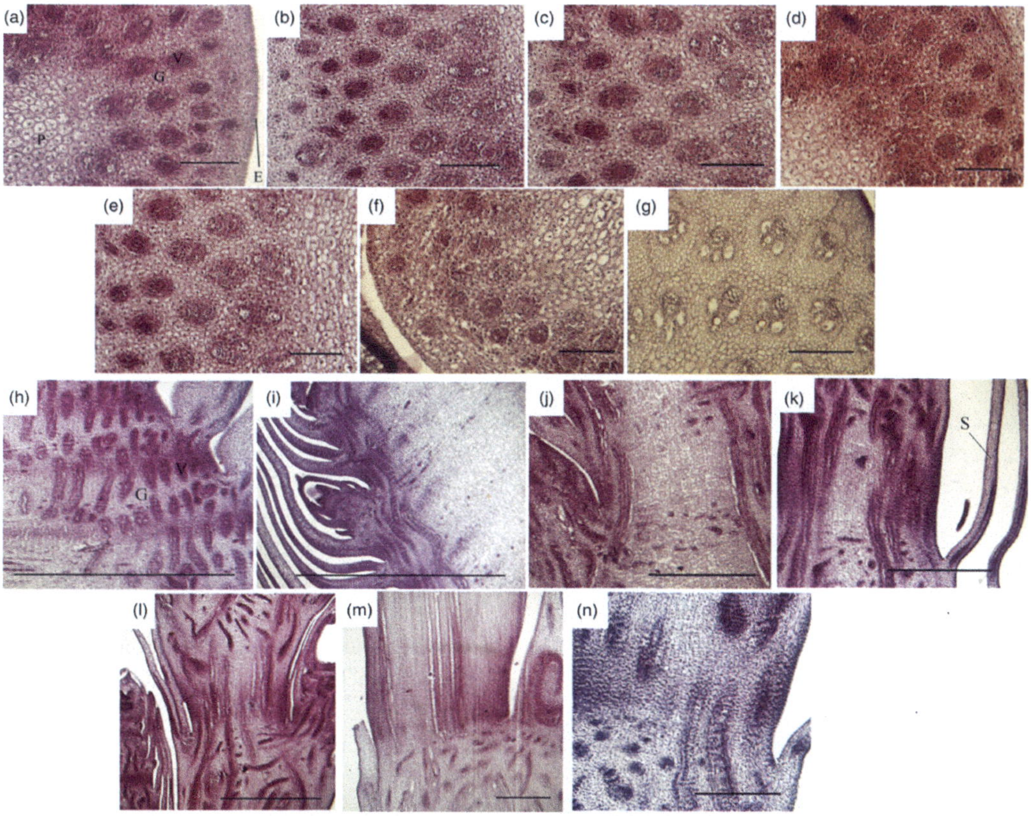

Fig. 3.5 Cross section and longitudinal section for tissue from emergence of shoots (the fifth internode) at seven growth stages (S1–S7). **a–g** Cross section for the fifth internode. Bars, 200 lm. **h–n** Longitudinal section for the fifth internode. Bars, 500 lm. E, G, P, S and V denote epidermis, ground tissues, pith, bamboo sheath and vascular bundles, separately (Li et al. 2018)

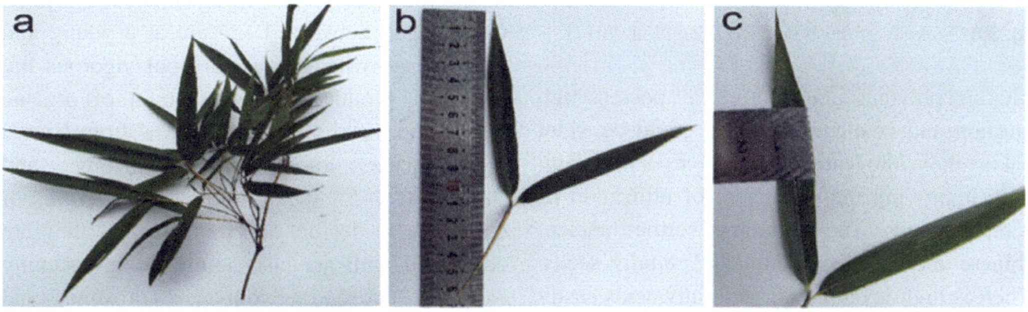

Fig. 3.6 Vegetative leaves of moso bamboo (Photograph courtesy of Yingchun Yu). **a**, Branchlets with leaves; **b**, leaf length; **c**, leaf width

viability upon being harvested (Cai et al. 2008). The germination percentage declined by 14% after 180 days storage treatment at 4 °C (Li et al. 2009).

Moso bamboo seeds belonged to starchy seeds, with high protein content, moderate fat content and rich amino acids. Particularly, seeds of *P. edulis* which contained rich essential amino

Fig. 3.7 Flowering of moso bamboo (Photograph courtesy of Jian Gao)

Fig. 3.8 Spikelet of moso bamboo (Gao et al. 2014)

acids and polyunsaturated fatty acids possess high nutrition and health values. We could consider making them into other food additives or the health food, improving the coefficient of utilization of *P. edulis* seeds. Above findings verified ancient Chinese records about edible *P. edulis* seeds. Whereas high-level unsaturated fatty acids easily trigger seed deterioration, it should be paid more attention in seeds preservation. Overall utilization and processing technology for *P. edulis* seeds are related to seed production and economic benefits. Two-year storage had significant influence on nutrition ingredients and antioxidant enzyme activity in *P. edulis* seeds. Nutrition content of fresh *P. edulis* seeds was significantly higher than that of old seeds. The content of protein, fat, amino acids and fatty acids separately decreased by 1.73, 0.33, 1.6 and 0.32% following storage (Ge et al. 2016). The remarkable decline in antioxidant enzyme activity suggested the reduction of vitality and germination percentage in *P. edulis* seeds following 2-year storage at 4 °C. The scientific and reasonable storage for *P. edulis* seeds requires further research, to provide a foundation for germplasm resource conservation and *P. edulis* genetic breeding realization.

3.8 Bamboo Maturation and Growth Characteristics of Bamboo Stands

Moso bamboo is a perennial herbaceous plant whose culm only has vascular bundles but no cambiums. They stop to grow higher and larger in diameter once they are fully formed and start to extend branches and put out leaves, and thus, their volume will not continue increasing with age. Newly formed culms have a low tissue density, high moisture content and low dry matter content, meaning that the fibrosis of culm wall mostly occurs during their maturation phase. Once culms reach full formation, they have a leaf replacement in the next spring and subsequently replace leaves once every 2 years. With every replacement, their growth age increases by 1°. Culms aged between 1 -3° are at a young and strong stage where they carry out vigorous life activities, produce a large amount of organic matter and the rhizomes they grow from have a high capacity for developing rhizomes and shoots. At 4–5°, the physiological activities of culms are in decline and the rhizomes they are connected with are also aging with weakening capacity for sending out new rhizomes and shoots. Culms aged above 6° lose vitality, their branches and leaves are withered and the rhizomes they have grown from have completely lost the capacity for shooting and developing new rhizomes.

Moso bamboos prefer warm and humid climate. They grow well on thick, fertile and well-drained acid soil and dislike low-lying land with poor drainage. In general, moso bamboo stands

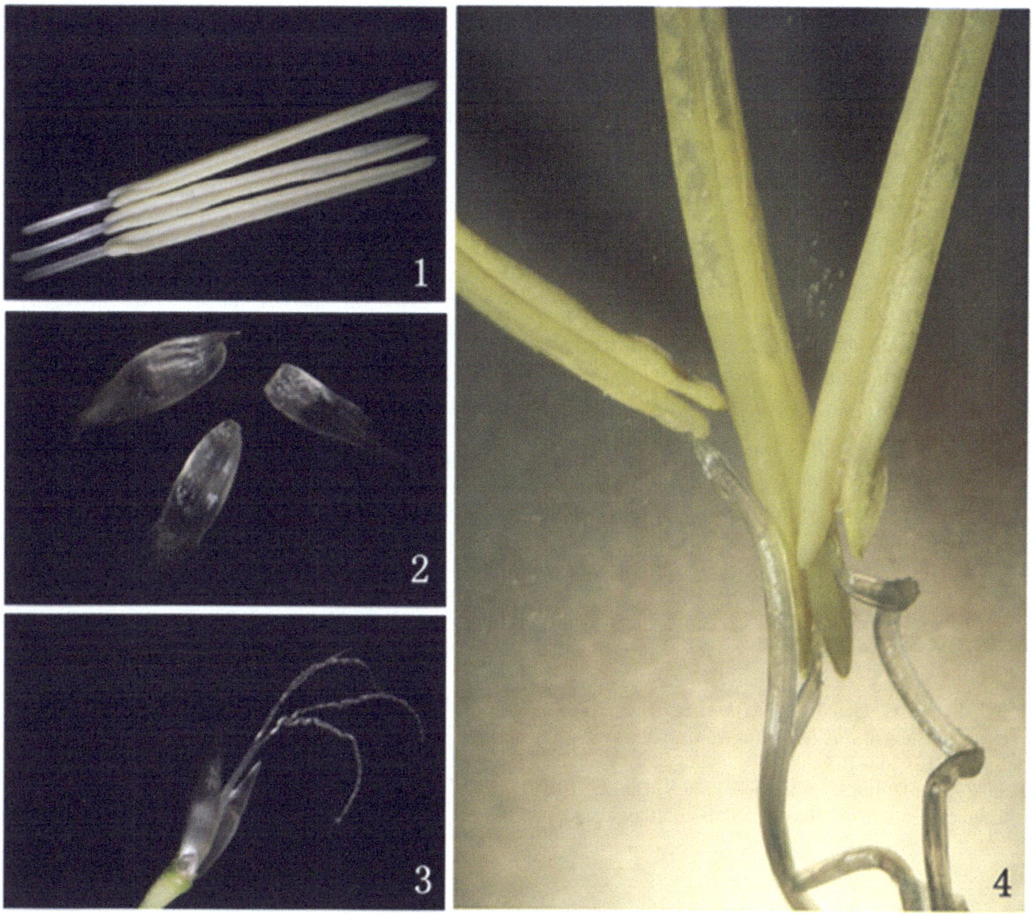

Fig. 3.9 The structure of moso bamboo flower. **1**: stamen (filament and anther); **2**: lodicules; **3**: pistil (style, dehiscence and ovary); **4**: matured stamen (filament and anther) (Gao et al. 2014)

Fig. 3.10 Moso bamboo seeds in shell. (Photograph courtesy of Zhanchao Cheng)

Fig. 3.11 Moso bamboo seeds without shell. (Photograph courtesy of Zhanchao Cheng)

exhibit distinct "on" and "off" years: in an "on" year, there is massive shoot emergence and bamboo formation, which will be followed by an "off" year when culms in the stand replace leaves and develop rhizomes while few shoots or even no shoot sprout out. The "on" year alternates with the "off" year in a cycle of 2 years, but pests, extreme climate events or human disturbance could affect this pattern. A bamboo stand that shows no obvious "on" and "off" years is called a constant stand, meaning that in a given time, some of the culms are in an "on" year while the others are in their "off" year. Constant stands produce a similar number of shoots and mature culms every year, leading to a stable annual yield, yet they also demand more sophisticated management measures.

References

Cai CJ, Peng ZH, Gao J et al (2008) Seed germination characteristics of *Phyllostachys edulis*. Chin Agric Sci Bull 24(12):163–167

Cheng Z, Hou D, Liu J, Li X, Xie L, Ma Y, Gao J (2018) Characterization of moso bamboo (Phyllostachys edulis) Dof transcription factors in floral development and abiotic stress responses. Genome 61(1):151-156

Cheng Z, Hou D, Ge W, Li X, Xie L, Zheng H, Cai M, Liu J, Gao J (2019) Integrated mRNA, MicroRNA Transcriptome and Degradome Analyses Provide Insights into Stamen Development in Moso Bamboo. Plant Cell Physiol 76:87

Gao J, Zhang Y, Zhang C, Li X, Mu S, Peng Z (2014) Characterization of the floral transcriptome of moso bamboo (Phyllostachys edulis) at different flowering developmental stages by transcriptome sequencing and RNA-Seq analysis. PLOS ONE 9(6):e98910. https://doi.org/10.1371/journal.pone.0098910

Ge W, Hu T, Gao J et al (2016) Difference of nutritional components between *Phyllostachys edulis* seeds and main grain crops. Turk J Biochem 41(5):331–337

Ge W, Zhang Y, Cheng Z, Hou D, Li XY, Gao J (2017) Main regulatory pathways, key genes and microRNAs involved in flower formation and development of moso bamboo (Phyllostachys edulis). Plant Biotechnol J 15(1):82–96

Li SZ (1596) Compendium of materia Medica. Chinese Classics Publishing House, Beijing

Li F, Hu M, Li M et al (1978) Into porcelain pillow. People's Literature Publishing House, Beijing

Li N, Jin QY, Peng HZ et al (2009) Effects of storage treatments on seed vigor of *Phyllostachys edulis*. J Bamboo Res 28(2):29–33

Qi F, Peng Z, Hu T, Gao J (2013) Changes of endogenous hormones in different organs during the flowering phase of moso bamboo. Forest Res 26(3):332–336

Moso Bamboo Genome

4

Xiangyu Li and Zhanchao Cheng

Abstract

Woody bamboo ranks among the foremost non-timber forest product across the world with fastest-growing and rarely flower character. The study of bamboo genome has carried out for a decade, bringing great advances in revealing the molecular mechanism underlying bamboo unique trait. The draft genome of *Phyllostachys edulis* (moso bamboo) has been updated to Version 2.0, greatly improve the quality and completeness of moso bamboo genome. The draft genome of *Raddia guianensis*, *Olyra latifolia*, *Bonia amplexicaulis,* and *Guadua angustifolia* issued latterly, taking a sight into the genome polyploidy of bamboos, and revealing the origin of woody bamboo. Comparative genomic research will provide new sight to study the unique traits of different bamboo species in further.

X. Li (✉) · Z. Cheng
Key Laboratory of Bamboo and Rattan Science and Technology, International Center for Bamboo and Rattan, State Forestry and Grassland Administration, Beijing 100102, China
e-mail: leerduo727@163.com

Z. Cheng
e-mail: czc@icbr.ac.cn

J. Gao (ed.), *The Moso Bamboo Genome*, Compendium of Plant Genomes,
https://doi.org/10.1007/978-3-030-80836-5_4

4.1 Introduction

Genome is a complete set of haploid genetic material in cells or organisms. After 1980, the development and successful application of several key technologies made it possible to sequence the whole-genome sequence, such as Sanger sequencing, pulsed-gel electrophoresis to separate large fragments of DNA, in vitro recombination of DNA and PCR technology. The current genomics comprise structural genomics and functional genomics. The former lays a foundation for the latter, which mainly includes three aspects: genetic map, physical map and genome sequencing. Its ultimate goal is to establish high-density genetic map, high-resolution physical map, and transcription map, and finally complete genome sequence determination and annotation. Genetic code is the basis of all biological life. The code in genome implies the origin and evolution of tellurian. Plants form the basis of the food chain that is responsible for maintaining the balance of the carbon cycles. The first sequencing of plant genome was *Arabidopsis* genome and completed in 2000, which had greatly invigorated and accelerated plant research. In recent years, many of plants have been sequenced. Although genome quality and completeness are different, all genomes offer unparalleled resources able to be utilized by diverse means.

There are nearly 1500 bamboo species from 119 genera of Bambusoideae are described worldwide (Guo et al. 2019; Peng et al. 2013). The Bambusoideae subfamily is divided into two categories: herbaceous and woody bamboo varieties (Zhou et al. 2017). Bamboo possesses many uses and efficacy, with great ecological benefit, economic value, and cultural importance. *Phyllostachys edulis* (moso bamboo) has extremely fast growth rates, a strong cloning regeneration ability, and a long vegetative period for decades prior to bloom. The study of moso bamboo genome has been last for years, bring great advances in molecular biology. In this chapter, the genome study of moso bamboo was reviewed together with the genome study of other four bamboo species.

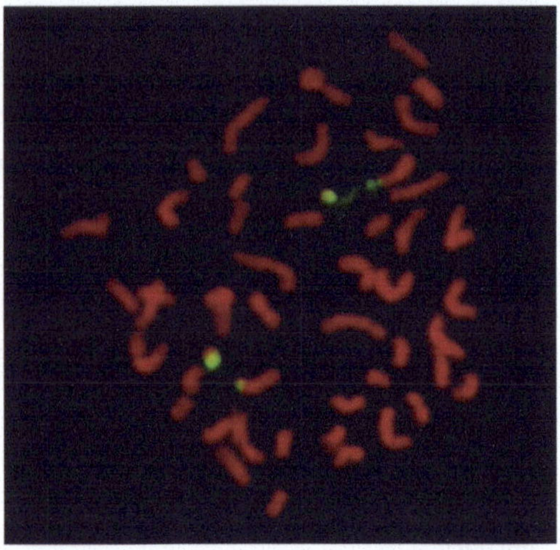

Fig. 4.1 Two copies of chromosome sets of moso bamboo (Peng et al. 2013)

4.2 Moso Bamboo Genome Size

Moso bamboo is allotetraploid woody bamboo. There are 48 chromosomes within moso bamboo (Fig. 4.1). Peng and colleagues (2013), Zhou and coworkers (2017) acquired 4.24 pg 2C DNA for moso bamboo by flow cytometry analysis (Peng et al. 2013; Zhou et al. 2017). It suggested that moso bamboo genome size is 2.075 Gb (Table 4.1) (Peng et al. 2013). Similar results were also reported in (Gui et al. 2007). Because of its peculiar cloning reproduction and rarely sexual reproduction, it seems almost impossible to produce the inbred line for moso bamboo or identify a variety. The study of moso bamboo takes the bamboo ecotype as research object. Peng and colleagues (2013) selected 5 plants from different ecotypes of moso bamboo rhizome to carry out genome-wide shotgun sequencing.

4.3 Moso Bamboo Draft Genome

The first draft genome in moso bamboo was released by International Center for Bamboo and Rattan, and Chinese Academy of Sciences in 2013 (Peng et al. 2013). It was assembled based on overlapping bacterial artificial chromosomes (BAC) clones which stand for the minimal tiling path in covering every chromosome arm. Ultimately, the draft genome of 2.05 Gigabase (Gb) containing Scaffold N50 of 328 Kilobase (Kb) was assembled at scaffolds level (Table 4.2). The first version genome was assembled with 295 Gb short-read sequencing data. In the first version genome, 31,987 protein-coding genes were predicted by RNA and cDNA sequencing data.

In the following years, Chinese researchers still work on the puzzle game of moso bamboo genome assembly. In 2018, a moso bamboo

Table 4.1 Genome sizes of several representative Poaceae plant (Arumuganathan and Earle 1991; Peng et al. 2013; Guo et al. 2019)

Scientific name	Common name	Haploid genome size/Mb
Oryza sativa	Rice	390
Sorghum bicolor	Sorghum	748,772
Triticum aestivum	Wheat	15,966
Saccharum robustum	Sugarcane	3151
Zea mays	Corn	2292–2716
Hordeum vulgare	Barley	4873
Phyllostachys edulis	Moso bamboo	2075
Bonia amplexicaulis		933
Olyra latifolia		692
Raddia guianensis		685
Guadua angustifolia		1775

Table 4.2 Statistics of assembly and annotation for moso bamboo genome (Peng et al. 2013)

Total length	2,051,719,643 bp
N50 length (contigs)	11,882 bp
N50 length (scaffolds)	328,698 bp
N80 length (scaffolds)	62,052 bp
Number of scaffolds (>N80 length)	5499
Largest scaffold	4,869,017 bp
GC content	43.9%
Number of protein-coding genes	31,987
Average length of protein-coding genes	3350 bp
Total size of transposable elements	1,210,862,930 bp
Content of transposable elements	59.0%

reference genome at chromosome level was reported by International Center for Bamboo and Rattan, and BGI Institute of Applied Agriculture (Zhao et al. 2018). For assembling the new version genome, ~603.3 Gb additional genomic data were obtained, and the read length was ranging from 76–250 bp. Different assembly strategies were performed to assemble moso bamboo genome, including WGS assembly and Hi-C assembly (Table 4.3). WGS assembly proceeded with ~154 Gb of newly obtained clean data and ~220 Gb clean data acquired in Peng's study. The WGS assembly strategies generated 1.91 Gb genome data with scaffold N50 reaching to 894 Kb. The WGS assembly obviously improved the statistics and quality of first version genome. Then, Hi-C assembly strategies were combined with draft assemblies obtained by WGS assembly strategies. A chromosome-level reference genome was generated, which was 1.91 Gb genome data with a scaffold N50 of 79.90 Mb, approximately 243 times longer than the first version (Fig. 4.2) (Zhao et al. 2018). There are 51,074 protein-coding genes that were predicted in this new version genome, in which 37,698 genes are annotated to 17,675 gene families. Moreover, around 3291 genes are coding transcription factors,

Table 4.3 Statistics of assembly for moso bamboo genome by various sequencing data (Zhao et al. 2018)

Statistics	WGS assembly		Hi-C assembly	
	Scaffold	Contig	Scaffold	Contig
Total number	19,285	76,900	19,684	84,758
Genome size (bp)	1,908,074,089	1,795,528,836	1,907,603,590	1,795,510,437
Gap number (bp)	112,545,253	0	112,093,153	0
Average length (bp)	98,940.84	23,348.88	96,911.38	21,183.96
N50 length (bp)	894,858	54,955	79,898,979	53,293
N90 length (bp)	115,487	11,757	44,603,463	10,445
Maximum length (bp)	5,406,526	738,589	137,299,170	738,589
Minimum length (bp)	926	157	318	1
GC content (%)	44.2	44.2	44.2	44.2

composing approximately 6.44% of protein-coding genes. 3866 non-coding RNA genes loci were predicted in new version genome, in which 349 conserved miRNA loci were found. All the information provides a reliable resource for molecular study of moso bamboo. Moreover, by massive transcriptomic sequencing, Zhao et al. provided an all-around AS profile after identifying 266,711 specific AS events in altogether 25,225 AS genes within 26 individual bamboo tissue samples through the Pacific Biosciences and Illumina sequencing platforms. Zhao and coworkers compared AS genes and orthologous ones of corresponding plant species; as a result, AS genes mostly distributed in genes with higher conservation levels were most likely to accumulate increased transcript abundances with lower tissue specificity (Fig. 4.3).

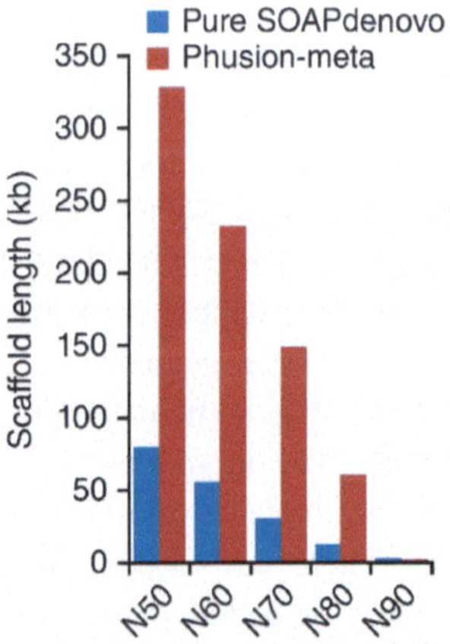

Fig. 4.2 Comparison for assembled scaffold length between pure SOAPdenovo method and Phusion-meta assembly method (Peng et al. 2013)

4.4 The Genome Database of Bamboos

With the development of bamboo research, the genome of bamboo was released. The Genome Database of Bamboos (BambooGDB) maintains by International Centre for Bamboo and Rattan (ICBR) aim to collect and display the bamboo genome resource and provide a website for users to download and align (Zhao et al. 2014; Zhao et al. 2018) (http://www.bamboogdb.org/#/). Lots of bamboo genomes will be upload and some powerful tools will be added with the Announcing the Genome Atlas of Bamboo and Rattan (GABR) program move award. Additionally, another genome database (http://forestry.fafu.edu.cn/db/PhePacBio/) of moso bamboo was established by Gu et al. (2017). In this genome database site, the raw sequence data of PacBio were collected from SRP093919 record, whereas the rest high throughput datasets of Illumina were obtained from GSM2405908-GSM2405916 records. RNA-seq was used in conjunction with PAS-seq to identify and quantify alternative polyadenylation (APA) and AS within the rhizome system in the whole genome.

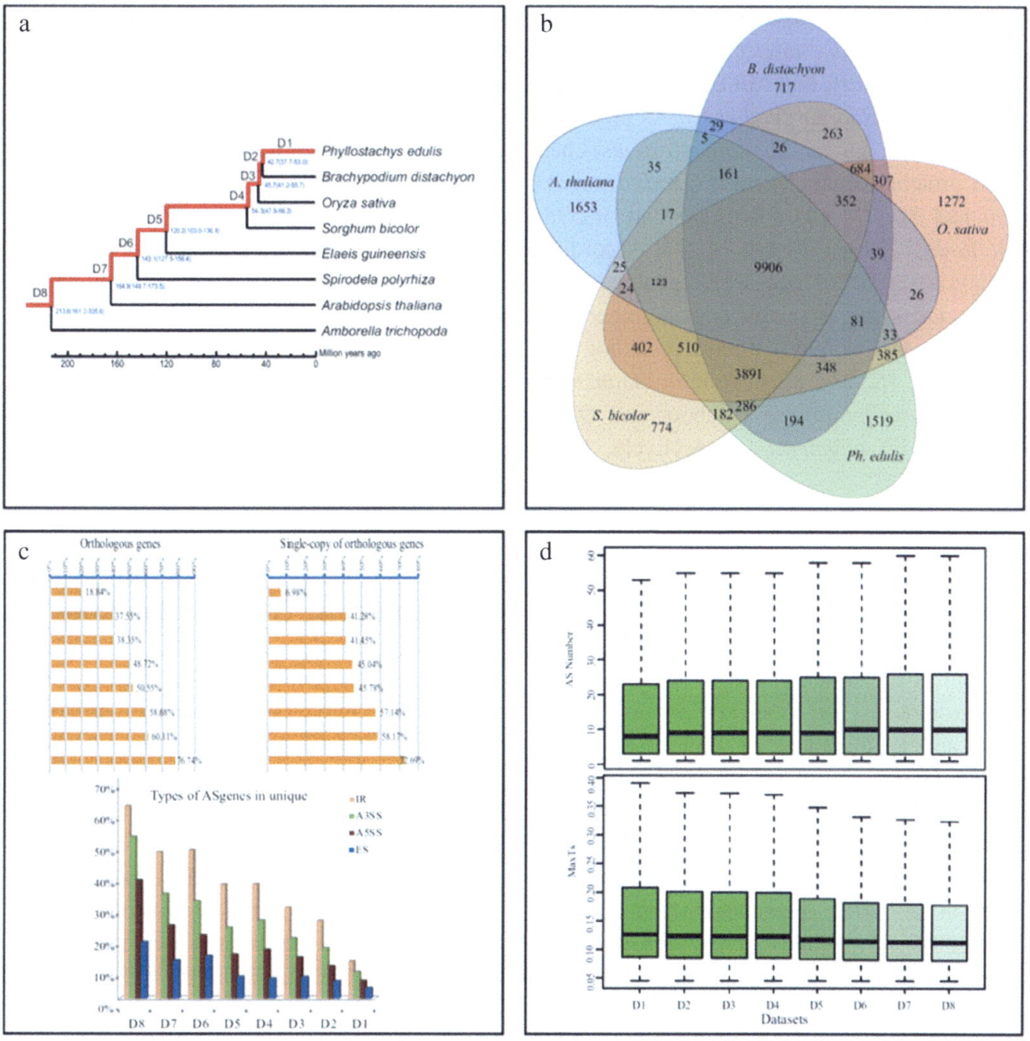

Fig. 4.3 Plant variety evolution across bamboo (Zhao et al. 2018). **a** phylogenetic relations among *Arabidopsis thaliana*, *Elaeis guineensis*, *Amborella trichopoda*, *Oryza sativa*, *Brachypodium distachyon*, *Spirodela polyrhiza*, *Ph. Edulis*, and *Sorghum bicolor*. Red line indicated the phylogenetic tree in those 8 selected plant varieties whose branches evolved into bamboo. The signal indicated those 8 specific orthologous gene datasets (D8-D1) selected in this work. **b** presents a Venn diagram regarding orthologous genes in those 8 plant varieties. **c** AS proportions and AS types in 2 out of those 8 orthologous gene datasets. **d** D8-D1 display growing AS abundance and diminishing tissues specificity

4.5 The Draft Genome of Other Bamboos

In recent study, Guo et al. (2019) studied two diploid herbaceous bamboo species' draft genomes (*Olyra latifolia*, 2n = 2x = 22; *Raddia guianensis*, 2n = 2x = 22), a neotropical tetraploid woody bamboo (*Guadua angustifolia*, 2n = 4x = 46), as well as a palaeotropical hexaploid woody bamboo (*Bonia amplexicaulis*, 2n = 6x = 72). The genome size of *Bonia amplexicaulis*, *Guadua angustifolia*, *Olyra latifolia*, and *Raddia guianensis* are 933 Mb, 1775 Mb, 692 Mb, and 685 Mb, respectively, estimated by flow cytometry analysis. In the draft genome sequences assembled, there were 1614 Mb of *Guadua angustifolia*, 848 Mb of *Bonia amplexicaulis*, 646 Mb of *Olyra latifolia*, and 626 Mb of *Raddia guianensis*. As predicted by genome annotations, *Bonia amplexicaulis* had 47,056 protein-encoding genes, *Guadua angustifolia* had 38,575, *Olyra latifolia* had 36,578, while *Raddia guianensis* had 24,275. According to analysis results of rice whole-genome gene synteny, the syntenic region copy number was verified by the bamboo ploidy, including the ratios of 1:3, 1:2, and 1:1 in *Raddia guianensis*, *Guadua angustifolia*/moso bamboo, and *Olyra latifolia*/*Raddia guianensis*, respectively, (Guo et al. 2019).

4.6 The Genome Polyploidy of Woody Bamboos

Polyploid is the result of the fusion of two or more chromosome sets (one via every ancestor or parent) and is ubiquitous in plants worldwide. Polyploidy represents a vital characteristic during plant species evolution, which has been identified to be the main force driving plant evolution closely associated with speciation and diversity (Bento et al. 2011; Fernández-Mazuecos and Glover 2017). Polyploidy generates the higher complex genotype and phenotype in bamboos. According to the distribution of bamboo species and the amount of basic sets for bamboo chromosomes, bamboo is further divided into temperate tetraploid woody bamboo (2n = 46–48), diploid herbaceous bamboo (2n = 20–24), palaeotropical hexaploid woody bamboo (2n = 70–72), and neotropical tetraploid woody bamboo (2n = 40–48).

The whole-genome duplication of Bambusoideae was studied in Guo et al.'s (2019) paper by analyzing the perfect syntenic orthologous genes, whose copies are in accord with ploidy levels. Progenitor genomes in woody bamboo are separated from herbaceous bamboos at ca. 42 million years ago (Ma) and then divided as 4 different lineages (A-D) at ca. about 32 Ma. During the independent evolution of around 10 million years, hybridization of ancestral B/C and C/D genomes occurred for generating 2 allotetraploid lineages, namely neotropical woody bamboo lineage: *Guadua angustifolia* (BBCC) and temperate tetraploid woody bamboo lineage: moso bamboo (CCDD). The palaeotropical hexaploid woody bamboo (*Bonia amplexicaulis*, AABBCC) was formed by the hybridization of ancestral A and B + C genomes about 19 Ma. C subgenome exists in overall extant woody bamboos and experiences more intrachromosome rearrangements and inversion than A and B (Guo et al. 2019), which may suggest that C subgenome could be critical for the origin of woody bamboos.

4.7 The Genome Clue to the Unique Biological Traits of Bamboos

The species differentiation of bamboos is driven by genome polyploidy. Genome polyploidy provides genes having new functions, which may generate genetic, and phenotypic novelty. In woody bamboo, absolute copy number of genes involved in cellulose and lignin biosynthesis pathways grew in polyploid woody bamboo species than diploid herbaceous species. Duplication for majority cellulose- and lignin-related genes had a Ks distance of 0.1–0.2, related to woody bamboo polyploidization, indicating the correlation between polyploidization and extra copies and woody trait origin of bamboo (Guo et al. 2019).

The orthologs of *OsSOC1* is an important gene for the activation and/or repression of flowering in dicots and monocots (Andersen et al. 2004; Tadege et al. 2003). The gene structures of *OsSOC1* orthologs are different in the woody and herbaceous bamboos. The *OsSOC1* orthologs displayed the same single copy number and identical gene structure as in rice. There were two *OsSOC1* orthologs found in *Bonia amplexicaulis*, *Guadua angustifolia*, and moso bamboo. The genovariation of 5' untranslated regions of these orthologs leads to the loss of MADS-box domains in coding proteins.

Changes of *OsSOC1* orthologs possibly decided the very long vegetative growth phase of woody bamboo (Guo et al. 2019).

References

Andersen CH, Jensen CS, Petersen K (2004) Similar genetic switch systems might integrate the floral inductive pathways in dicots and monocots. Trends Plant Sci 9(3):105–107

Arumuganathan K, Earle ED (1991) Nuclear DNA content of some important plant species. Plant Mol Biol Report 9(3):208–218

Bento M, Gustafson JP, Viegas W et al (2011) Size matters in Triticeae polyploids: larger genomes have higher remodeling. Genome 54(3):175–183

Fernández-Mazuecos M, Glover BJ (2017) The evo-devo of plant speciation. Nature Ecology & Evolution 1 (4):0110

Gui Y, Wang S, Quan L et al (2007) Genome size and sequence composition of moso bamboo: a comparative study. Sci China, Ser C Life Sci 50(5):700–705

Guo Z-H, Ma P-F, Yang G-Q et al (2019) Genome sequences provide insights into the reticulate origin and unique traits of woody bamboos. Mol Plant 12(10):1353–1365

Peng Z, Lu Y, Li L et al (2013) The draft genome of the fast-growing non-timber forest species moso bamboo (Phyllostachys heterocycla). Nat Genet 45(4):456

Tadege M, Sheldon CC, Helliwell CA et al (2003) Reciprocal control of flowering time by OsSOC1 in transgenic Arabidopsis and by FLC in transgenic rice. Plant Biotechnol J 1(5):361–369

Zhao H, Peng Z, Fei B, et al (2014) BambooGDB: a bamboo genome database with functional annotation and an analysis platform. Database j biolo databases curation bau006

Zhao H, Gao Z, Wang L, et al (2018) Chromosome-level reference genome and alternative splicing atlas of moso bamboo (moso bamboo). GigaScience 7(10), giy115

Zhou M, Xu C, Shen L et al (2017) Evolution of genome sizes in Chinese Bambusoideae (Poaceae) in relation to karyotype. Trees 31(1):41–48

Moso Bamboo Transposon

5

Dan Hou

Abstract

Transposable elements (TEs, also called transposons and jumping genes) represent the congenital genetic elements moving around higher plant genomes, which are generally identified to be the hallmarks of evolution. It is still unclear about their exact activities in genome, and TEs have been suggested to spontaneously produce genomic variations through transposition. In the latest studies on molecular diagnosis, TEs have been applied as the tools to investigate evolution, genomic variations, silencing of target gene, and gene tagging, and they contribute to crop improvement. TE levels in bamboos are related to 6 bamboo genome sizes, which suggests that TE proliferation potentially serves as a main factor to derive the evolution of bamboo species. TEs occupy 63.3% in the genome of moso bamboo (*Phyllostachys edulis*). Typically, TE insertions can be usually detected within flanking regions in 26,366 genes, accounting for 51.6% of all annotated bamboo genome genes. LTR retroelements are located far away from genes, which display potent transcriptional activity. On the other hand, retrotransposons take up 45.7% of the whole genome, and DNA transposons rank the second place (10.4%), while the rest 7.2% remain unclassified yet. 43.9% of retroelements are identified as the long terminal repeat elements (LTREs), 1.7% as the long interspersed nuclear elements (LINEs), whereas 0.03% as the short interspersed nuclear elements (SINEs). Repeat insertions have been reported previously to be detected within the regulatory or genic regions in many *FPI*- or *CONSTANS* (*CO*)-encoding homologs, thus inducing the down-regulated gene levels within the floral tissues. Downstream flowering genes that carry the TE deletions are up-regulated compared with those carrying TE insertions. It suggests that the TEs can affect target gene expression within the genome of moso bamboo and might be involved in bamboo flowering time regulation. Additionally, a LTR transposon, PHRE5's copies, and insertion polymorphism were analyzed in eight *Ph. edulis* cultivars. Compared with moso bamboo, the copy number was increased in these cultivars, which was verified by the polymorphism. In summary, TEs can serve as the strong regulating factors for moso bamboo genome genes with significantly different epigenetic and genetic structures. Since many TEs can be controlled by specific environmental stimuli, they can be adopted as the attractive agents for crop breeding and used

D. Hou (✉)
State Key Laboratory of Subtropical Silviculture, Zhejiang Agricultural and Forestry University, No. 666, Wu Su street, Hangzhou, Zhejiang 311300, China
e-mail: 20184007@zafu.edu.cn

J. Gao (ed.), *The Moso Bamboo Genome*, Compendium of Plant Genomes,
https://doi.org/10.1007/978-3-030-80836-5_5

for bamboo genome manipulation and for extending the genetic variation pool. Further knowledge on TEs can shed more lights on bamboo genetic improvement.

5.1 Introduction

Transposable elements (TEs) have been identified as the DNA fragments with the ability to migrate to diverse positions in the single-cell genome. TEs can be divided into 2 main types according to the transposition mechanism. Of them, retroelements (class I) adopt the RNA intermediate to generate DNA copy under the action of a reverse transcriptase and to insert into host genome under the action of an integrase; on the other hand, DNA transposons (class II) can directly migrate, which catalyzes the joining and cutting of DNA under the action of a replicase or a transposase (Feschotte et al. 2002). TEs take up 63.3% in the genome of moso bamboo (Zhao et al. 2018), close to that of 62% in sorghum genome (Paterson et al. 2009), but slightly decreased compared with that of 85% in maize genome (Schnable et al. 2009). Nonetheless, TEs occupy an increased proportion in the genome of moso bamboo than that in rice (40%) (Goff et al. 2002) as well as in *Brachypodium* (28%) (Initiative, 2010).

As reported by Zhao and coworkers (2018), TEs were completely searched within the genome of moso bamboo. As a result, a high TE insertion frequency was detected surrounding flanking regions in 26,366 genes, accounting for 51.6% of all annotated genes within the genome of moso bamboo. As for TE insertions surrounding the bamboo genes, their overall length was about 46 Mb, and genes with high conservation degrees were related to a greater number of TEs. It was interesting that whole-genome analysis indicated the far distance between LTR retroelements and genes, which showed markedly potent transcriptional activity. Furthermore, for TE insertions, the divergence rate was predicted to be 30%, indicating their ancestral origin. Across the diverse TEs classes detected from the genome of moso bamboo, retrotransposons occupied 45.7% of whole genome, followed by DNA transposons (10.4%), while the rest 7.2% remained undefined so far. In the meantime, 43.9% of all retroelements were LTREs, 1.7% were LINEs, whereas 0.03% were SINEs. In addition, for 31,987 genes in moso bamboo genome, about 85.7% of their promoter regions within the 2000 bp upstream sequences were located in TE sequences. Besides, 3 kinds of TEs, including miniature inverted-repeat elements (MITEs), hAT-like transposons, and SINEs tend to have insertions within promoter regions (Zhou et al. 2018b).

5.2 Long Terminal Repeat Retroelements (LTR Retroelements) from Moso Bamboo

LTR retroelements represent the single greatest components in the genomes of many plants, which have great influences on genome from different aspects. The structural method was used combined with homology-based methods to originally analyze 982 LTR-retroelement families that comprised 2,004,644 related sequences, taking up over 40% of genomic genes in moso bamboo (Table 5.1, Zhou et al. 2017b). As indicated by additional analyzes, the solo LTRs-to-intact elements (S/I) ratio (about 0.28:1) markedly decreased in moso bamboo, which suggested the potentially lower illegitimate/unequal recombination frequencies in bamboo LTR retroelements. Further, according to phylogenetic analysis, 5 Ty3-gypsy and 4 Ty1-copia evolutionary lineages existed prior to the divergence of monocot and eudicot species, yet there were markedly different time frames and scales in proliferation among different lineages and families. The insertion time was also estimated, which discovered the application of LTR retroelements for about 0–3 million years, and they had prolonged activity time period compared with *Arabidopsis* and rice. The above results indicated that LTR retroelements levels were related to the great genome size in the host during moso bamboo evolution.

Table 5.1 LTR-retroelements distribution within the genome of moso bamboo (Zhou et al. 2017b)

Superfamily	Lineage	Family[a]	Structure	Number	Total length (bp)	Percentage of sequences (100%[b])
Ty1-copia	Tork	48	Gag-PR-INT-RT-RH	132,836	57,946,850	2.82
	Retrofit	249	Gag-PR-INT-RT-RH	330,726	108,542,148	5.29
	Sire	60	Gag-PR-INT-RT-RH-env	235,066	125,428,928	6.11
	Oryco	101	Gag-PR-INT-RT-RH	130,970	45,135,364	2.2
	Total	458		829,598	337,053,290	16.43
Ty3-gypsy	Del	63	Gag-PR-RT-RH-INT-CHR	443,394	195,466,907	9.53
	Reina	181	Gag-PR-RT-RH-INT-CHR	117,128	36,722,443	1.79
	Crm	71	Gag-PR-RT-RH-INT	158,181	49,677,331	2.42
	Tat	186	Gag-PR-RT-RH-INT	406,315	198,348,797	9.67
	Total	501		1,125,018	480,215,478	23.41
Unknown		23		50,028	10,618,889	0.52
Sum		982		2,004,644	827,887,657	40.35

[a]Family number in the lineage
[b]LTR retroelement proportion within the genome of moso bamboo (2,051,719,643 bp)

A previous study identifies an intact LTR retrotransposon with full length from the genome of moso bamboo, which is named PHRE2 (*Phyllostachys edulis* retrotransposon 2) (Zhou et al. 2018a). There are diverse domains essential for transposition in PHRE2, including pr, gag, rt, int, and rh, whose coding frames are also intact with no distinct mutation. Besides, there is one prime binding site (PBS) in PHRE2, along with two characteristic LTR retrotransposon sequences and one polypurine tract (PPT). As revealed by the whole-genome scan, the genome of moso bamboo contains just a single full-length PHYRE2 sequence. When it is transferred into *Arabidopsis thaliana*, T3 plants have a greater PHRE2 copy number than T2 plants. Following growing within tissue culture or plant hormone/irradiation treatment, the PHRE2 copy number markedly elevated in the seedlings of moso bamboo. The above results suggested the possibility of PHRE2 transposition triggered by environmental factors.

5.3 MITEs from Moso Bamboo

MITEs have been extensively presented within plants and animals. Numerous transposable elements are contained within the genome of moso bamboo. 362 MITE families involving altogether 489,592 MITE-associated sequences have been previously discovered in moso bamboo, which takes up 4.74% of its genome (Table 5.2, Zhou et al. 2016b). These 362 identified MITE families can be divided as 6 known and 1 novel superfamilies. According to our results, MITEs in moso bamboo tended to exist within or around genes related to regulating gene expression in the host. Three out of those 7 superfamilies experienced 2 main expansion events at 8–11 and 22–28 million years ago (MYA), separately;

Table 5.2 Classification and statistics of MITEs from the genome of moso bamboo (Zhou et al. 2016b)

Superfamily	Family	Intact MITEs[a]	Approx. Size (bp)	Truanted MITEs[b]	Intact/truanted (%)[c]	Length of MITEs (bp)[d]	Content (100%)[e]	TSD (bp)	TIR (bp)
Stowaway like	31(PhSt1-PhSt31)	7221	100–330	16,811	43	3,967,317	0.19	TA	9–26
Tourist-like	57(PhTo1-PhTo57)	19,298	90–880	37,123	52	12,483,078	0.61	TWA	8–22
hAT-like	79(PhhA1-PhhA79)	16,928	90–890	44,090	38	13,424,296	0.65	2–10	8–18
Micron-like	3(Phmi1-Phmi3)	1976	120–400	5164	38	955,469	0.05	2	8–14
Mutator-like	43(PhMu1-PhMu43)	8977	100–570	29,254	31	6,736,386	0.33	2–9	9–18
CACTAlike	1(PhCA1)	110	600–660	306	36	143,024	0.01	2–3	57
Unknown	148(PhUn1-PhUn148)	75,158	100–780	227,176	33	59,570,270	2.91	2–9	8–47
Total	362	129,668	90–890	359,924	36	97,279,840	4.74		

[a]Intact MITEs number with complete TIRs
[b]Truanted MITEs number with no complete TIRs
[c]Ratio of intact MITEs proportion in truanted MITEs number
[d]MITEs length (bp)
[e]MITEs proportion within the genome of moso bamboo

meanwhile, 2 underwent main expansion events at 6–13 MYA. Nearly 1/3 of the small RNAs were possibly originated from MITE sequences. In addition, certain MITE families mostly produced small RNAs from terminals, whereas the remaining ones were mostly from central region.

MITEs have great copy number, MITEs are preferentially inserted in gene regions, most miRNAs and siRNAs originate from MITE sequences; as a result, MITEs are possibly associated with bamboo differentiation and genomic diversity. The PhTst-3 (*Phyllostachys edulis* Tourist-like element 3), one of the Tourist-like MITE family members with high conservation degree, was investigated in moso bamboo (Zhou et al. 2017a). Amplification polymorphisms were detected in the PhTst-3 copy (PhTst-3–55), and it was localized within the PH01002699G0010 intron. PhTst-3–55 carrying insertion significantly affected gene expression, as suggested through fluorescence qRT-PCR. Compared with PhTst-3–55 insertion, non-PhTst-3–55 insertion increased the gene level by 4 folds. Such result indicated that PhTst-3, which is localized within the intron, participated in regulating gene expression. For examining how PhTst-3–55 affected the surrounding genes, PhTst-3–55 insertion was completed between 2 promoter sequences in the pxk7S2D promoter analysis vector. As suggested by transient expression regulated by *Agrobacterium*, PhTst-3–55 insertion down-regulated GUS (upstream) and GFP (downstream) gene expression, respectively. Therefore, PhTst-3–55 possibly had knockdown effect through the bidirectional inhibition of gene expression.

5.4 Transposable Mariner-Like Elements (MLE) in Moso Bamboo

In moso bamboo, DNA transposons have been extensively investigated across TEs (Zhong et al. 2010; Zhou et al. 2010a, b, 2011, 2012). DNA transposons can be divided as 3 superfamilies, including MLEs, Pong-like elements, and P instability factor (PIF)-like elements. Typically, *Ppmar1* has the length of 3472 base pairs (bp), its perfect inverted terminal repeats (ITRs) are

28 bp, while the open reading frame (ORF) encoding transposase is 1986 bp; by contrast, *Ppmar2* has the length of 2679 bp, its perfect ITRs are 27 bp, while its ORF encoding transposase is 1877 bp. For the sake of investigating the above two transposons for their transposition activities, a heterologous transposition system was prepared in *Arabidopsis thaliana*. As a result, transposition occurred in the above two transposons of *A. thaliana*. Removing Ppmar1 or Ppmar2 was generally unsuitable because 1–4 bp lengths of these two transposons were located at the excision sites. Ppmar1/Ppmar2 reinsertions were observed in the TA dinucleotides, with the preferential insertion in TA-rich regions. Besides, disperse and unconnected insertion sites were found. Here, the 2 active transposons in moso bamboo may be used for constructing the mutant bamboo libraries and for gene tagging based on additional plant transposons (Fig. 5.1, Zhou et al. 2016a).

Ppmar1 and *Ppmar2* have the same representative duplications at TA/AT target site and representative ITR consensus sequences in plant MLEs (5′-CTC CCT CCR T-3′, with R being A/G). According to phylogenetic analysis, *Ppmar1* is in clade A2 of plant MLE subfamilies, while *Ppmar2* is in clade C. As for those 2 encoded transposases, they possess one DD39D catalytic domain and complete DNA-binding motifs; besides, most residues important for transposase activity are conserved within these two transposons. These findings suggest that the two transposons may be active naturally. Cloning these 2 full-length MLEs can lay certain basis for investigating MLEs transposition activities in moso bamboo. Furthermore, the above 2 transposons contain the nuclear export signal (NES) domain, with no nuclear localization signal (NLS). NES exerts a vital role in regulating the transposase nuclear export within the Ppmar elements, whereas NES domain mutations may promote or suppress export signal transduction (Fig. 5.2, Ramakrishnan et al. 2019). As a result, NESs possibly modulate MLEs transposition activities in moso bamboo for maintaining their genome integrity.

5.5 Transposable Elements and Moso Bamboo Flowering

TF genes, which are the OsMADS1/2/3/14 homologues of rice, are found to participate in the floral meristem identity (FMI), a process converting vegetative meristems into the flowering state. Nonetheless, genes are related to the specific flowering-promoting pathways (like genes associated with gibberellins, photoperiod, light-quality, and ambient-temperature pathways) as well as the floral pathway integrator (FPI) genes showed high expression levels within such bamboo floral tissues (Peng et al. 2013). Repeat insertions can be detected within the regulatory or genic region in many homologs that encoded *FPI* and *CO* genes, giving rise to their down-regulation within floral tissues. *FPI* and *CO* genes form an important network connecting the flowering-promoting pathways with FMI involved in the flowering gene network. Down-regulation of *FPI* and *CO* genes, whereas up-regulation of genes related to FMI revealed that FMI was activated mostly based on such flowering-promoting pathways. To investigate whether the transposable elements are involved in moso bamboo flowering regulation, we chose those promoter sequences in 4 FPI gene homologs harboring transposons in later experiments (Figs. 5.3, 5.4, 5.5, and 5.6, Zhou et al. 2018b). According to results, relative to TE insertions, those downstream flowering genes were up-regulated in the presence of TE deletions. It suggests that the TEs can affect target gene expression within the genome of moso bamboo and might be involved in bamboo flowering time regulation.

5.6 Transposable Elements and Moso Bamboo Variants

Moso bamboo generates many variants during the cultivation, such as the stem shape, stem color, leaf color, and so on, but these variants cannot be inherited steadily and will have back mutation when the environment changes. A LTR

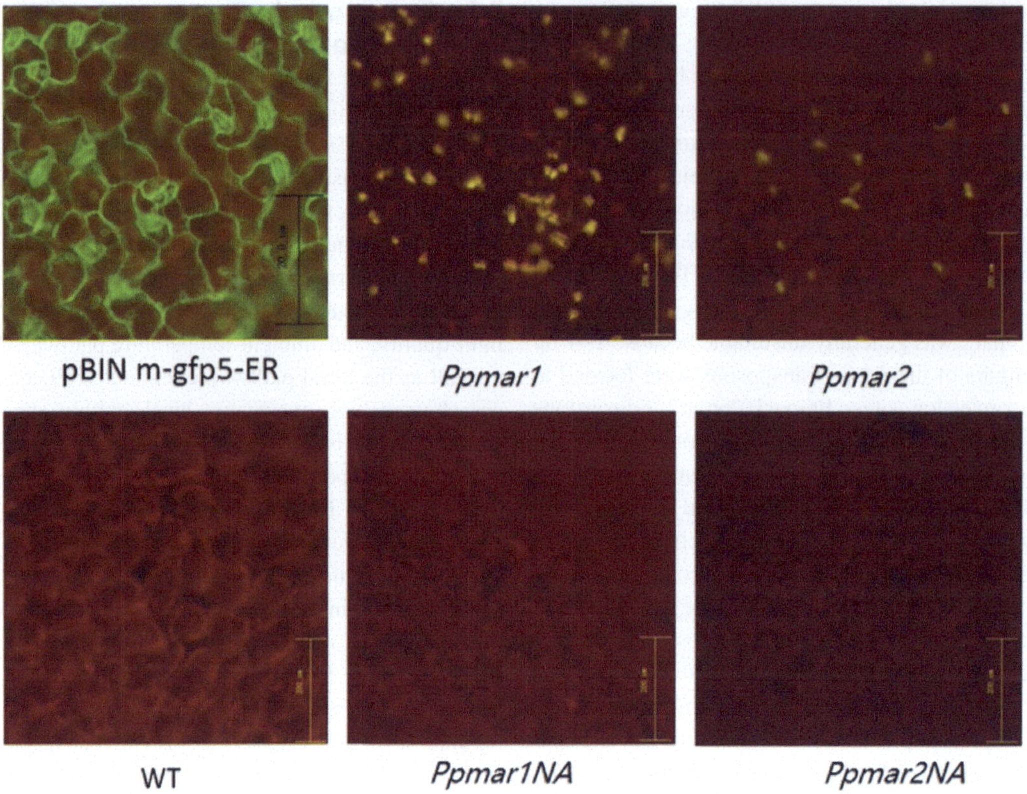

Fig. 5.1 Observation of green and red fluorescence signals by GFP and chlorophyll in Arabidopsis leaves. pBinmgfp5-ER: Arabidopsis leaves exposed to pBinmgfp5-ER transfection; Ppmarl: Arabidopsis leaves exposed to Ppamrl transfection in pBINm-gfp5-ER; Ppmar2: Arabidopsis leaves exposed to Ppmar2 transfection in pBINm-gfp5-ER; PpmarNA1: Arabidopsis leaves exposed to PpmarNA1 transfection in pBINm-gfp5-ER; PpmarNA2: Arabidopsis leaves exposed to PpmarNA2 transfection in pBINmgfp5-ER; WT: wild-type Arabidopsis (Zhou et al. 2016a)

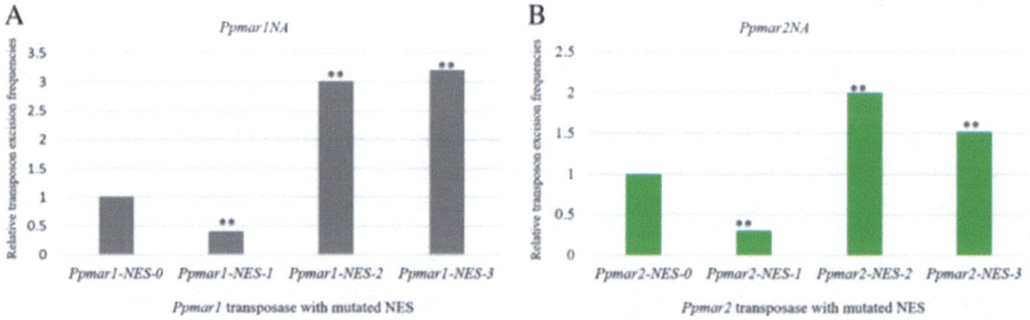

Fig. 5.2 **a** Ppmar1NA transposition frequency under the catalysis of Ppmarl transposase in the presence of NES mutation. **b** Ppmar2NA transposition frequency under the catalysis of Ppmar2 transposase in the presence of NES mutation. Y-axis stands for the ratio of average excision frequency under the catalysis of Ppmarl/Ppmar2 transposase mutants from 6 individual assays, to that under the catalysis of WT Ppmarl/Ppmar2 transposases. ** $P < 0.01$ (Ramakrishnan et al. 2019)

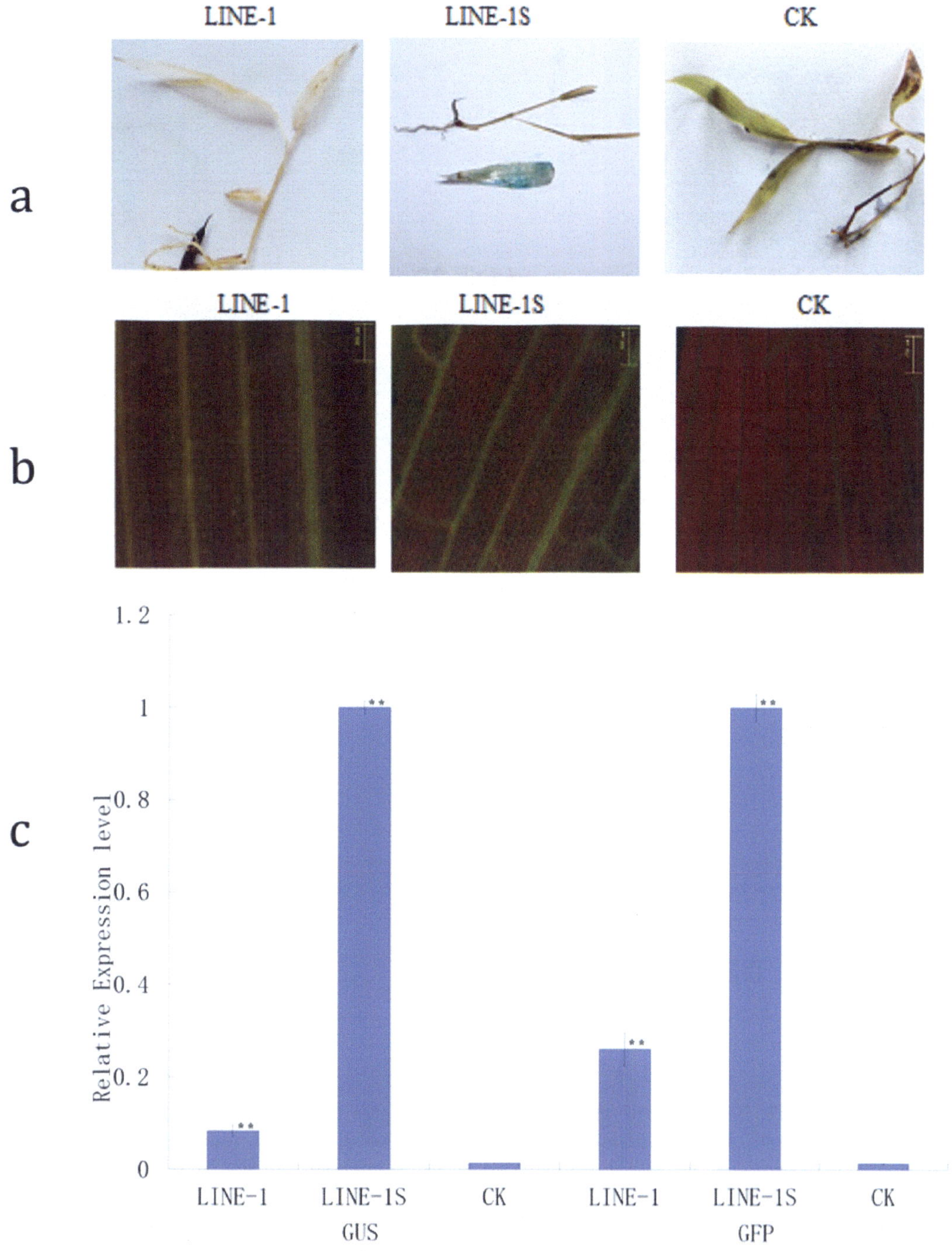

Fig. 5.3 Role of TE insertions into FPI homolog promoter sequences within PH01000216 in downstream reporter gene expression. **a** Observation under GUS staining, **b** detection of green fluorescence signal, **c** relative GUS/GFP gene expression levels. LINE-1S, promoted by a truncated Pro-Line1 sequence in the absence of LINE-like sequence. LINE-1, promoted by a complete Pro-Line1 sequence. CK, absence of promoter sequence. $*p < 0.05$, $**p < 0.001$ (Zhou et al. 2018b)

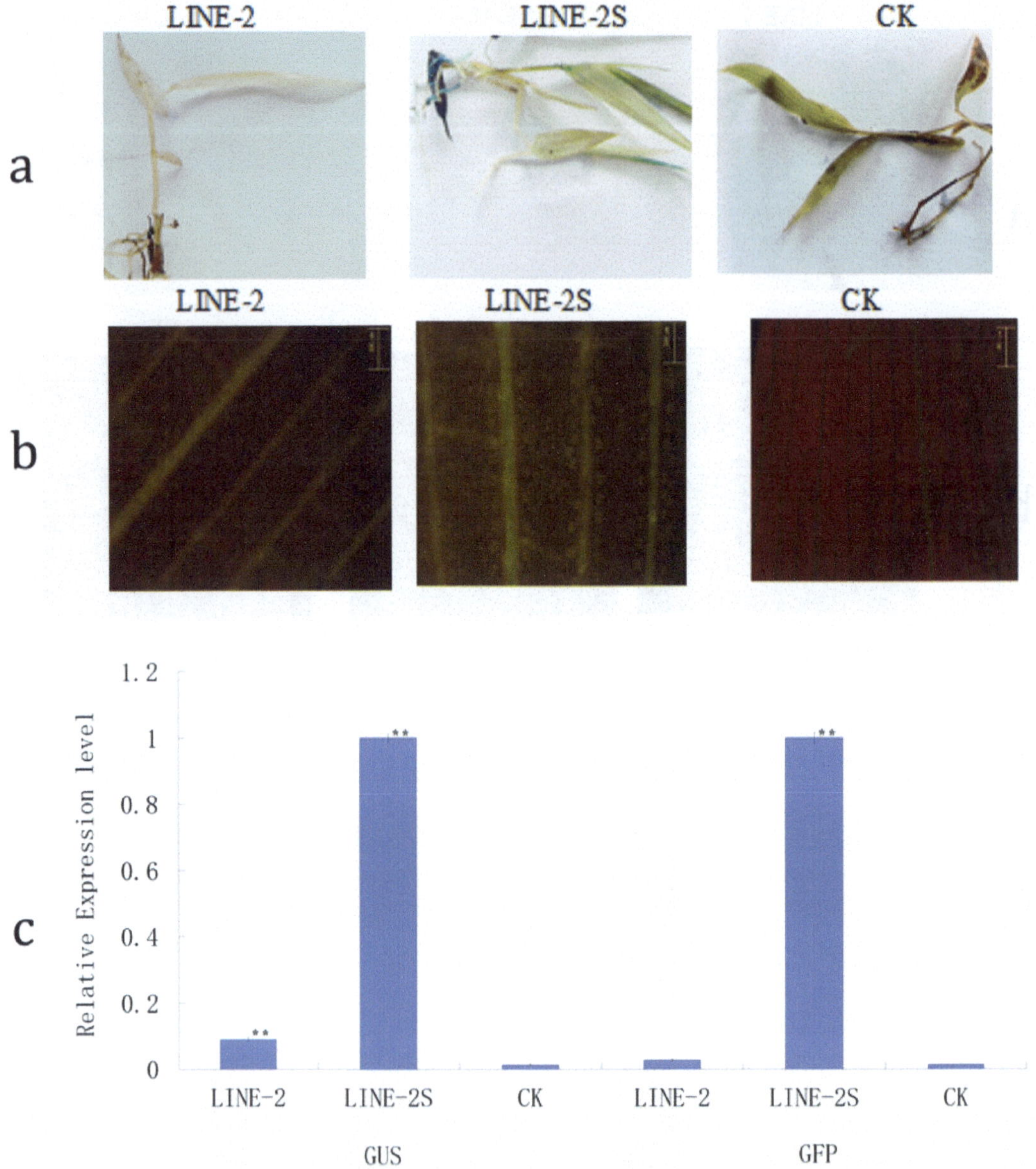

Fig. 5.4 Role of TE insertions into FPI homolog promoter sequences within PH01000265 in downstream reporter gene expression. **a** Observation under GUS staining, **b** detection of green fluorescence signal, **c** relative GUS/GFP gene expression levels. LINE-2, promoted by a complete Pro-Line2 sequence. LINE-2S, promoted by a truncated Pro-Line2 sequence in the absence of LINE-like sequence. CK, absence of promoter sequence. $*p < 0.05$, $**p < 0.001$ (Zhou et al. 2018b)

transposon, PHRE5's copies, and insertion polymorphism were analyzed in eight *Ph. edulis* cultivars. Compared with moso bamboo, the copy number was increased in these cultivars, which was verified by the polymorphism (Figs. 5.7 and 5.8, Liu 2017). However, more research is needed about whether transposition is directly related to phenotypic changes in moso bamboo.

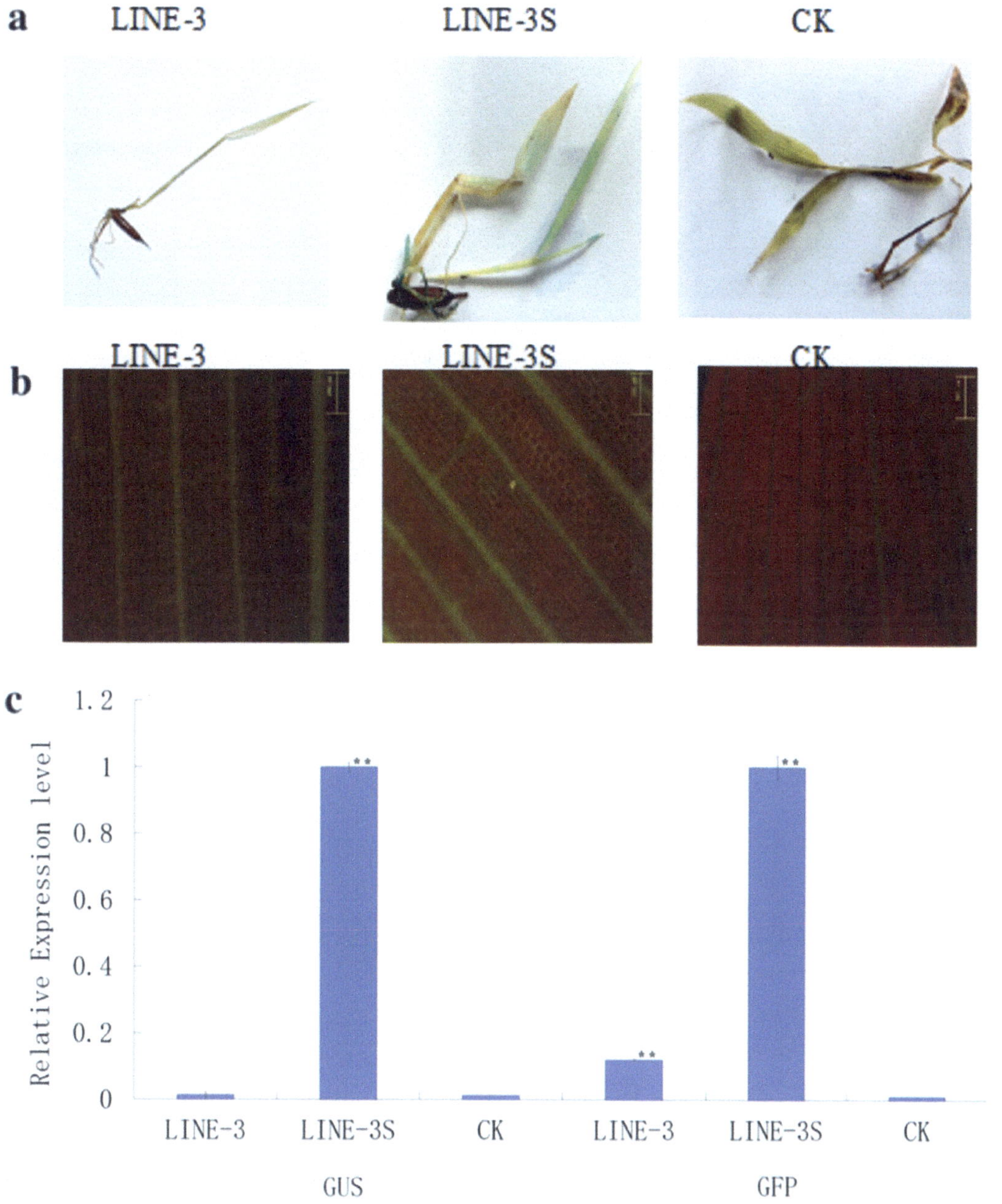

Fig. 5.5 Role of TE insertions into FPI homolog promoter sequences within PH01001953 in downstream reporter gene expression. **a** Observation under GUS staining, **b** detection of green fluorescence signal, **c** relative GUS/GFP gene expression levels. LINE-3, promoted by a complete Pro-Line3 sequence. LINE-3S, promoted by a truncated Pro-Line3 sequence in the absence of LINE-like sequence. CK, absence of promoter sequence. $*p < 0.05$, $**p < 0.001$ (Zhou et al. 2018b)

Fig. 5.6 Role of TE insertions into FPI homolog promoter sequences within PH01004268 in downstream reporter gene expression. **a** Observation under GUS staining, **b** detection of green fluorescence signal, **c** relative GUS/GFP gene expression levels. MuDR-1, promoted by a complete Pro-MuDR1 sequence. MuDR-1S, promoted by a truncated Pro-MuDR1 sequence in the absence of MuDR-like sequence. CK, absence of promoter sequence. $*p < 0.05$, $**p < 0.001$ (Zhou et al. 2018b)

Fig. 5.7 Phenotypic Changes of Bamboo Varieties. 1–8: *Ph. edulis cv. Pubescens; Ph. edulis cv. Luteosulcata; Ph. edulis cv. Ventricosa; Ph. edulis cv. Edulis; Ph. edulis cv. Tubaeformis; Ph. edulis cv. Viridisulcata; Ph. edulis cv.* [DT3] *Obliquinoda; Ph.edulis* (Liu 2017)

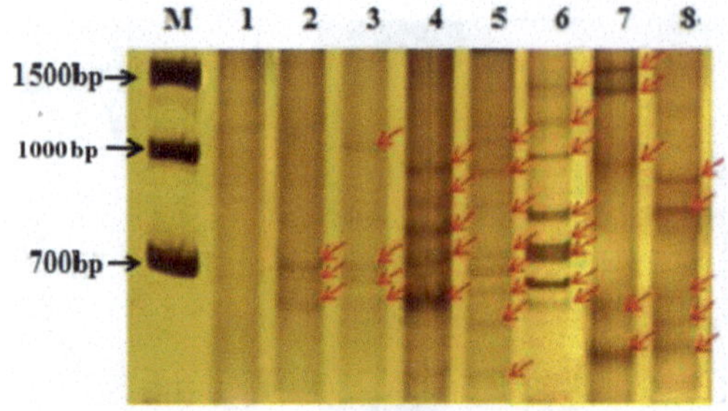

Fig. 5.8 Polymorphism in different *Ph. edulis* of PHRE5. M: Marker, lane1: *Ph.edulis*; 2–8: *Ph. edulis cv. Pubescens; Ph. edulis cv. Luteosulcata; Ph. edulis cv. Ventricosa; Ph. edulis cv. Edulis; Ph. edulis cv. Tubaeformis; Ph. edulis cv. Viridisulcata; Ph. edulis cv.* [DT7] *Obliquinoda;* (Liu 2017)

References

Feschotte C, Jiang N, Wessler SR (2002) Plant transposable elements: where genetics meets genomics. Nat Rev Genet 3:329–341

Goff SA, Ricke D, Lan TH et al (2002) A draft sequence of the rice genome (*Oryza sativa* L. *ssp. japonica*). Science 296:92–100

Initiative IB (2010) Genome sequencing and analysis of the model grass *Brachypodium distachyon*. Nature 463:763–768

Liu (2017) The Distribution and evolution of three full-length LTR retrotransposons, and characterization and functional analysis of PHRE5 from *Phyllostachys edulis*.

Paterson AH, Bowers JE et al (2009) The Sorghum bicolor genome and the diversification of grasses. Nature 457:551

Peng, Z., Lu, Y., Li, L., et al. (2013). The draft genome of the fast-growing non-timber forest species moso bamboo (*Phyllostachys heterocycla*). Nat Genet 45 (4), 456–461, 461e451–452. doi: DOI https://doi.org/10.1038/ng.2569.

Ramakrishnan M , Yrjl K , Vinod K , et al. Genetics and genomics of moso bamboo (*Phyllostachys edulis*): Current status, future challenges, and biotechnological opportunities toward a sustainable bamboo industry [J]. Food and Energy Security, 2020.

Ramakrishnan M, Zhou MB, Pan CF et al (2019) Nuclear export signal (NES) of transposases affects the transposition activity of mariner-like elements Ppmar1 and Ppmar2 of moso bamboo. Mob DNA 10:35

Schnable PS, Ware D, Fulton RS et al (2009) The B73 maize genome: Complexity, diversity, and dynamics. Science 326:1112–1115

Zhao, H., Gao, Z., Wang, L. E., et al . (2018). Chromosome-level reference genome and alternative splicing atlas of moso bamboo (*Phyllostachys edulis*). GigaScience, 7, giy115.

Zhong H, Zhou M, Xu C et al (2010) Diversity and evolution of Pong-like elements in Bambusoideae subfamily. Biochem Syst Ecol 38:750–758

Zhou, M., Chen, A., Zhou, Q., et al. (2017a). A moso bamboo (*Phyllostachys edulis*) miniature inverted-repeat transposable element (MITE): the possible role of a suppressor. Tree Genetics & Genomes 13(6). doi: https://doi.org/10.1007/s11295-017-1210-4.

Zhou, M., Hu, B., and Zhu, Y. (2017b). Genome-wide characterization and evolution analysis of long terminal repeat retroelements in moso bamboo (*Phyllostachys edulis*). Tree Genetics & Genomes 13(2). doi: https://doi.org/10.1007/s11295-017-1114-3.

Zhou M, Hu H, Liu Z et al (2016) Two active bamboomariner-like transposable elements (Ppmar1- and Ppmar2) identified as the transposon-based genetic tools for mutagenesis. Mol Breeding 36 (12):163

Zhou M, Liang L, Hanninen H (2018) A transposition-active *Phyllostachys edulis* long terminal repeat (LTR) retrotransposon. J Plant Res 131(2):203–210. https://doi.org/10.1007/s10265-017-0983-8

Zhou M, Liu X, Tang D (2011) Transposable elements in *Phyllostachys pubescens* (Poaceae) genome survey sequences and the full-length cDNA sequences, and their association with simple-sequence repeats. Genet Mol Res 10:3026–3037

Zhou M, Liu X, Tang D (2012) PpPIF-1: First isolated fulllength PIF-like element from the bamboo *Phyllostachys pubescens*. Genet Mol Res 11:810–820

Zhou M-B, Lu J-J, Zhong H et al (2010) Distribution and diversity of PIF-like transposable elements in the Bambusoideae subfamily. Plant Sci 179:257–266

Zhou M-B, Lu J-J, Zhong H et al (2010) Distribution and polymorphism of mariner-like elements in the Bambusoideae subfamily. Plant Syst Evol 289:1–11

Zhou M, Tao G, Pi P et al (2016) Genome-wide characterization and evolution analysis of miniature inverted-repeat transposable elements (MITEs) in moso bamboo (*Phyllostachys heterocycla*). Planta 244(4):775–787. https://doi.org/10.1007/s00425-016-2544-0

Zhou M, Zhou Q, Hänninen H (2018) The distribution of transposable elements (TEs) in the promoter regions of moso bamboo genes and its influence on downstream genes. Trees 32(2):525–537. https://doi.org/10.1007/s00468-017-1650-3

Transcriptome of Moso Bamboo

6

Jian Gao, Zhanchao Cheng, Long Li,
Ying Zhang, and Juan Li

Abstract

With the development of biotechnology, transcriptome approach (RNA-Seq) has become an important means in identifying key genes. The next-generation sequencing technologies (NGS) feature unprecedented lower costs as well as greater depth. This chapter introduces the transcriptome information of bamboo shoot development, flower development, and seed germination by NGS. Transcriptome analysis is used to reveal the dynamic changes of development process and explore the regulatory mechanism of development, and to identify potential candidate genes for molecular breeding of moso bamboo.

J. Gao (✉) · Z. Cheng · J. Li
Key Laboratory of Bamboo and Rattan Science and Technology, International Center for Bamboo and Rattan, State Forestry and Grassland Administration, Beijing 100102, China
e-mail: gaojian@icbr.ac.cn

J. Li
e-mail: lijuan@icbr.ac.cn

L. Li
Northwest Agriculture and Forestry University, No.3 Taicheng Road, Yangling 712100, Shanxi, China

Y. Zhang
Beijing Academy of Agricultural and Forestry Sciences, No. 11 Middle Road, Dawning Garden, Haidian District, Beijing 100097, China

6.1 Global Transcriptome Analysis in the Growth of Moso Bamboo Shoots

6.1.1 Introduction

Moso bamboo (*Phyllostachys edulis*) contains rich lignocellulose, characterized by substantial ecological, economic, and cultural values. Moso bamboo has the greatest growth rate of overall plants (Lin et al. 2010). Under proper conditions at springtime, shoot growth rate possibly reaches 1 m/day at peak, and shoots can grow up to 20 m in 45–60 days. Due to the special strength and superb growth ratio, moso bamboo has produced 5 billion US dollars in the fields of nutrition and timber (Peng et al. 2013). As revealed by previous transcriptomic sequencing, the hormone network, cell wall metabolism, and cell cycle control play important roles in rapid moso bamboo shoot development (Peng et al. 2013; He et al. 2013). Plant hormones- and cell wall development-associated genes played important roles in the initial thickening procedure of underground moso shoots (Wei et al. 2016). In growing bamboo shoots, the size of SAM determined the biosynthesis of cytokinin and auxin and the expression of hormone signalling genes. A smaller shoot apical meristem (SAM) also results in a thinner and shorter culm (Wang et al. 2019). The reduction of SAM causes the reduction of hormone level, including

J. Gao (ed.), *The Moso Bamboo Genome*, Compendium of Plant Genomes,
https://doi.org/10.1007/978-3-030-80836-5_6

auxin and cytokinin, while decreasing the hormone signal transduction and down-regulating downstream gene expression, including the metabolic genes, eventually dwarfing culm and diameter, and lowering biomass. Comparative transcriptome analysis between *Pseudosasa japonica* and the variant *P. japonica* var. *Tsutsumiana* with slow growth rate recognized some cell growth-linked genes having significant down-regulation in the variant, including those ones associated with vesicle transport, auxin, cytoskeleton organization, as well as cell wall modification. This demonstrated that the variant indeed had thinner cell wall and lower cellulose content. Combined with mechanic force exerted by extrusion pressure in adjacent fast-growth cells as well as weight pressure above growing cells, it possibly leads to the radial and irregular growth in variant cell (Wei et al. 2017). In growing bamboo internode, sugar significantly promotes the rapid growth of bamboo internodes by inhibiting BmSnf1. Mechanical stress engages in the triggering of internode growth decline by activating the formation of reactive oxygen species (ROS) through CaMs up-regulation (Wei et al. 2019). NAC/NAM/ATAF TFs, basic region/leucine zipper TFs, basic helix-loop-helix TFs, WRKY TFs and MYB TFs played a vital role in the postharvest senescence of bamboo shoots, among them NACs were the most abundant (Li et al. 2019).

Here, we introduced the spatio-temporal variation of transcriptome during moso bamboo shoots growth and development, and explained how hormone signaling genes participated in moso bamboo shoot growth.

6.1.2 Overview of Gene Expression in Each Shoot Growth Stage

Comparison between sample and CK proved that 10,344 genes had differential expression in shoot growth during one growth stage at least. Particularly, up-regulated genes progressively reached 3,089 in quantity at S7. At S1–S4, down-regulated genes ranged between 3544 and 3787 in quantity and decreased after S4 (Fig. 6.2a). By analyzing the DGEs intersected across diverse growth periods, there were more overlapped DEGs between two adjacent periods than those non-adjacent ones. For instance, S2 and S3 shared more DGEs (2360 with up-regulation, while 3245 with down-regulation) relative to S2 and S6 (2083 with up-regulation, while 2472 with down-regulation; Fig. 6.1b, c). Merely 1365 genes had up-regulation, while 947 genes had down-regulation at S1–S7. Briefly, there were more DGEs at early growth stage than late growth stage. Likewise, the intercalary meristem showed strong division ability at early stages, but it weakened in later stages.

6.1.3 Functional Analysis on DGEs

For further investigating changes of transcriptomic landscape in shoot elongation, this study carried out gene ontology (GO) analysis. According to sequence homology, we obtained 10,344 genes that were categorized into three GO terms, including biological processes (BPs), cellular components (CCs), and molecular functions (MFs) (Fig. 6.2). In terms of BPs, 'hormone-regulated signal transduction pathway,' 'redox process,' and 'transmembrane receptor protein serine/threonine kinase signal transduction pathway' were the most significant terms; meanwhile, 'monocarboxylic acid biosynthesis' and 'protein autophosphorylation' were also significantly enriched (Fig. 6.2). Of the MF terms, 'activity of transmembrane receptor protein serine/threonine kinase,' 'heme binding,' and 'activity of sequence-specific DNA binding TF,' ranked the top among the enriched terms with significance. Across the CC terms, 'plasmodesmata,' 'extracellular region,' and 'cytoplasmic membrane-bounded vesicle' were the three most significant ones. GO analysis indicated that hormone signaling genes were critical to moso bamboo shoot growth.

For analyzing phytohormone interaction, the network consisted of hormone signaling genes with differential expression was established (Fig. 6.3). Of the top 15 most closely associated

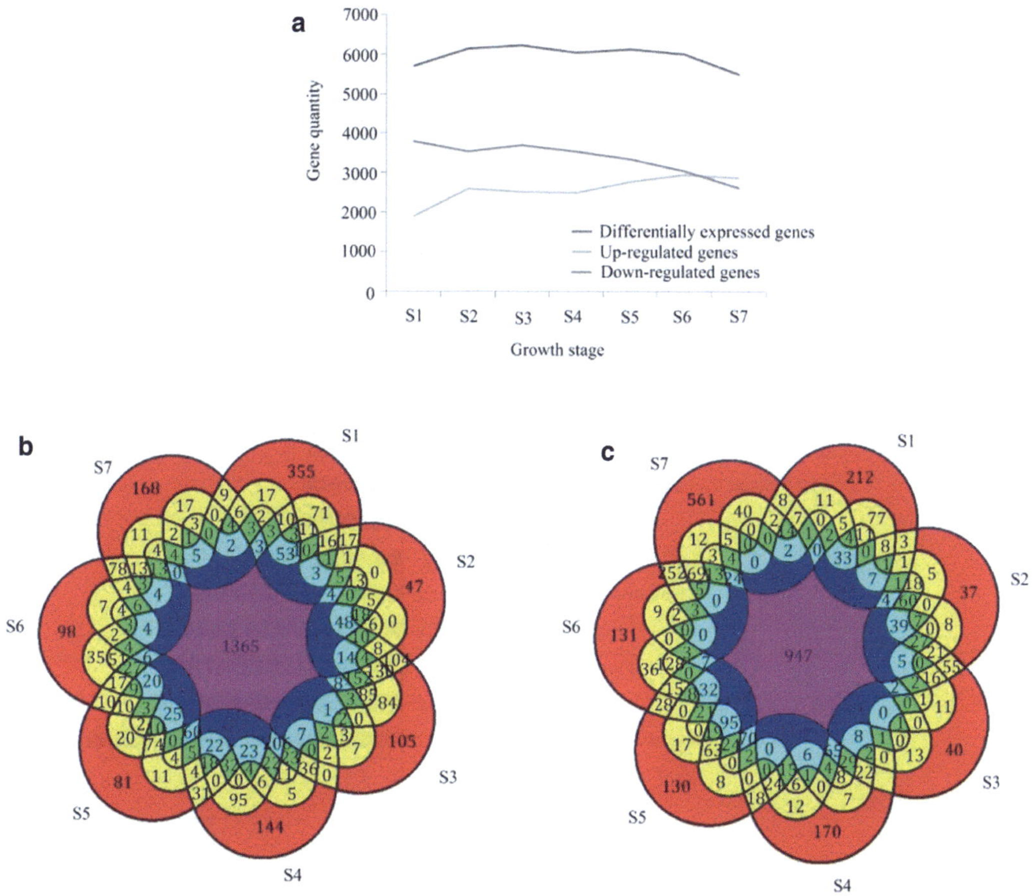

Fig. 6.1 a Transcriptome relations of samples. **b** Gene expression in eight samples. The x-axis and y-axis of every chart denote number of genes and eight growth stages, separately. From Li et al. (2018)

(DGEs $\geq$ 7) of network (hub genes), 5 were related to auxin signal transduction, whereas 4 belonged to the AUX/IAA family. The auxin signaling pathway has more hub genes compared with other hormone signaling pathways, revealing the importance of auxin signal transduction, particularly AUX/IAA, in the growth of shoots. Among these, *PheIAA1* with 13 edges was related to jasmonic acid (JA), brassinosteroid (BR), and cytokinin signal transduction genes, indicating that *PheIAA1* exerted vital parts in the communication of auxin signal transduction with additional signals, which could be seen from *PheIAA2* and more AUX/IAA genes. Furthermore, 4 AUX/IAA members displayed close relationship with the BR signaling, whereas AUX/IAA-triggered BR signals were detected within *Arabidopsis* (Ayako et al. 2003). Cytokinins modulated the activities of auxin efflux carriers to affect the transport of polar auxin (five had up-regulation in the dataset, Table 6.1) (Su et al. 2011). Therefore, increasing auxin efflux carrier activity possibly promotes elongation of shoots. Numerous latest reports indicate that ARF families engage in auxin signaling, and also mediate cytokinin (Kobayashi et al. 2012; Cheng et al. 2013) as well as gibberellin signaling (Maaike et al. 2011; Elena et al. 2015). The proposed gene interaction network comes to an alike conclusion. For instance, *PH01000057G1420* interacts with AUX/IAA, cyclin D3, and BAK1 genes that get involved in

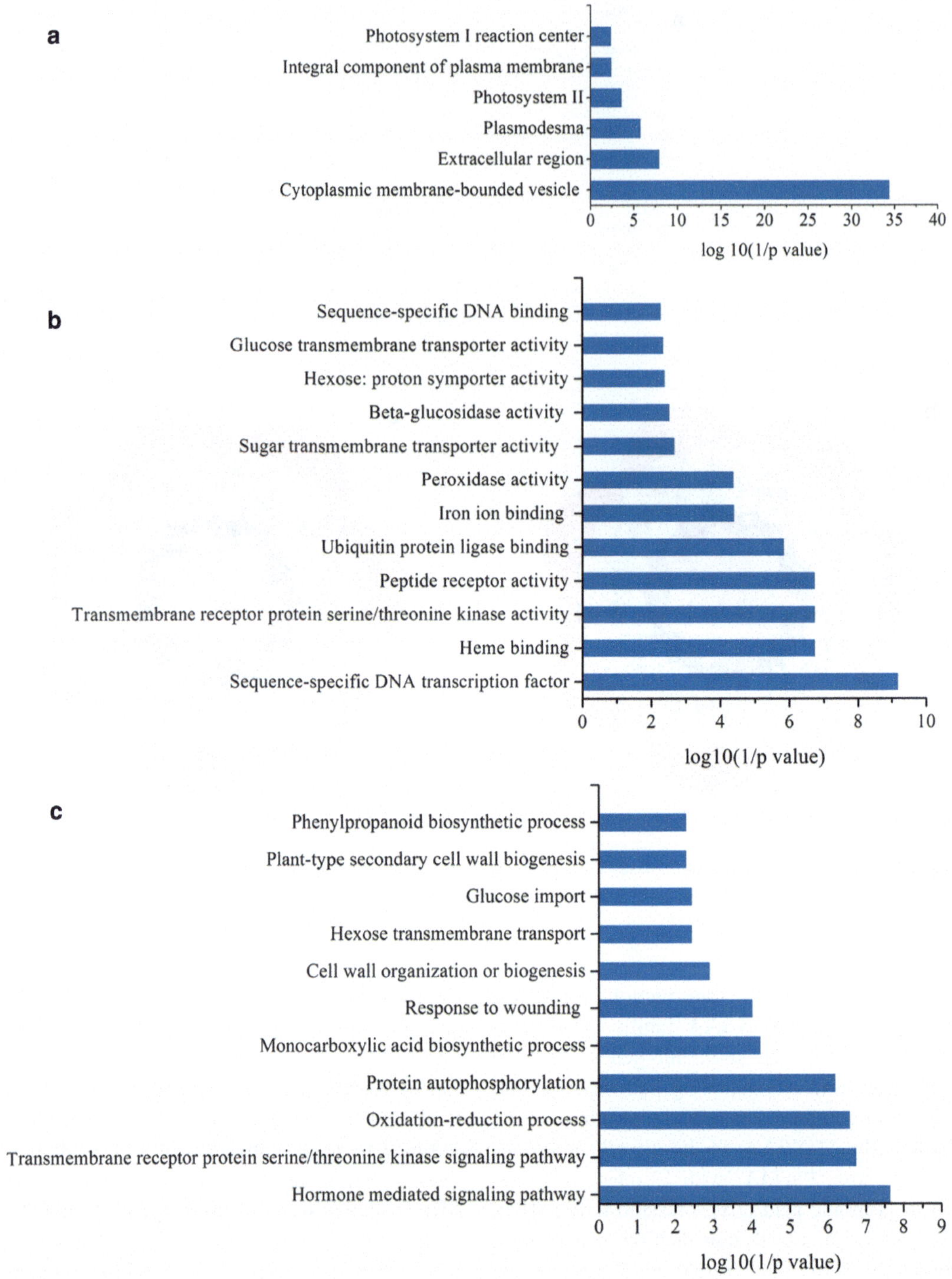

Fig. 6.2 GO terms with significant abundance (FDR <0.05) in DGEs **a** GO component, **b** GO function, **c** GO biological process. From Li et al. (2018)

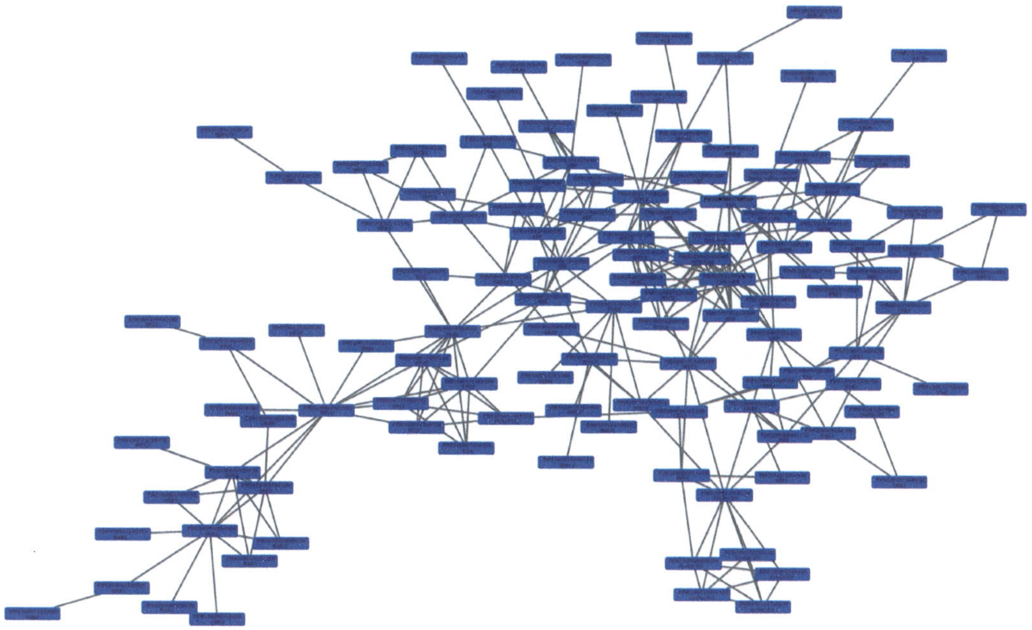

Fig. 6.3 The co-expression network for hormone signaling genes with differential expression in moso bamboos. From Li et al. (2018)

cytokinin signaling and BR signaling, separately. In summary, auxin signaling genes play a vital role in moso bamboo shoot development by triggering or coordinating other hormone signaling genes as hub genes.

Four auxin signaling genes experiencing significant up-regulation in growth of bamboo shoots had been further examined using in-situ hybridization. In S1, *PheIAA1* was up-regulated within bamboo sheaths and ground tissues relative to pith meristems. In S4 and S6 periods, *PheIAA1* was up-regulated within vascular bundles, epidermis, and lateral buds (Fig. 6.4a). PheARF1 was expressed within the epidermis, apical buds, and lateral bud apical growth cones, yet it was not detected within intercalary meristem or bamboo sheath (Fig. 6.4b). The above four genes were highly expressed within apical buds and lateral bud apical growth cones (Fig. 6.4). Additionally, *PheIAA1*, *PheIAA2*, and *PheLAX1* were expressed in epidermis and bamboo sheaths (Fig. 6.6a, c, d). In maize, *RUM1* modulates lateral roots as well as seminal initiation. This study histologically analyzed the *rum1* mutant roots and discovered the damaged differentiation of xylem-surrounding pith cells (Zhang et al. 2014). *PheIAA1* and *PheIAA2* were highly expressed within lateral buds, intercalary meristems, and vascular bundles, indicating that these genes exerted important roles in tissue cell differentiation and division. Mutations of *iaa3* and *iaa7* in *Arabidopsis* induced the disappearance of apical dominance (Tian and Reed 1999). Therefore, *PheIAA1/PheIAA2* up-regulation in apical buds possibly suggests the vital effect of *AUX/IAA* on keeping apical dominance for moso bamboo shoot development. In-situ hybridization outcomes indicate the multiple functions of auxin signaling genes in bamboo shoot growth.

6.1.4 Importance of Cell Division and Elongation to Shoot Elongation

Plentiful cell nuclei could be seen from ground tissues and vascular tissues in winter and early growth stages, suggesting the existence of meristematic tissues. With shoot growth in

Table 6.1 The genes associated with transcription factor families

Transcription factor families	Gene numbers
MADS-Box gene (MADS)	38
Heat shock transcription factor 20 (HSP20)	28
Basic helix-loop-helix factor (bHLH)	34
Basic leucine zipper (bZIP)	6
NAM-ATAF-CUC family (NAC)	61
Heat shock transcription factor 70 (HSP70)	1
WRKY transcription factor (WRKY)	110
MYB transcription factor (MYB)	143
DNA-binding with-one finger (Dof)	15
AP2/EREBP transcription factor gene (AP2/EREBP)	32
Heat shock transcription factor (HSF)	28
High-mobility group box (HMG)	4
MADS-Box gene (MADS)	38

subsequent stages (8 m–12 m), nuclei of intercalary meristem reduced, and merely a few nuclei were able to be found at 12 m. Such tendency conformed to expression of cell cycle-linked genes (cyclin A, cyclin B, and cyclin D) and division (Fig. 6.5). As gene expression progressively decreased at S6–CK, cell division activity simultaneously reduced. Additionally, many genes were labeled as 'monocarboxylic acid biosynthesis,' 'oxidation–reduction process,' along with 'phenylpropanoid biosynthesis'; at the same time, the remaining BPs had statistical significance after up-regulation (Fig. 6.5). It was possibly associated with sustained cell proliferation of meristem (Sun et al. 2014; Gao et al. 2015). Moreover, multiple mitochondria could be seen from internode cells, suggesting the substantial energy demands for shoot cell proliferation.

Aquaporin genes are related to cell elongation and water absorption for auxin-mediated development (Teale et al. 2006). Therefore, rich aquaporins provide the necessary conditions for cell differentiation in moso bamboo shoot growth (Table 6.1). Almost two-thirds of kinesins had significant up-regulation in shoot growth. Former reports suggested the role of kinesin family in regulating cytokinesis progression (Hirokawa et al. 2009; Tanaka et al. 2004) and forming cell plate (Michiko et al. 2015) for Arabidopsis. Therefore, the increased levels of kinesins are related to chromosome segregation as well as organelle transport during bamboo shoot development. Cyclin and chromosome transmission fidelity protein relate to chromosome replication, and they both showed an extensive up-regulation trend in the research (Table 6.1), suggesting the mitosis process experienced by parenchyma and fiber cells in shoot elongation. To sum up, cell division-associated genes had up-regulated levels, which promoted meristems to keep vibrant at wintertime and early growth stages. Whereas, there was a great difference in terms of growth pattern between early and winter growth periods. In winter growth stage, meristem cell proliferation led to primary thickening growth in bamboo shoots, but no change could be seen in bamboo shoot height (Wei et al. 2016). However, early meristem cell proliferation led to rapid internode elongation, finally causing the longitudinal growth in bamboo shoots.

Factors and enzymes related to the expansion of cell wall, including cellulose synthase, hydrolase, or xyloglucan endotransglucosylase, showed high levels in late development (Fig. 6.5), where intercalary meristem had reduced cell division ability and cell elongation substituted constant bamboo shoot development

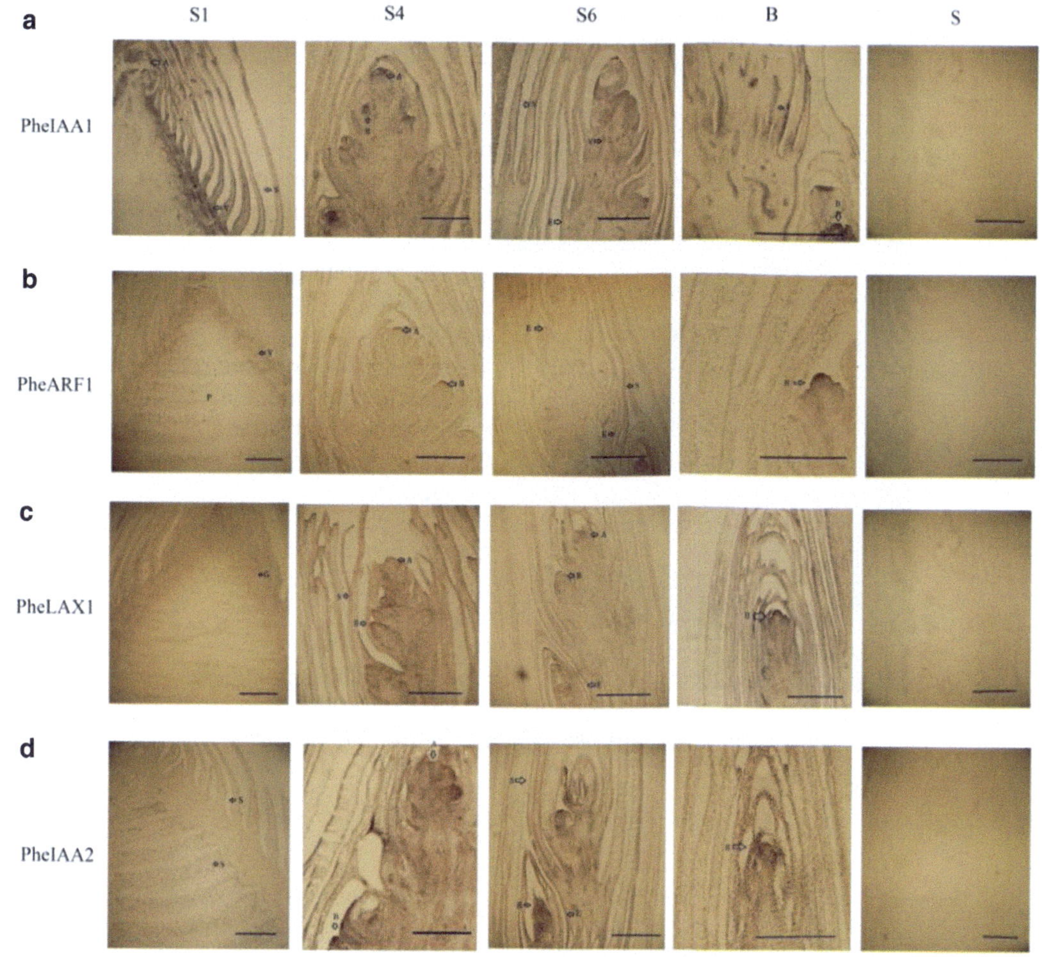

Fig. 6.4 Auxin signaling-associated gene expression in moso bamboo shoots at different development periods. S1, S4, S6. B stands for bamboo shoots (height, 3/9 m, with lateral buds) during wintertime; S stands for expression of sense probe detected by ISH. Arrowheads indicate the transcript abundance. A stands for apical buds, B for lateral buds, E for epidermis, G for ground tissues, P for pith, S for bamboo sheaths, and V for vascular bundles. Bars = 2 mm. **a** PH01000025G1600 (AUX/IAA), **b** PH01000046G0220 (ARF), **c** H01000373G0290 (AUX/LAX), **d** PH01001098G0440 (AUX/IAA). From Li et al. (2018)

(Cui et al. 2012). In the case of auxin-mediated growth, the great expression quantity modulates the plasticity of cell wall via a loosening process (Paque et al. 2014).

6.1.5 Spatio-Temporal Dynamic Changes in Moso Bamboo Shoots

Moso bamboo shoots' biological characteristics have gained thorough exploration, though most studies concentrate on the early growth of shoots (Xu et al. 2008; Wei et al. 2016). Internodes with sequential elongation on the top and bottom have also been extensively studied (Cui et al. 2012). As confirmed by relevant proteomic studies, several metabolic processes relate to cell wall structure and rapid bamboo growth (Cui et al. 2012). Whereas, the spatio-temporal change in moso bamboo shoots in growth stage remains unknown. In winter growth stage, bamboo shoots finished primary thickening growth under the cover of a thin layer of soil (Wei et al. 2016). Till

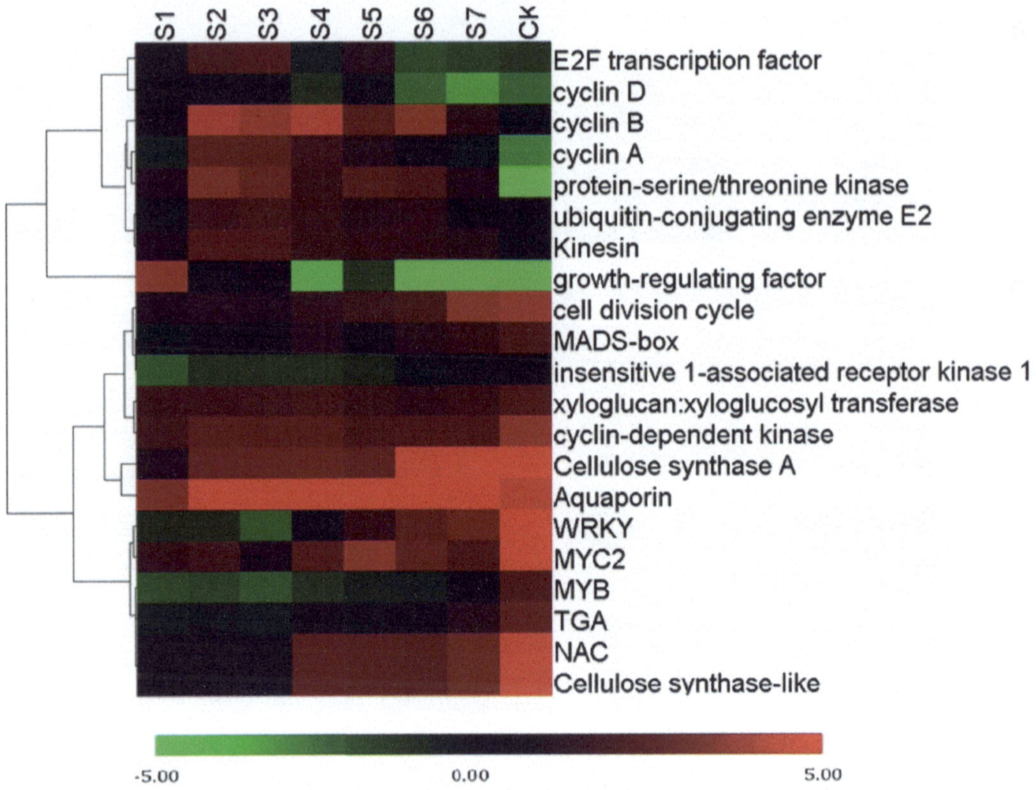

Fig. 6.5 Heat map produced from average expression of families related to shoot growth. From Li et al. (2018)

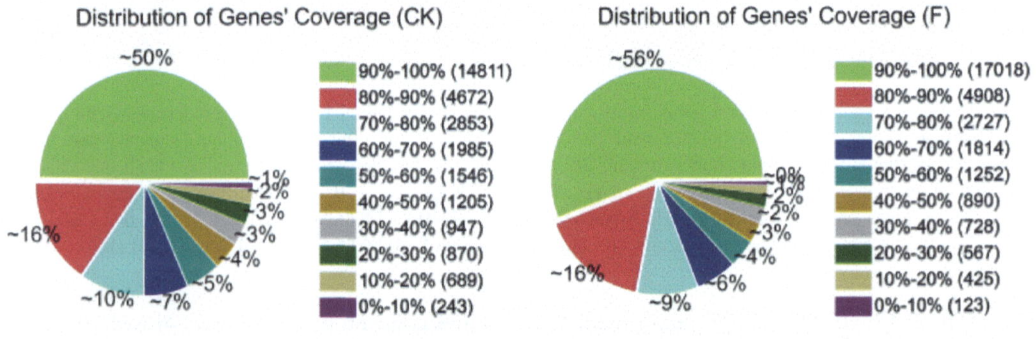

Fig. 6.6 Distribution of gene coverage identification of DEGs

spring, shoots burst out of soil, and lateral bud appeared, suggesting the initiation of early growth stage. Intercalary meristem at internodes from shoot tips developed vibrantly at early growth stage, whereas decreased in late growth stage. Such tendency coincided with middle internodes, though growth decreased later (Cui et al. 2012). Besides, AS changed at regular intervals in terms of its number during shoot development. Typically, 1-year culm (CK) had the lowest AS frequency, which in early/late development periods ranked the second place, whereas that in bamboo shoot was the highest (Li et al. 2016). Likewise, the number of DEGs in early growth stage was more than that in late growth stage.

6.1.6 Conclusion

Growth of bamboo shoots comprises three stages, namely winter, early, and late growth stages according to anatomical structure and gene level. In early development period, lateral bud sprouting was seen from bamboo shoot tip. Intercalary meristems developed vibrantly in winter and early growth stages, while mitosis of intercalary meristems reduced at late growth stage. The early growth stage had more expression of cell cycle-related genes and DGEs compared with late growth, possibly due to hormonal concentration. Gene expression analysis reveals the importance of hormone signaling genes to shoot growth, in which auxin signaling genes have core effects. In situ hybridization analysis demonstrates the regulating role of signaling genes in apical dominance, meristem maintenance, and lateral bud growth.

6.2 Global Transcriptome Analysis in Moso Bamboo Flower Development

6.2.1 Quality Control and Filtering of Sequence Data

For gaining in-depth insights into transcriptome of moso bamboos, cDNA libraries were produced using non-flowering plant leaves (short for CK) and blended them with panicle RNA in four diverse flowering stages (short for F) through the Illumina platform-based paired-end sequencing. To sum up, over 67 million reads were obtained in CK and F samples. When short reads were mapped to reference genome, approximately 70% reads were only mapped into the moso bamboo genome. Besides, the proportion of reads-covered genes was also determined to measure gene coverage. 15,497 reads were searched from F samples, whereas 167,767 from CK samples and corresponding gene coverage was 90–100% (Fig. 6.8).

6.2.2 Reads Mapped to the Reference Genome Dataset

For identifying genes that correspond to above clean reads from libraries, these clean reads should be mapped into the moso bamboo reference genomic genes. As confirmed by the mapping analysis, 39,386,902 and 39,026,185 reads from each library had a perfect match with reference genome, and approximately 22,824,394 and 23,064,627 reads had perfect match with reference genes (Table 6.2). 47,847,557 and 47,709,687 reads had unique match with reference genome, while 30,640,507 and 30,868,269 reads had match with reference genes in the two libraries. 53,035,796 and 52,489,347 reads were perfectly mapped into reference genome, whereas 32,346,195 and 31,843,985 were mapped into reference genes. In addition, the Illumina platform achieved marked sequencing depth, and the genome of moso bamboo was incompletely annotated, so 14,582,070 and 15,815,789 reads were not mapped into reference genome, while 35,271,671 and 36,461,151 reads were not mapped into reference genes in each library. Table 6.2 presents the alignment statistical results for reads in CK and F libraries.

6.2.3 Putative Moso Bamboo Floral Development Transcription Factors

Moso bamboo is featured by lengthy vegetative growth period and uncertain flowering time before flowering. For comparing flowering and vegetative tissues in gene expression pattern, we obtained vegetative leaf and panicle samples from the moso bamboo not at flowering period to carry out Illumina sequencing. The 714 genes related to floral growth were up-regulated within panicle samples (fold change FC expression between leaf and panicle samples was twofold), covering 476 flowering genes under transcription factor genes (Table 6.1). Transcriptional regulation can be mediated via the association between TFs and certain regulation regions within the genome, causing gene activation or other

Table 6.2 Alignment statistics results in CK and F libraries

Tissue	F	CK
Total reads	68,305,136	67,617,866
Total basepairs	6,147,462,240	6,085,607,940
Map to genome		
Total mapped reads	52,489,347	53,035,796
Perfect match	39,026,185	39,386,902
≤ 5 bp mismatch	13,463,162	13,648,894
Unique match	47,847,557	47,709,687
Multi-position match	4,641,790	5,326,109
Total unmapped reads	16,730,549	15,499,315
Map to gene		
Total mapped reads	31,843,985	32,346,195
Perfect match	22,824,394	23,064,627
≤ 5 bp mismatch	9,019,591	9,281,568
Unique match	30,640,507	30,868,269
Multi-position match	1,203,478	1,477,926
Total unmapped reads	38,062,793	36,891,402

variations (Zhang et al. 2008). Thus, several transcription factors are determined as master regulators, because their action triggers downstream effects, like activating other transcription factors, and finally leads to extensive variations of organism (as (Nadgauda et al. 1990)). This research was concerned with such critical regulatory genes during moso bamboo flower developmental process. Altogether, 476 putative transcription factor families could be recognized from moso bamboo panicle tissues, WRKY (110), MADS-Box (38), bHLH (34), Dof (15), NAC (61), HSF (4), HMG (4), bZIP (6), HSP20 (28), HSP70 (1), MYB (143), and AP2/EREBP (32) included.

6.2.4 Detection of Putative Genes Associated with Flowering Time Control and Flower Development

Former reports concerning model plant species proved how floral induction timing is under the control of complicated regulatory networks monitoring environmental variations, including the ambient temperature pathway, autonomous pathway, age pathway, gibberellins pathway, and photoperiod/circadian clock pathway(Simpson and Dean 2002; Fornara et al. 2010). The uncommon and infrequent flowering of bamboos has aroused concerns from scientists and laypeople for hundreds of years (Janzen 1976). Whereas, few studies have been made on bamboo flowering molecular mechanism because of the difficulty in flowering sample collection. By comparing moso bamboo genes of the research and NCBI and Uniprot databases, it was found that 238 genes were homologous to existing flowering-associated genes of other plants. These genes were compared with the flowering ones in Hordeum vulgare, Arabidopsis thaliana, Oryza sativa, Sorghum bicolor, and *Brachypodium distachyon*. Whereas, these genes were proved to function in representative flowering promotion pathways (like autonomous pathways, ambient-temperature) and floral pathway integrator (*FPI*) genes (as (Fornara et al. 2010; Ehrenreich et al. 2009)) had no high expression in floral tissues of bamboos. Just the *CONSTANS* (*CO*), *phytochrome-interacting factor 3* (*PIF3*), *late elongated hypocotyl* (*LHY*), *GIGANTEA ()*, and *crytochrome-1* (*CRY1*) genes were detected from the photoperiod/circadian clock pathway in moso bamboos. There was no gene similar to components in the above pathway, including pseudo response regulators (*PRR5*, *PRR7*, and *PRR9*), *early flowering 3* and *4* (*ELF3*, *ELF4*), *timing of CAB1* (*TOC1*), *phtyochromes* (*PHYA*, *PHYB*), and *circadian clock-associated 1* (*CCA1*) (Mouradov et al. 2002). The *CO* gene is important to the integration of light and temporal information, and also determination of flowering time (Paul et al. 2004). There are two zinc fingers in CO protein, including the B-domain at the N-terminal that mediates inter-protein interaction as well as CCT domain at C-terminal related to nuclear localization (Robson et al. 2001). The TF *CO* in Arabidopsis contributes to flowering by promoting the expression of *flowering locus T* (*FT*) (Samach et al. 2000). Rice orthologues in Arabidopsis genes *CO* and *FT* are considered *Hd1* and *Hd3a*, separately; and *Hd1* controls rice flowering through *Hd3a* regulation ((Li et al. 2009)). In current research, 41 putative *CO* transcription factors had been found to engage in flower development. Whereas, they mostly had down-regulation in moso bamboo floral tissues. The upstream floral suppressors, including *PIF3* and *LHY*, were up-regulated, while *FPI* genes (*CRY1* and *GI*) were down-regulated, causing the decreased levels of *CO* genes. Low-expression *CO* genes indicated flower initiation's independence from conventional promotion pathways during bamboo flowering.

Gibberellins pathway genes are highly expressed, and this pathway is considered a promotion pathway for complicated regulatory networks. GA can regulate different developmental stages in plant growth, including seed germination, leaf expansion, stem elongation, and floral initiation (Richards et al. 2001). Putative homologs in genes in GA-signaling pathway had high expression, gibberellin response modulator (GAMYB) (Tsuji et al. 2006), and gibberellin receptors (GID1, GID2) included (Alexandrov et al. 2009). Gibberellin initiates degradation of DELLAs (including GAI, RGA, and RGL78-80) while up-regulating *MYB21*, *MYB24*, and *MYB57*, and these are essentially required by stamen filament growth ((Peng 2009)). These high-expression genes demonstrated a relation of GA with bamboo flowering.

Unlike genes associated with the flowering pathway, over 252 stress-responsive genes (like drought or oxidative stress OS) were up-regulated. Some FMI-related genes, together with the associated upstream drought response genes showed high expression in flowering, including *Dof3*, *Dof5*, *Dof7*, *Dof10*, *Dof12*, and *bZIP* transcription factors. Putative homologs in FMI genes (Bowman et al. 1993; Jeon et al. 2000; Dreni et al. 2011), including *AP1*, *AP2*, *MADS*1, *MADS14*, and *MADS15*, had high expression. High expression of such genes suggested the underlying correlation between drought or additional environmental stresses and bamboo flowering.

6.2.5 Change of Gene Expression Profile in Varying Flower Developmental Stages

DGEs have absolute expression, which prohibits multiple intrinsic constraints in microarrays (as (Tang et al. 2011)). We adopted DGEs to interpret gene expression in four flower development stages of moso bamboos according to floral organ anatomical structure, namely F1, F2, F3, and F4 stages (Fig. 6.7). Further, four cDNA libraries (F1-F4) were sequenced, and 10–11 million reads with high quality were acquired. The above read sequences should be matched with reference moso bamboo transcriptome database. Around 8,412,664–9,367,437 reads should be mapped to reference genome database-derived genes, with about 80% transcriptome sequences being identified from the genome of moso bamboo. About 5,185,207–5,820,078 reads were matched with reference gene database-derived genes, and around 79% transcriptome sequences were perfectly matched with genes in moso bamboo. Additionally, Figure S2 demonstrates quality evaluation for reads, sequencing saturation analysis, randomness evaluation, and gene coverage. DEGs in samples

Fig. 6.7 Floral tissue sampling. A1-A2: Typical flowers collected to generate F1 samples (floral bud formation period). B1-B2: Typical flowers collected to generate F2 samples (inflorescence development period). C1-C2: Typical flowers collected to generate F3 samples (florescence, flowers with both pistils, stamens emergence from glumes). D1-D3: representative flowers gathered for producing samples for F4 (embryo formation stage)

were recognized by the algorithm of Audic and Claverie (Audic and Claverie 1997). As predicted, most gene expression changes were seen in CK and F1/F2/F3/F4 samples, accompanied by growing up-regulated genes (Fig. 6.8). Furthermore, about 1% genes with up-regulation were in the orphan sequences with no annotations from Gene Ontology database (http://www.Geneontology.org/), while there was significant difference in gene expression level. Therefore, it can be assumed that such genes possibly involve in additional special processes or pathways in the flowering period of moso bamboo.

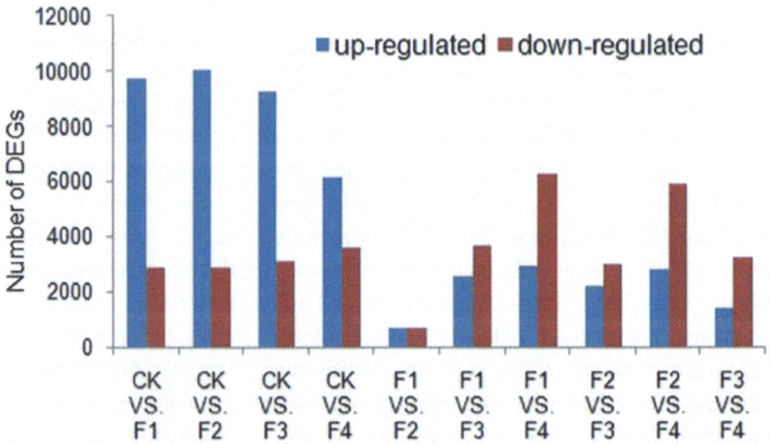

Fig. 6.8 Changes of gene expression profile in various flowering periods. Up-/down-regulated DEG counts in CK vs F1/F2/F3/F4; F1 vs F2/F3/F4; F2 vs F3/F4; F3 vs F4. DEGs: differentially expressed genes

6.2.6 Expression Level of Putative Genes During Flower Development

80 genes had been enriched into the flowering regulation pathways in the transcriptome of moso bamboo (Table S10). For finding if such genes had putative involvement in flowering regulating pathway, a comparison was made between such genes and CK samples in expression profile using DGEs (shown in Fig. 6.9). At CK–F1 and CK–F2, MADS, Dof, GID1, GID2, LHY, and MYB genes all had up-regulation. The expression of several genes declined with time. Dof genes' expression remarkably declined during anthesis and embryo formation stages, indicating the greater effects of Dof genes on early flower developmental stage flowering inhibitors, like LHY and PIF3 genes, had up-regulation in flower developmental stages, which possibly led to CO genes' down-regulated expression in floral tissues. The decreased expression of CO genes, whereas increased expression of FMI and stress-responsive genes suggested FMI activation' independence from conventional flowering regulating pathway during the flowering of moso bamboos.

6.2.7 Gene Validation and Expression Analysis

For confirming expression of genes in clean reads and carrying out experimental

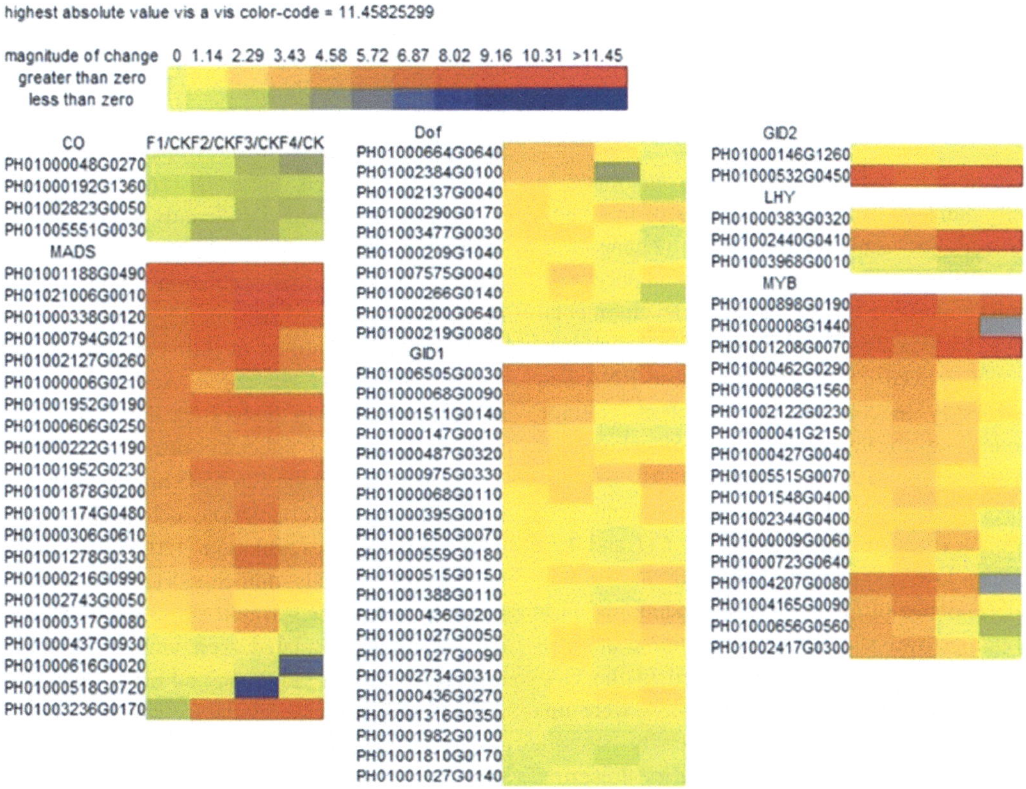

Fig. 6.9 Gene expression profile clustering during 4 flowering periods of moso bamboo

computational analysis, we enrolled eight candidate flower development-related genes for RT-PCR and qRT-PCR analyses (Fig. 6.3 and Table S11). As revealed by RT-PCR results, all selected genes were positive for PCR, and one individual band was detected in the computation size (data not shown). According to qRT-PCR analysis (Fig. 6.10 and Table 6.4), *CO* genes were down-regulated within moso bamboo floral tissues; besides, *PIF3* and *LHY* were up-regulated during the flowering period as the upstream floral suppressors. Drought-responsive genes such as *Dof* were highly expressed within floral tissues, especially in early flowering period; *GID1* showed similar expression profile to *Dof* within floral tissues. Expression of *MADS-14*, *MADS-15*, and *MYB* increased within floral tissues relative to non-flowering leaf samples. qRT-PCR results conformed to the functions predicted from Illumina analysis.

6.2.8 Conclusions

Rare genetic and genomic studies are made to study bamboo flowering regardless of few bamboo varieties' economic values (as (Zhang et al. 2012)). Sequences and transcript abundance level across moso bamboo flower developmental stages were analyzed via Illumina GAII platform. Results proved it was feasible to use RNA-Seq technology in combination with transcriptome sequencing to investigate floral development-related genes of moso bamboos. All the 714 genes related to bamboo floral development were highly expressed within panicles. *CO* and *FPI* genes' low expression indicated the higher independence of FMI from aforementioned promotion pathways in the flowering period of moso bamboo. In addition, genes related to the GA-signal and stress-responsive pathways were up-regulated, which suggested adversity stress's relation with GA and bamboo flowering (Fig. 6.11). The dataset provides a major resource for subsequent genetic or genome study on bamboo species and helps better learn about moso bamboo flower developmental mechanism.

6.3 Global Transcriptome Analysis in Germinating Moso Bamboo Seeds

6.3.1 Global Transcriptome Analysis

For having a whole picture of seed germination at the level of transcription, seeds from the three germination stages were chosen for further RNA-Seq (CK, PS, MS, FS). After high-throughput RNA-Seq, 44,405,150 to 54,486,954 total reads and 32,125,623 to 39,155,754 mapped reads for each library were obtained, and 59.74–71.16% unique reads were mapped to the moso bamboo genome (Table 6.3). Pearson's correlation coefficient (PCC) of all samples was determined to verify reproducibility. A total of 5174, 4697, and 6853 DEGs had been recognized during PS, MS, and FS stages, respectively, compared with CK. During seed development, 3828, 3266, and 4972 DEGs were up-regulated in PS, MS, and FS stages, and 1346, 1431, and 1881 DEGs were down-regulated in the respective PS, MS, and FS stages. Of these DEGs, 2353 DEGs were persistently and significantly shifted during seed germination, whereas 1221, 725, and 2279 DEGs had differential regulation uniquely only in respective PS and MS, or FS stage.

The genes identified at different seed germination stages were compared. According to the subcategories during the biological process, these genes fell into 14 subcategories, in which ell matrix and membrane (26,654) ranked first, followed by cell differentiation, cell proliferation, gene expression regulation (20,574), energy conversion of cells–mitochondria, chloroplasts (18,519), ribosome, protein (15,404), biochemical component-related enzymes (13,041), response stimulus (11,724), and other biological processes with fewer regulatory genes.

To identify biological processes associated with seed germination, the GO enrichment analysis of DEGs was carried out. A total of 2302, 1999, and 2900 DEGs were differentially expressed and assigned to 123,128, and 143 pathways between CK-versus-PS, CK-versus-

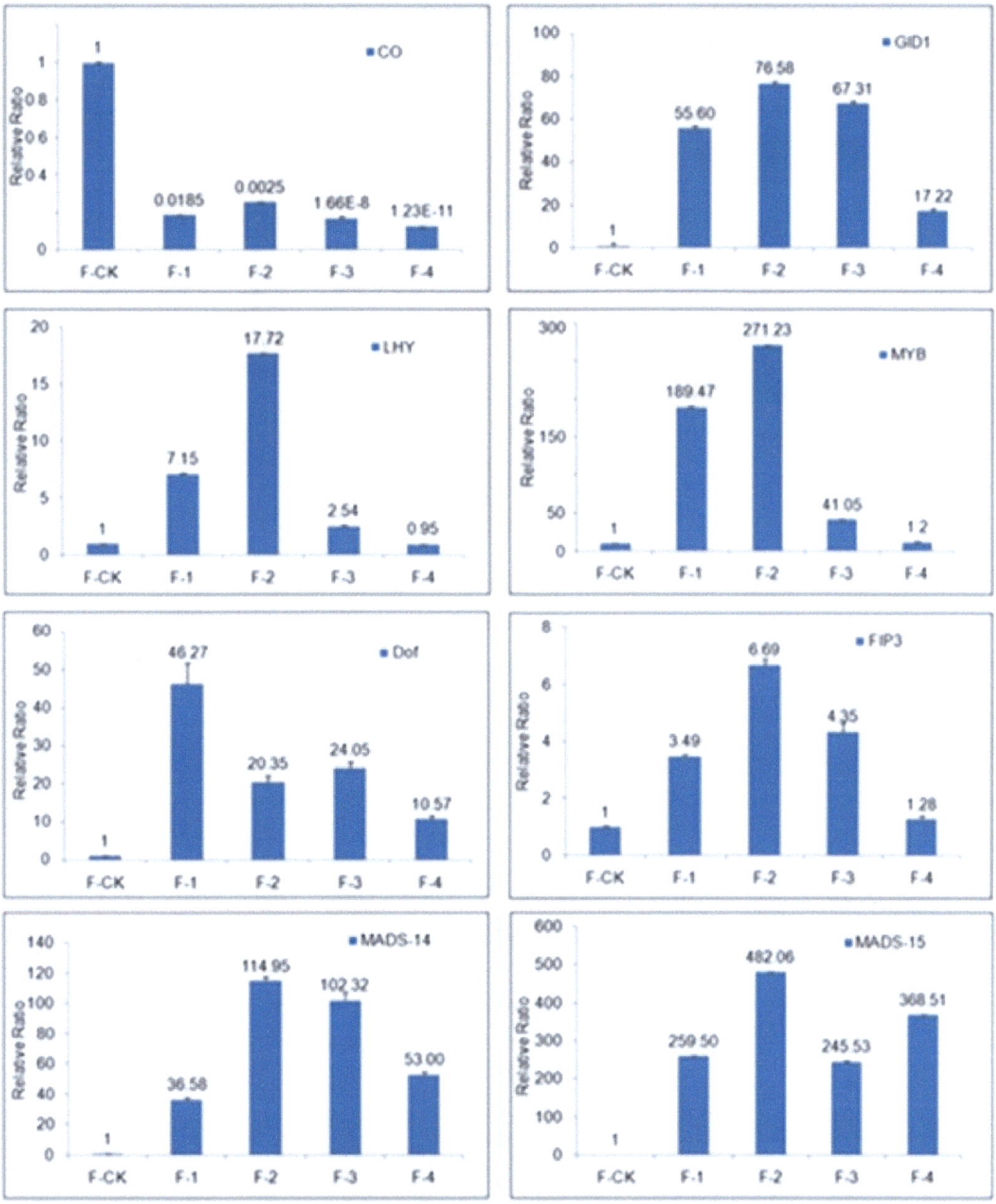

Fig. 6.10 The expression profile for eight chosen genes in flowering tissues during diverse flowering period and in moso bamboo leaf samples in non-flowering periods. Transcript abundances were standardized based on TIP41 (tonoplast intrinsic protein 41), while the control individual gene expression was deemed as 1.0. Error bars stand for SD from 3 individual assays. LHY, late elongated hypocotyl; CO, CONSTANS; DOF, Dof zinc finger protein 12; GID1, gibberellin receptor GID1; MYB, MYB TF; FIP3, phytochrome-interacting factor 3; MADS-14, MADS-box gene-14; MADS-15, MADS-box gene-15;

MS, and CK-versus-FS, respectively. Among all clusters, C1, C2, and C3 contained the greatest number of enriched DEGs related to cell differentiation and cell cycle, including cell movement and subcellular components, microtubule-based processes, cell proliferation,

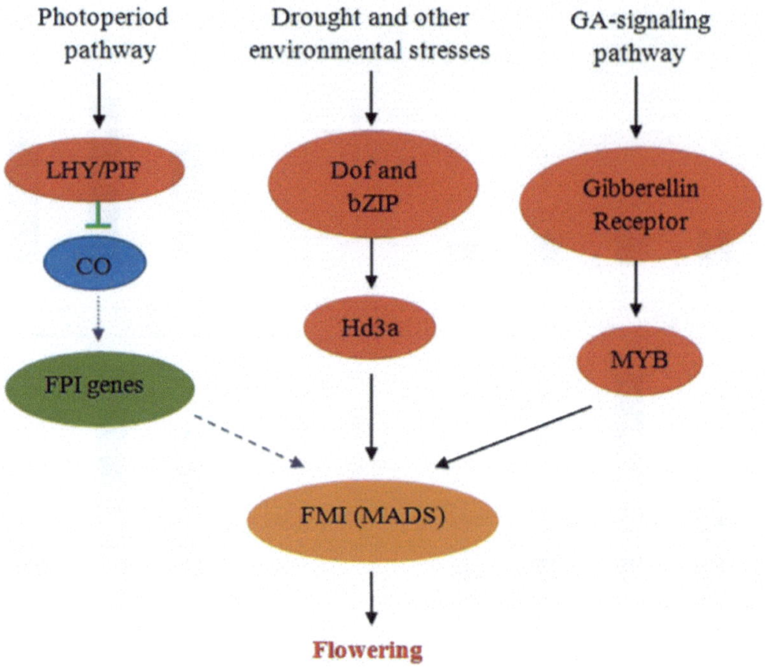

Fig. 6.11 Hypothesized pathway in moso bamboo flowering regulation. Yellow and red ovals suggest high expression of genes from floral tissues, and blue oval suggests gene down-regulation, and green oval means the non-activation of genes. Purple dashed arrows denote non-used pathways in flowering. Black arrow denotes promotion. Green bar denotes suppression

Table 6.3 Summary statistics of RNA-Seq results for seed development of moso bamboo

Sample name	Replicate	Total reads	Mapped reads	Uniquely mapped reads	Uniquely mapped read rate
CK1	R1	5129,9128	3688,3458	3487,4296	67.98
CK2	R2	4440,5150	3212,5623	3023,8058	68.10
CK3	R3	4900,1540	3324,5201	3200,1542	69.12
PS1	R1	5086,2112	3727,9985	3619,5519	71.16
PS2	R2	5106,5294	3915,5754	3520,9440	68.95
PS3	R3	5048,7519	4014,5135	3415,2167	70.33
MS1	R1	5071,7276	3624,2824	3526,1368	69.53
MS2	R2	5052,7104	3583,9322	3165,8725	62.66
MS3	R3	5106,5294	3915,5754	3520,9440	68.95
FS1	R1	4439,7580	3187,4765	2652,2929	59.74
FS2	R2	5448,6952	3891,0224	3666,7842	67.30
FS3	R3	5448,6954	3779,3454	3663,4542	66.35

cell division, cytokinesis, mitotic cell cycle process, cytoskeleton-dependent cytokinesis, cell cycle, cytokinetic processes, mitotic cytokinetic process, cell cycle regulation, and cell cycle G2/M phase transition.

Regardless of the identified exclusiveness, such exclusive gene sets shared similarities in remarkably enriched categories. In this study, movement of cell or subcellular components, microtubules, microtubule-based processes, and carbohydrate metabolic process were enriched throughout seed germination, indicating the positive effect of associated genes on the regulation of seed germination.

6.3.2 Germination Respiration Pathway

Seeds germination is related to respiratory pathways. The main respiration pathways are the glycolytic pathway (EMP), TCA, and PPP (Khan, 1971). By comparing the number of up-regulated DEGs in different respiratory pathways, it was found that EMP > TCA > PPP, and the down-regulated DEGs in different respiratory pathways followed the pattern EMP > PPP > TCA. There were fewer up-regulated DEGs in the EMP pathway and TCA pathway in CK versus MS than in CK versus PS or CK versus FS (Fig. 6.12a). 54 DEGs associated with seed respiratory had been recognized, including 21 glucose DEGs, 1 phosphohexose DEG, 3 phosphoenolpyruvate DEGs, 12 pyruvate DEGs, 9 acetyl-CoA DEGs, 7 citrate DEGs, 1 succinate DEG, and 1 fructose DEG (Fig. 6.12b).

In glucose, *PH01003291G0100* and *PH01001968G0050* DEGs associated with glucose metabolic processes exhibited continuous up-regulated patterns from CK to FS, which revealed that these genes might be responsible for glucose metabolism during seed germination. *PH01007114G0040* DEG annotated golgi stack, *PH01001359G0510* DEG annotated ammonia assimilation cycle, and *PH01001245G0400* DEG annotated carbohydrate transport and metabolism showed lower expression levels during the germination process.

In pyruvate, the *PH01000820G0110* DEG annotated to mitochondrial isocitrate lyase was

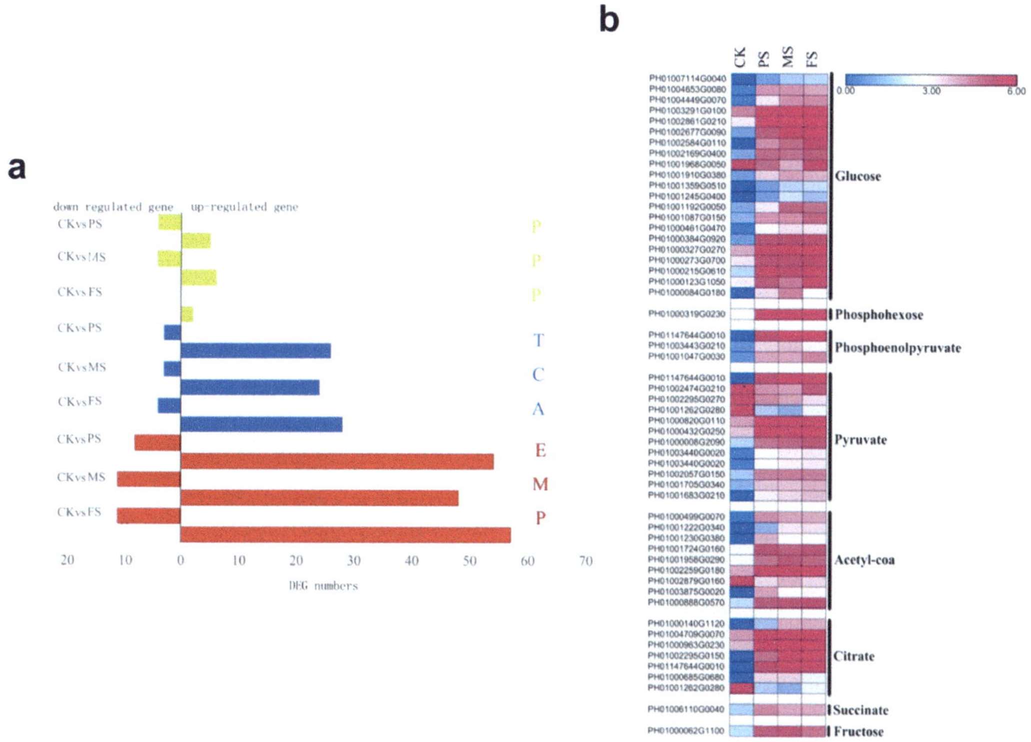

Fig. 6.12 (**a**) Up-regulated and down-regulated amounts of DEGs. The x-axis represents the number of DEGs, and the y-axis represents germination stage. (**b**) Main DEGs related to respiratory pathways during seed germination. The $-\log10P$ value and adjusted P values $\leq$ 0.01 are included to indicate significance

expressed highly from PS to FS; however, *PH01001262G0280DEG* annotated to the mitochondrial matrix was up-regulated in CK but remained at a low level from PS to MS. This phenomenon indicated these DEGs regulate seed respiration in different stages. The *PH01000432G0250* DEG related to energy production and conversion had high expression throughout seed germination.

In citrate, the DEGs associated with energy production and conversion (*PH01004709G0070, PH01000963G0230, PH01002295G0150, and PH01147644G0010*) were up-regulated significantly from PS to FS, but the expression of the PH01001262G0280 was much lower from PS to FS than in CK. DEGs in succinate (*PH01006110G0040*) and fructose (*PH01000062G1100*) related to carbohydrate transport and metabolism showed the same expression patterns: they were down-regulated in CK and up-regulated from PS to FS.

References

Alexandrov NN, Brover VV, Freidin S et al (2009) Insights into corn genes derived from large-scale cDNA sequencing. Plant Mol Biol 69:179–194

Ap RT (1980) Assessment of the contribution of metabolic pathways to plant respiration. In: Davies DD (ed) The biochemistry of plants, vol 2. Academic Press, New Yourk, p 129

Ap RT (1985) The organization of glycolysis and the oxidative pentose phosphate pathway. In: Douce R, Day DA (eds) Encyclopedia of plant physiology, vol 18. Springer, New York, pp 391–417

Audic S, Claverie JM (1997) The significance of digital gene expression profiles. Genome Res 7(10):986–995

Ayako N, Kanako H, Hideki G et al (2003) Brassinolide induces IAA5, IAA19, and DR5, a synthetic auxin response element in *Arabidopsis*, implying a cross talk point of brassinosteroid and auxin signaling. Plant Physiol 133:1843–1853

Bai QS, Hou D, Li L, et al (2016) Genome-wide analysis and expression characteristics of small auxin-up RNA (*SAUR*) genes in Moso bamboo (*Phyllostachys edulis*). Genome https://doi.org/10.1139/gen-2016-0097

Bewley JD (1997) Seed germination and dormancy. Plant Cell 9(7):1055–1066

Botha FC, O'Kennedy MM (1989) Characterization of the cytosolic aldolase from germinating *Phaseolus vulgaris* seeds. J Plant Physiol 135:433–438

Botha FC, Small JGC, Grobbelaar N (1983) The effect of water stress on the respiration and some aspects of respiratory metabolism of *Citrullus lantus* seeds. S Afr J Bot 2(3):247

Botha FC, Potgieter GP, Botha AM (1992) Respiratory metabolism and gene expression during seed germination. Plant Growth Regul 11:211–224

Bowman JL, Alvarez J, Weigel D et al (1993) Control of flower development in Arabidopsis thaliana by APETALA1 and interacting genes. Development 119:721–743

Cai CJ, Cao BH (2013) Physiological and biochemical changes of moso bamboo seed in artificial aging. Scientia Silvae Sinicae 49(8):29–34

Cai CJ, Liu F, Peng ZH et al (2009) Effect of storage temperature and moisture content on seed vitality of moso bamboo (*Phyllostachys edulis*). J Anhui Agric Univ 36:607–611

Cai CJ, Liu F, Guo QR et al (2010) A study on responses of moisture contents to seed germplasm preservation of moso baomboo (*Phyllostachys edulis*). Acta Agric Univ Jiangxiensis 32(2):0312–0317

Campbell JJN (1985) Bamboo flowering patterns: a global view with special reference to East Asia. J Am Bamboo Soc 6:17–35

Carrie C, Murcha MW, Giraud E et al (2013) How do plants make mitochondria? Planta 237(2):429–439

Cheng ZJ, Wang L, Sun W et al (2013) Pattern of auxin and cytokinin responses for shoot meristem induction results from the regulation of cytokinin biosynthesis by Auxin response factor 3. Plant Physiol 161:240–251

Chia TYP, Pike MJ, Rawsthorne S (2005) Storage oil breakdown during embryo development of *Brassica napus* (L.). J Exp Bot 56(415): 1285–1296

Cui K, He CY, Zhang JG et al (2012) Temporal and spatial profiling of internode elongation-associated protein expression in rapidly growing culms of bamboo. J Proteome Res 11:2492–2507

Dreni L, Pilatone A, Yun D et al (2011) Functional analysis of all AGAMOUS subfamily members in rice reveals their roles in reproductive organ identity determination and meristem determinacy. Plant Cell 23(8):2850–2863

Ehrenreich IM, Hanzawa Y, Chou L et al (2009) Candidate gene association mapping of Arabidopsis flowering time. Genetics 183:325–335

Ehrenshaft M, Brambl R (1990) Respiration and mitochondrial biogenesis in germinating embryos of maize. Plant Physiol 93:295–304

Elena P, Laura P, Sabrina S (2015) Plant hormone cross-talk: the pivot of root growth. J Exp Bot 66:1113–1121

Fornara F, de Montaigu A, Coupland G (2010) SnapShot: control of flowering in Arabidopsis. Cell 141:550, 550e1–550e2

Gahan PB, Dawson AL, Black M et al (1986) Localization of glucose-6-phosphatedehydrogenase activity in seeds and its possible involvement in dormancy breakage. Ann Bot 57:791–799

Gao J, Ge W, Zhang Y et al (2015) Identification and characterization of microRNAs at different flowering developmental stages in Moso bamboo (*Phyllostachys edulis*) by high-throughput sequencing. Mol Genet Genomics 290:2335–2353

Gonzalez DO, Vodkin LO (2007) Specific elements of the glyoxylate pathway play a significant role in the functional transition of the soybean cotyledon during seedling development. BMC Genomics 9(18):468

Gosling PC, Ross JD (1979) Characterization of glucose-6-phosphate dehydrogenase and 6-phosphogluconic dehydrogenase from hazel cotyledons. Phytochemistry 8:1441–1445

He CY, Cui K, Zhang JG et al (2013) Next-generation sequencing-based mRNA and microRNA expression profiling analysis revealed pathways involved in the rapid growth of developing culms in Moso bamboo. BMC Plant Biol 13:119

Hirokawa N, Noda Y, Tanaka Y et al (2009) Kinesin superfamily motor proteins and intracellular transport. Nat Rev Mol Cell Biol 10:682–696

Hou X, Ding L, Hao Y (2013) Crosstalk between GA and JA signaling mediates plant growth and defense. Plant Cell Rep 32:1067–1074

Howell KA, Millar AH, Whelan J (2006) Ordered assembly of mitochondria during rice germination begins with promitochondrial structures rich in components of the protein import apparatus. Plant Mol Biol 60:201–223

Hu HN, Gary AC (1990) Activities of catalase and pathway dehydrogenases release in *Nectarine* seed pentose phosphate during dormancy. J Am Soc Hortic Sci 115(6):987–990

Janzen DH (1976) Why bamboos wait so long to flower. Ann Rev Ecol Syst 7:347–391

Jeon JS, Jang S, Lee S et al (2000) Leafy hull sterilel is a homeotic mutation in a rice MADS box gene affecting rice flower development. Plant Cell 12(6):871–884

Jessica ASB, João HFC, David BM et al (2017) Autophagy deficiency compromises alternative pathways of respiration following energy deprivation in *Arabidopsis thaliana*. Plant Physiol 175:62–76

Jiang ZH (2002) The flowering and fruiting of bamboo and bamboo seedling breeding. In: Liu H, Luan SL, eds, Bamboo and rattan in the world, vol 17, 1st edn. Liaoning Science and Technology publishing house, Shenyang, pp 124–125

Jiang FL, Bo LP, Xu JJ et al (2018) Changes in respiration and structure of non-heading Chinese cabbage seeds during gradual artificial aging. Sci Hortic 238:14–22

Kazan K, John M (2013) MYC2: the master in action. Mol Plant 6:686–703

Khan AA (1971) Cytokinins: permissive role in seed germination. Science 171:353–359

Kobayashi K, Baba S, Obayashi T et al (2012) Regulation of root greening by light and auxin/cytokinin signaling in Arabidopsis. Plant Cell 24:1081

Li XW (2013) The law of material and energy metabolism regulated by the key enzyme of respiration in the process of seed germination. MA. Thesis. Lanzhou university, Lanzhou

Li DJ, Yang CH, Li XB et al (2009) Functional characterization of rice OsDof12. Planta 229:1159–1169

Li ZG, Chen HW, Li QT (2015) Three SAUR proteins SAUR76, SAUR77 and SAUR78 promote plant growth in Arabidopsis. Sci Rep 5:12477

Li L, Hu T, Li XP et al (2016) Genome-wide analysis of shoot growth-associated alternative splicing in moso bamboo. Mol Genet Genomics 291:1–20

Li XY, Xie LH, Zheng HF et al (2019a) Transcriptome profiling of postharvest shoots identifies PheNAP2- and PheNAP3-promoted shoot senescence. Tree Physiol 39:2027–2044

Li J, Zhang BS, Gao HY et al (2019b) Anatomical structural changes and correlation between water content and germination characteristics during seed germination of *Phyllostachy edulis* (CARRIÈRE) J. HOUZ. Propag Ornamental Plants 5:3–8

Lin XC, Chow TY, Chen HH et al (2010) Understanding bamboo flowering based on large-scale analysis of expressed sequence tags. Genet Mol Res 9:1085–1093

Liu M (2014) Changes of enzyme activity and gene expression related to respiratory metabolism during Shannon17 wheat seed germination. MA. Thesis. Shandong agriculture university, Taian

Liu F, Cao BH, Cai CJ et al (2009) Study on the change of physiology and biochemistry of *Phyllostachys edulis* seed during germination. Seed 28(2):12–14

Logan DC, Leaver CJ (2000) Mitochondria-targeted GFP high lights the heterogeneity of mitochondrial shape, size and movement within living plant cells. J Exp Bot 51:865–871

Maaike DJ, Mieke WA, Jose LG-M et al (2011) The *Solanum lycopersicum* AUXIN response factor 7 (SlARF7) mediates cross-talk between auxin and gibberellin signaling during tomato fruit set and development. J Exp Bot 62:617–626

Makita A (1998) The significance of the mode of clonal growth in the life history of bamboos. Plant Spec Biol 13:85–92

Mendes CR, Moraes DMD, Lima MDGDS et al (2009) Respiratory activity for the differentiation of vigor on soybean seeds lots. Revista Brasileira Sem 31(2):171–176

Michiko S, Nanako I, Tsuyoshi H et al (2015) The carboxyl-terminal tail of the stalk of Arabidopsis NACK1/HINKEL kinesin is required for its localization to the cell plate formation site. Plant Res 128:327–336

Møller IM (2001) Plant mitochondria and oxidative stress: electron transport, NADPH turnover, and metabolism of reactive oxygen species. Annu Rev Plant Physiol Plant Mol Biol 52:561–591

Morohashi Y (1980) Development of mitochondrial activity in pea cotyledons following imbibition; influence of the embryonic axis. J Exp Bot 31(3):805–812

Mouradov A, Cremer F, Coupland G (2002) Control of flowering time: interacting pathways as a basis for diversity. Plant Cell 14:S111–S130

Nadgauda RS, Parasharami VA, Mascarenhas AF (1990) Precocious flowering and seeding behaviour in tissue-cultured bamboos. Nature 344:335–336

Navrot N, Rouhier N, Gelhaye E et al (2007) Reactive oxygen species generation and antioxidant systems in plant mitochondria. Physiol Plant 129:185–195

Nawa Y, Asahi T (1971) Rapid development of mitochondria in pea cotyledons during the early stage of germination. Plant Physiol 48(6):671–674

Nemhauser JL, Hong FX, Chory J (2006) Different plant hormones regulate similar processes through largely nonoverlapping transcriptional responses. Cell 126:467–475

Pan RZ (2008) Respiratory metabolic pathways in plants. In: Li GY, Meng L, eds, Plant physiology, 6th edn. Higher education press, Beijing, pp 106–123

Paque S, Mouille G, Grandont L (2014) AUXIN binding protein1 links cell wall remodeling, auxin signaling, and cell expansion in Arabidopsis. Plant Cell 26:280–295

Paul KB, Ruth MB, Joshua SM et al (2004) Multiple pathways in the decision to flower: enabling, promoting, and resetting. Plant Cell 16:S18–S31

Peng J (2009) Gibberellin and jasmonate crosstalk during stamen development. J Integr Plant Biol 51:1064–1070

Peng ZH, Zhang CL, Zhang Y, et al (2013) Transcriptome sequencing and analysis of the fast-growing shoots of Moso bamboo (*Phyllostachys edulis*). PLoS ONE 8: e78944

Ren SJ (2010) Study on the cell cycle and physiological characteristics of pinus massoniana seed germination. MA. Thesis. Nanjing forestry university, Nanjing

Richards DE, King KE, Ait-ali T et al (2001) How gibberellin regulates plant growth and development: a molecular genetic analysis of gibberellin signaling. Annu Rev Plant Physiol Plant Mol Biol 52:67–88

Robson F, Costa MMR, Hepworth SR et al (2001) Functional importance of conserved domains in the flowering-time gene CONSTANS demonstrated by analysis of mutant alleles and transgenic plants. Plant J 28(6):619–631

Rodríguez JL, De Diego JG, Rodríguez FD et al (2015) Mitochondrial structures during seed germination and early seedling development in Arabidopsis thaliana. Biologia 70(8):1019–1025

Salon C, Raymond P, Pradet A (1988) Quantification of carbon fluxes through the tricarboxylic acid cycle in early germinating lettuce embryos. J Biol Chem 263:12278–12287

Samach A, Onouchi H, Gold SE et al (2000) Distinct roles of CONSTANS target genes in reproductive development of Arabidopsis. Science 288:1613–1616

Sarkar PK, Kumar PR, Singh AK, et al (2019) Effect of priming treatments on seed germination and seedling growth in bamboo [Dendrocalamus strictus (Roxb.) Nees]. Acta Ecologica Sinica. https://doi.org/10.1016/j.chnaes.2018.11.004

Simpson GG, Dean C (2002) Arabidopsis, the rosetta stone of flowering time? Science 296:285–289

Smiri M, Chaoui A, Ferjani EE (2009) Respiratory metabolism in the embryonic axis of germinating pea seed exposed to cadmium. J Plant Physiol 166 (3):259–269

Staehelin LA, Newcomb EH (2000) Membrane structure and membranous organelles. In: Buchanan BB, Gruissem W, Jones RL (eds) Biochemistry and molecular biology of plants. American Society of Plant Physiologists, Rockville, pp 1–50

Su YH, Liu YB, Zhang XS (2011) Auxin-cytokinin interaction regulates meristem development. Mol Plant 4:616–625

Sun Y, Tan XF, Luo M et al (2014) The sequencing analysis of transcriptome of *Vernicia fordii* flower buds at two development stages. Scientia Silvae Sinicae 50:70–74

Swamy PM, Sandhyarani CK (1986) Contribution of the pentose phosphate pathway and glycolytic pathway to dormancy breakage and germination of peanut *Arachishypogaea* seeds. I Exp Bot 37:80–88

Tanaka H, Ishikawa M, Kitamura S et al (2004) The AtNACK1/HINKEL and STUD/TETRASPORE/AtNACK2 genes, which encode functionally redundant kinesins, are essential for cytokinesis in Arabidopsis. Genes Cells 9:1199–1211

Tang Q, Ma XJ, Mo CM et al (2011) An efficient approach to finding Siraitia grosvenorii triterpene biosynthetic genes by RNAseq and digital gene expression analysis. BMC Genomics 12:343

Teale WD, Paponov IA, Palme K (2006) Auxin in action: signaling, transport and the control of plant growth and development. Nat Rev Mol Cell Biol 7:847–859

Tian Q, Reed J (1999) Control of auxin-regulated root development by the *Arabidopsis thaliana SHY2/IAA3* gene. Development 126:711–721

Tsuji H, Aya K, Ueguchi-Tanaka M et al (2006) MGAMYB controls different sets of genes and is differentially regulated by microRNA in aleurone cells and anthers. Plant J 47(3):427–444

Wang JY (2002) Glycolysis. In: Biochemistry, 3rd edn. Higher education press, Beijing, pp 63–90

Wang XY, Fang SZ (2007) Storage substance content and corresponding enzyme activity during the stratification of *Cyclocarya paliurus* seeds. J Nanjing for Univ (nat Sci) 31(1):111–113

Wang XX, Gao J, Zhou LR (2010) Research of POD isoenzyme and PAPD molecular detection on moso bamboo 60Co-γirradiation. Acta Laser Biology Sinica 19(6):758–763

Wang Y, Li Y, Xue H, et al (2015) Reactive oxygen species-provoked mitochondria-dependent cell death during ageing of elm (Ulmus pumila L.) seeds. Plant J 81(3):438–452

Wang YJ, Sun XP, Ding YL et al (2019) Cellular and molecular characterization of a thick-walled variant reveal a pivotal role of shoot apical meristem in transverse development of bamboo culm. J Exp Bot 70:3911–3926

Wei Q, Jiao C, Ding YL et al. Cellular and molecular characterizations of a slow-growth variant provide insights into the fast growth of bamboo. Tree Physiol 4:1–14

Wei Q, Jiao C, Guo L et al (2016) Exploring key cellular processes and candidate genes regulating the primary thickening growth of moso underground shoots. New Phytol 214:81–96

Wei Q, Guo L, Jiao C et al (2019) Characterization of the developmental dynamics of the elongation of a bamboo internode during the fast growth stage. Tree Physiol 39:1201–1214

Weitbrecht K, Müller K, Leubner-Metzger G (2011) First off the mark: early seed germination. J Exp Bot 62:3289–3309

Xu YM, Hao PY, Fei BH (2008) Dynamic change of anatomical structure and chemical constitute of bamboo shoot for *Phyuostachys pubescence* at development stage. J Northeast for Univ 36:8–16

Xu QT, Yang L, Zhou ZQ et al (2013) Process of aerenchyma formation and reactive oxygen species induced by water logging in wheat seminal roots. Planta 238:969–982

Yan Q (2001) Theory and determination of seed viability. In: Seed, vol 6. China agricultural press, Beijing, pp 103–106, 304

Yang WF, Zhang JL, Lü WZ et al (2014) Study on the differential genes expression in maize embryo treated by a controlled deterioration treatment. Sci Agric Sin 47(10):1878–1893

Yao Z, Tian F, Cao X, et al. (2016) Global transcriptomic analysis reveals the mechanism of *Phelipanche aegyptiaca* seed germination. Int J Mol Sci 17(7):1139

Yin XJ, He DL, Gupta R et al (2015) Physiological and proteomic analyses on artificially aged *Brassica napus* seed. Front Plant Sci 6:112

Yu YL, Guo GF, Lv DW et al (2014) Transcriptome analysis during seed germination of elite Chinese bread wheat cultivar Jimai 20. BMC Plant Biol 14:20

Zaynab M, Pan D, Fatima M et al (2018) Transcriptomic approach to address low germination rate in *Cyclobalnopsis gilva* seeds. S Afr J Bot 119:286–294

Zhang DF, Li B, Jia GQ, et al. (2008) Isolation and characterization of genes encoding GRF transcription factors and GIF transcriptional coactivators in Maize (Zea mays L.). Plant Sci 175(6):809–817

Zhang XM, Zhao L, Larson-Rabin Z, et al. (2012) De novo sequencing and characterization of the floral transcriptome of Dendrocalamus latiflorus (Poaceae: Bambusoideae). PLoS One 7(8):e42082

Zhang YX, Anja P, Caroline M et al (2014) The Aux/IAA gene rum1 involved in seminal and lateral root formation controls vascular patterning in maize (*Zea mays L.*) primary roots. J Exp Bot 65:4919–4930

Zhao YH, Yang SL, Liu HQ et al (2001) Relationship between phosphopentose pathway and seed dormancy releasing of *Panax quinque folius*. Chin Tradit Herbal Drugs 32(3):259–261

Zhou FC (1998) The flowering and fruiting of bamboo. In: Editor Board of Bamboo research (eds) Bamboo cultivation. Nanjing Forestry University Printing Company, Nanjing, pp 93–100

Alternative Splicing and Alternative Polyadenylation in Moso Bamboo

7

Lianfeng Gu, Hangxiao Zhang, and Jian Gao

Abstract

The processes of gene expression can happen at diverse levels and utilizes multiple molecular mechanisms. Alternative polyadenylation (APA) and alternative splicing (AS) account for two important regulatory events for generating distinct alternatively spliced isoforms or mRNA transcripts with different 3' termini at post-transcription level. Therefore, AS and APA increase diversity and complexity of transcriptome and proteome. Moso bamboo is non-timber forest species which present unique fast growing and late flowering characteristic. AS and APA are prevalent in bamboo to regulate development. Therefore, understanding the post-transcriptional mechanisms involving unique characteristic in moso bamboo will provide key insights into gene regulation on development. We review here the profile of AS and APA from latest research progresses which will help to understand underlying role of AS and APA during the diverse processes in moso bamboo.

L. Gu (✉) · H. Zhang
College of Forestry, Fujian Agriculture and Forestry University, Fuzhou, Fujian 350002, China
e-mail: lfgu@fafu.edu.cn

H. Zhang
e-mail: hangxiaozhang@126.com

J. Gao
Key Laboratory of Bamboo and Rattan Science and Technology, International Center for Bamboo and Rattan, State Forestry and Grassland Administration, Beijing 100102, China
e-mail: gaojian@icbr.ac.cn

7.1 Introduction

7.1.1 AS and APA in Plant

The discoveries of alternative splicing (AS) in 1978 are the key milestone in understanding RNA procession (Gilbert 1978). Most of precursor-mRNAs in plants include introns which must be spliced out during transcription to form mature mRNA before export from the nucleus to the cytoplasm. This splicing machine can generate multiple isoforms from a single gene including multiple introns by differential usage of splicing sites to increase the diversity of transcription (Reddy 2007). On the base of prevenient research finding, AS exerts a vital part in regulation at post-transcription level and represents larger proportion (Pan et al. 2008; Wang et al. 2008b). Partial AS regulation mechanisms have been revealed (Bartel 2004; Jones-Rhoades and Bartel 2004; Barbazuk et al. 2008). Apart from AS, alternative polyadenylation (APA) acts as another post-transcriptional level of regulation. APA is defined as multiple poly(A) sites can be generated from a gene to form distinct isoforms with different 3' termini (Tian and Manley 2017). The usage of poly(A) sites can be

J. Gao (ed.), *The Moso Bamboo Genome*, Compendium of Plant Genomes,
https://doi.org/10.1007/978-3-030-80836-5_7

altered during development (Shen et al. 2011) and hormone response (Hong et al. 2018). In arabidopsis *thaliana, Oryza sativa, Chlamydomonas reinhardtii,* and *Medicago truncatula,* the APA sites have been hosted in PlantAPAdb (Wu et al. 2016), which shown that APA is most prevalent events in plant.

Bamboo (*Poaceae: Bambusoideae*) belongs to the family Gramineae. Bamboo forest ecosystem has important role in carbon cycling due to its huge biomass. The moso bamboo genome sequences (Peng et al. 2013), herbaceous species as well as several woody bamboos (Guo et al. 2019) contribute to investigating functional evolution and molecular mechanisms for genes related with fast growing and late flowering characteristic. The purpose of this review was to characterize AS and APA atlas of moso bamboo. We anticipate that much broader studies in future will advance our understanding of post-transcriptional regulation in moso bamboo.

7.1.2 The Tissue-Specific AS and APA Pattern in Moso Bamboo

Multiple groups have identified many AS genes in plant many years ago (Haas et al. 2003; Iida et al. 2004; Kitagawa et al. 2005, Nagasaki et al. 2006, Wang and Brendel 2006; Chen et al. 2007; Wang et al. 2008a). With the development of height throughtput sequencing, recent studies in moso bamboo have shed some light on this important post-transcriptional regulation. In total, AS is suggested to regulate about 29,731 AS events from 36% of moso bamboo genes in samples of shoots, mature culms, flowering samples, and leaf using paired-end sequencing from Illumina HiSeq TM ^{2500} platform (Li et al. 2018). AS events within different tissues are compared, which reveals that they tissue specificity and most of AS events do not show overlap in all samples (Li et al. 2018). Using more tissues in another report, 266,711 typical AS events have been identified from 25,225 AS genes in 26 representative bamboo tissues by both single-molecule long-read sequencing from single-molecule real-time (SMRT) long-read isoform sequencing (Iso-Seq) by Pacific BioSciences (PacBio) and RNA-seq from Illumina sequencing (Zhao et al. 2018). Overlapped analysis between PacBio and Illumina reveals that about 81% AS events from Iso-Seq are also detected in RNA-Seq (Zhao et al. 2018). In total, this study identifies 885 AS genes, which can be detected in all samples and are obviously enriched into the mRNA processing, RNA metabolic processing, and RNA splicing together with RNA processing (Zhao et al. 2018). The conserved genes tend to have less tissue specificity. Especially, in moso bamboo, the lignin biosynthesis-associated genes present abundant AS (Zhao et al. 2018). In additional to above studies, PacBio Iso-Seq is used combined with RNA-Seq to reveal 21,254 AS events which are identified from leaf, lateral bud of rhizome, tip of underground rhizome, root initiating from rhizome, and tip of new shoot (Wang et al. 2017). Moreover, pair-wise comparison of different tissues reveals several hundred differential AS events (Wang et al. 2017). As reported by the systemic view of the percentage of AS type in moso bamboo, intron retention is dominant among the AS events (Wang et al. 2017; Li et al. 2018; Zhao et al. 2018). In addition to AS events in different tissues, GA treatment in seedling of moso bamboo also induces the differential AS events, which include 74 alternative donor (AltD), 135 alternative acceptor (AltA), 120 exon skipping (ES), and 113 intron retention (IR) (Zhang et al. 2018). These differential AS events generate from genes, which are involved in hormone signaling pathway and splicing factors (Zhang et al. 2018).

In addition to AS, APA has also been investigated using PacBio Iso-Seq and strand-specific polyadenylation site sequencing (PAS-Seq) in moso bamboo (Wang et al. 2017). In total, five tissues generate 6311 APA genes with two or more polyadenylation sites, which include 11 cellulose synthaselike (*Csl*), 11 cellulose synthase (*CesA*), and two lignin genes using PacBio full-length reads, which can avoid the artifacts because of possible endogenous priming (Wang et al. 2017). Case study from one of the most well-known flowing gene FPA shows that, within the final intron, one proximal poly(A) site is included, which can be validated by 3' RACE (Wang et al. 2017). Strand-specific PAS-seq reveals 1224 differential poly(A) genes, which show enrichment in diversity categories (such as transcription regulator, molecular transducer, or translation transporter and regulator) (Wang et al. 2017). At present, only limited tissues have been sequenced in moso bamboo. Thus, AS and APA events of moso bamboo are still far from saturation. With the development of sequencing technologies, more AS and APA information from spatiotemporal transcriptome will be added in moso bamboo.

7.2 The Database of AS and APA in Moso Bamboo

Since AS and APA are most prevalent, corresponding databases can provide comprehensive information for exploration and visualization of AS and APA. Splice variants information in plant are stored at different databases (Kim et al. 2005, Holste et al. 2006, Stamm et al. 2006, Thierry-Mieg and Thierry-Mieg 2006, Bhasi et al. 2007, de la Grange et al. 2007, Kim et al. 2007, Rambaldi et al. 2007, Ryan et al. 2008). At present, there are several database resources for moso bamboo. The genome databases are provided by BambooGDB (http://www.bamboogdb.org) and genobank (http://www.genobank.org/bamboo), which are wonderful resource to deposit genome resource of moso bamboo (Zhao et al. 2014) and diploid herbaceous and three polyploid woody lineages (Guo et al. 2019).

With the release of post-transcriptional information from moso bamboo, there are two databases including PhePacBio (Wang et al. 2017) and ASmiR (Wang et al. 2019a) to deposit post-transcriptional regulation information about moso bamboo. PhePacBio (http://forestry.fafu.edu.cn/db/PhePacBio) contains the genome-wide profile for AS and APA in the rhizome system using SMRT Iso-Seq (Wang et al. 2017). For quantification of APA, the database also contains differential APA sites in rhizome system using PAS-seq. In addition to the post-transcriptional profile of bamboo, PhePacBio also contains a series of tools for query, analysis, and visualization such as BLAST, sequence extraction, and coverage plots from bigwiggle for gene expression levels. The tracks of genome browser include multiple samples, such as flower, leaf, rhizome system, and seedling in response to different hormone treatments. ASmiR (http://forestry.fafu.edu.cn/bioinfor/db/ASmiR) is the other database, which mainly identifies and deposited miRNA targets located in AS region of circular or linear RNAs. This database deposits detail coordinate information of 12,981 highly reliable AS events which includes 5331 miRNA targets located in alternatively spliced regions events in moso bamboo (Wang et al. 2019a).

In future, post-transcriptional information from more tissues and treatment will be generated in moso bamboo. Thus, these databases should be continuously maintained when new AS or APA events are published. At the same time, most of datasets use different methods for libraries construction or different sequencing platform. It will be very useful to integrate these datasets from different resource and provide uniform database interface for end-user to investigate the potential molecular mechanism and biological function from these post-transcriptional regulations in moso bamboo.

7.3 Methods for Identification of AS and APA

7.3.1 Experiment and Sequencing

The methods for AS identification included probe-based method, full-length cDNA clones, RT-PCR, RNA-Seq, and PacBio Iso-Seq, respectively. Microarray is a traditional method to quantify gene expression by hybridization-based method. Several splicing alteration observes by investigation of multiple probe sets of the same gene (Pearson et al. 2008). Furthermore, based on the Affymetrix exon junction array, it is reported that MADS+ can discover differential splicing (Shen et al. 2009). Exon junction microarray application shows that it is powerful in animal systems to detect known splicing variants (Johnson et al. 2003). However, exon junction array relies on reference annotation of splice junction and cannot detect new splice isoforms. Exon array is also used to reveal the hnRNP L (splicing regulator) targets (Hung et al. 2008), many splicing variants changes (Kwan et al. 2008), and splicing factor polypyrimidine tract binding protein (PTB) (Xing et al. 2008). For example, exon array of SMN-deficient mice identifies hundreds of genes with alteration of mRNA splicing (Zhang et al. 2008). However, high-uncertainty and low validation rate depress the application of exon array on detecting AS in plant. At the same time, exon array cannot be used to detect IR which is the most prevalent phenomena in plant. Thus, probe-based method for AS identification has not been reported in moso bamboo.

Full-length cDNA clone is another wonderful method for detecting AS (Kikuchi et al. 2003). This method has been used in moso bamboo to generate 10,608 putative full-length cDNA sequences and 38,000 expressed sequence tags (EST) which revealed the AS in 304 cDNA groups (Peng et al. 2010). EST from young leaf (Gao et al. 2011) and flowering sample (Lin et al. 2010) is also reported in bamboo. But it is costly and non-quantitative to sequence ESTs and cDNA (Wang et al. 2008a,2008b). Thus, recently studies seldom used this technology to detect AS in moso bamboo.

After Illumina high-throughput sequencing is developed, RNA-Seq turned into the gold standard to identify AS in primates and plants. About 90% genes with AS in human have been detected from different tissue types (Wang et al. 2008b), and lineage-specific and sex-specific alteration of AS is revealed (Blekhman et al. 2009) by RNA-Seq. With this techonology, ES is identified as the most prevalent form of AS in human (Sultan et al. 2008). In arabidopsis, RNA-Seq was used to detect the AS change under different stress treatments, and about 40% multiexon genes are identified as AS genes (Filichkin et al. 2009). This is the first publication using RNA-Seq to investigate AS in plant species. Then AS in rice also is analyzed using this method (Zhang et al. 2010). Since RNA-Seq for quantitative measurements of splicing isoforms is possible (Pan et al. 2008), analyzing differential splicing by this technology is possible, which can provide unbiased isoform determination and avoid signal-to-noise problem (Moore and Silver 2008). Due to this advantage, RNA-Seq presently provides most of the transcriptome-wide quantification of AS in moso bamboo (Wang et al. 2017; Li et al. 2018; Zhang et al. 2018).

Comparison with RNA-Seq based on PacBio, Illumina platform, together with Oxford Nanopore Technologies (ONT), provides more advantages with regard to sequence length to identify AS and APA (Zhao et al. 2019). The ONT and PacBio derived long reads are more advantageous in read length. At present, two distinct platforms including PacBio (Wang et al. 2017; Zhao et al. 2018) and Illumina (Wang et al. 2017; Li et al. 2018; Zhao et al. 2018) were used for identification of AS in moso bamboo.

The AS events based on high-throughput sequencing technology need validation by RT-PCR which can provide quantitative measurements using exon junction-specific primers. For example, 37 splicing isoforms for Isochorismate synthase (ICS) gene are detected by sequencing of 184 RT-PCR *Populus* clones (Yuan et al. 2009). Hundreds of significant changes in splicing isoforms are detected in CBC mutants using this way (Raczynska et al. 2009). Differential AS events in moso bamboo were validated by RT-PCR (Wang et al. 2017). However, RT-PCR method is low throughput, and it is quite impractical to detect AS at a genome-wide level.

At present, the gold standard for quantitative profiling of APA is poly(A) tag sequencing (PAT-Seq) approach, which has been used in Arabidopsis (Wu et al. 2011; Zhang et al. 2015), rice (Fu et al. 2016), and moso bamboo (Wang et al. 2017). But, PAT-Seq based on short reads has internal priming issue, which can be overcome by long reads from PacBio and ONT. At present, the strategies adopted in moso bamboo is a combination the advantage of long reads for qualitative evaluation and short reads for quantification (Wang et al. 2017).

7.3.2 Bioinformatics Analysis

For genome-wide RNA-Seq analysis of AS, rMATS is one statistical method for the AS identification and quantification and yields high validation rate (Shen et al. 2014). Several studies in moso bamboo use this method to analyze differential AS events. For PAT-Seq analysis of APA, differential poly(A)-clustering (DPAC) is the pipeline to identify differential poly(A)-cluster usage (Routh 2019). For PacBio long reads, the transcriptome analysis pipeline for isoform sequencing (TAPIS) (Abdel-Ghany et al. 2016) and post-transcriptional regulation analysis pipeline for Iso-Seq (PRAPI) (Gao et al. 2018) are available tools to identify AS and APA. PRAPI also can be used for other analysis, such as circular RNAs (circRNAs), alternative transcription initiation (ATI), and natural antisense transcripts (NAT) (Gao et al. 2018). For example, NAT from moso bamboo seedling is identified by PRAPI (Zhang et al. 2018).

7.4 The Potential Functions of AS and APA

7.4.1 Alteration in Protein Domain by AS and APA

One outstanding question is how AS and APA contribute to biological functionality. The most intuitive function of AS and APA can generate protein isoforms. Mass spectrometry data in moso bamboo provides direct evidences that different protein isoforms from the same locus can be translated (Yu et al. 2019). In total, more than 64% AS events are located in open reading frame (ORFs) in moso bamboo (Li et al. 2018), which is consistent with that AS favored alteration in coding regions rather that random distribution (Kriventseva et al. 2003). Previous study about comparing protein domains between variants identifies changes in 475 genes (Loraine et al. 2003). Furthermore, 554 protein domains are identified to be altered by AS in human (Resch et al. 2004). Another interesting example in arabidopsis is JAZ which contain two conserved domain (ZIM and Jas) (Chico et al. 2008; Staswick 2008). One isoform JAZ10.4 deriving from AltD produces one frame-shift protein at the C-terminal and deletes Jas domain (Chung and Howe 2009). In addition, JAZ10.4 with no Jas domain will be degraded through binding to JAZ proteins like MYC2, but not COI1 (Chung and Howe 2009). By this way, AS enhances JA-mediated responses through generating the isoform of splicing-out domain that is not destroyed via SCF^{COI1}, the ubiquitin E3 ligase (Chung and Howe 2009).

In moso bamboo, mass spectrometry identified four different types for domain alteration of protein isoforms by label-free high-throughput quantitative proteomic analysis (Yu et al. 2019). In the protein domains, the dramatically changed protein isoforms exert an important part during large-scale evolution (Tress et al. 2007). Mass spectrometry in moso bamboo reveals that the same gene can generate different protein isoforms with completely different domains by post-transcriptional regulation in a total of 134 genes (Yu et al. 2019). More evidences show that splicing isoforms produce novel proteins and have important function. For example, isoform SR45.1 has important function in flower petal, whereas isoform SR45.2 pays major function in root growth (Zhang and Mount 2009). In future, more experimental evidences are essential to explore biological function of distinct protein isoforms related with fast growing and late flowering characteristic in moso bamboo.

7.4.2 Nonsense-Mediated mRNA Decay (NMD) Regulation by AS and APA

NMD can degrade isoforms containing premature termination codons (PTCs), and this indicates a mechanism of quality control at post-transcription level (Chang et al. 2007). Splicing isoform with NMD has biological function rather than stochastic noise and can regulate the transcription level of normal transcription (Spellman et al. 2007; Schoning et al. 2008). For example, HnRNP L can bind to conserved CA-rich motif of its own pre-mRNA and an analogous hnRNP LL gene (Rossbach et al. 2009). Including exon 6A in the final splicing isoform leads to NMD and down regulates the level of normal transcript by negative feedback. By this way, hnRNP L can autoregulate its own mRNA and cross-regulate of an analogous hnRNP LL gene (Rossbach et al. 2009). In addition to the role of regulator using NMD, 601 splicing isoforms with PTCs could be detected by mass spectrometry in moso bamboo (Yu et al. 2019), which was consistence with previous report that even isoforms with frame-shift are linked with polyribosomes, suggesting these isoforms can exported to cytoplasm and are translated into proteins (Ner-Gaon et al. 2004).

7.4.3 Alteration of miRNA Target Sites by AS and APA

MicroRNAs (miRNAs) were short single-strand RNAs that play role in suppressing mRNA translation (Brodersen et al. 2008) and degradation (Henderson and Jacobsen 2008). In plant, fold back structures in primary miRNA (pri-miRNA) is trimmed to miRNA precursors (pre-miRNAs) further processed into the mature miRNA in nucleus (Han et al. 2004; Yang et al. 2006; Matranga and Zamore 2007) by DCL1 that contained additional cofactors, like HYL1 (Han et al. 2004; Vazquez et al. 2004), SE (Yang et al. 2006), CBP20, and CBP80 (Gregory et al. 2008; Kim et al. 2008; Laubinger et al. 2008). Then mature miRNAs associates with different AGO proteins define RNA silencing complexes and are involved in miRNA regulatory pathway (Vaucheret 2008). Target mimics show that miRNA in arabidopsis can regulate morphology and development (Todesco et al. 2010). The miRNA target happens as early as embryos (Nodine and Bartel 2010) and leads to the mRNA-destabilization both in plant and animal (Guo et al. 2010). Investigating the impact of AS on miRNAs target in plant is amusing because most plant target sites locate in coding regions, which provides one chance for miRNAs to regulate splice variants in a qualitative fashion. In total, 5,331 miRNA targets locating in alternatively spliced regions are identified in moso bamboo (Wang et al. 2019a). However, the impact of the interplay between miRNA and AS should be further explored in bamboo, so validating the miRNA-targeted functional splicing variants in future is of great importance.

7.4.4 Diverse Back-Splicing Landscape: Insights into Circular RNAs in Moso Bamboo

The circular RNA is produced by back-splicing of downstream 3' acceptor site and the upstream 5' donor site. Unlike linear transcripts, circular RNA presents closed circular structure and is not affected by exonuclease. The expression of circular RNA is relatively stable and suitable as biomarker (Chen et al. 2019; Vo et al. 2019). In 1979, circular RNA is observed with the aid of electron microscopy (Hsu and Coca-Prados 1979). After that, circular RNA is generally regarded as by-product of splicing process for more than thirty years. Circular RNA does not include typical poly(A) structure. Thus, circular RNA cannot be detected by conventional transcriptome library which is constructed through the process of poly(A) selection. Thus, circular RNAs or other non-coding RNAs are greatly underestimated due to the method for library construction (Szabo and Salzman 2016). After library construction and high-throughput sequencing are developed, circular RNA is widespread in many species through non-poly(A) selection transcriptome sequencing methods (Jeck et al. 2013; Memczak et al. 2013). Plant circRNAs are detected from rice (Lu et al. 2015), arabidopsis (Ye et al. 2015; Zhang et al., 2019), tomato (Zuo et al. 2016), cotton (Zhao et al. 2017), and maize (Chen et al. 2018). Circular RNA presents a very old mechanism in evolution (Wang et al. 2014).

In total, 175 circRNAs in lariat intron and 720 circRNAs in moso bamboo produced by exon circularization have been identified from eight different heights in shoots using rRNA depleted libraries (Wang et al. 2019b). circRNAs subgroups show high expression levels in the early shoot elongation period (Wang et al. 2019b). Interestingly, 18 candidate 5'- and 26 candidate 3'-back-spliced circRNA isoforms are identified in moso bamboo (Wang et al. 2019b), which conform to the candidate back-splicing patterns previously identified from circRNAs (Zhang et al. 2016).

7.5 Future Prospects

The expression of isoform shows tissue and development stage specificity. The tight regulation is actualized by the reciprocity of diverse cis-regulatory factors and trans-splicing factors. Two well-characterized regulatory factors (SR protein and hnRNPs) regulate not only splicing, but also translation (Blaustein et al. 2009). SR proteins have been suggested to bind to the splicing enhancers in exons (Blencowe 2000; Graveley 2000; Lam and Hertel 2002) and recruit the splicing component by RS domain and/or by antagonizing the action of nearby silencer elements (Shen and Green 2004; Shen et al. 2004). In addition to splicing factor, transcription factor can have indirect effects on AS through affecting RNA polymerase II elongation efficiency (Ip et al. 2011). Thus, alteration in splicing factor as well as change in transcription factor can cause the alteration in splicing patterns. Cis-acting regulatory motifs such as 3' and 5' splicing sites, or branch site play indispensable role in recognizing the boundaries of introns/exons or the generation of functional splicing variants (Cartegni et al. 2002). Since prediction of splicing regulatory motif is less accurate, high-throughput sequencing for RNAs extracted through cross-linking immunoprecipitation (CLIP-seq or HITS-CLIP) (Licatalosi et al. 2008; Sanford et al. 2008) is the most appropriate high-throughput techniques to detect functional splicing regulatory element. Using this technology, global poly (A) usage has been identified in HLP1 mutation, an hnRNP A/B protein (Zhang et al. 2015). We expect new knowledge about post-transcriptional regulation using this technology in moso bamboo in future.

APA also was regulated by cis-regulatory elements or polyadenylation factor (RNA binding proteins) to determine the usage of distal or proximal PAS (Tian and Manley 2017). Interestingly, N^6-methyladenosine (m^6A) is highly enriched in 3' UTR. For the three m^6A methylase complexes, knockdown of their components altered the proximal poly(A) site use (Ke et al. 2015). VIRMA recruiting METTL3/METTL14/WTAP is related to the polyadenylation cleavage factors, which facilitates the selection of proximal polyadenylation (Yue et al. 2018). m^6A-level and isoform-characterization sequencing (m^6A-LAIC-seq) also revealed bias of methylated and nonmethylated transcript isoforms coupled with proximal APA and distal sites, respectively (Molinie et al. 2016). At present, the interplay between RNA modification and post-transcriptional regulation in moso bamboo remains unknown.

Since AS and APA are the major contribution for producing protein diversity, it is necessitous to explore the function of different isoforms at a large scale. Recent studies reveal numerous APA and AS variants within moso bamboo based on experimental and bioinformatics methods; however, it remains challenging to investigate the function impact of these isoforms. The speed of experimental identification of isoforms function is asymmetrical with the speed of detecting isoforms using high-throughput sequencing. Hence, it is still important to elucidate the diverse functions to understand the important roles of APA and AS. Investigating biological function of specifically isoform is time-consuming because there is no high-through experimental method to identify whether isoforms own specific function in vivo. The technology of systematic overexpression system (Qiao et al. 2014; Ye et al. 2017) along with genome editing (Ye et al. 2019) in bamboo will be useful tools to investigate the molecular mechanism for these isoforms in bamboo. We expect that the knowledge about the regulation of AS and APA in moso bamboo will continue to grow rapidly with the development of different technologies from both wet-lab experiments and dry computational methods (Fig. 7.1 and Tables 7.1 and 7.2).

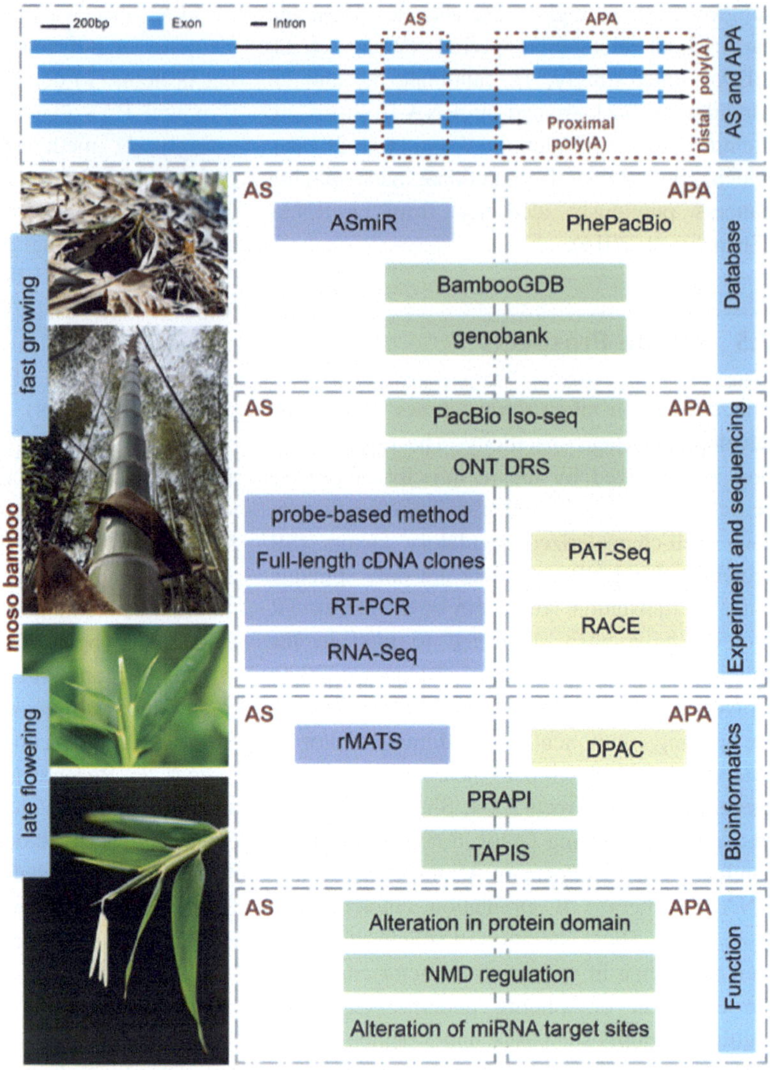

Fig. 7.1 AS and APA within moso bamboo. Top panel presented one case genes including both AS and APA within moso bamboo. Left panel shown fast growing shoots and flowering bamboo, which were two unique characteristics in moso bamboo. The right panel summarized available experimental and bioinformatics method and databases to identify and deposit AS and APA within moso bamboo. Bottom panel presented multiple biological functions of AS and APA within moso bamboo

Table 7.1 List of all publications related with alternative splicing and alternative polyadenylation in moso bamboo

Types of tissue	Sequencig technology	Results	References
Shoots, mature culms, flowering samples, and leaf	Paired-end sequencing (Illumina HiSeq ™ 2500 platform)	29,731 AS events	Li et al. (2018)
Rhizome, root, shoot, leaf, sheath, bud(different developmental stages)	SMRT long-read isoform sequencing (Iso-Seq) and RNA-seq (Illumina sequencing)	266,711 AS events	Zhao et al. (2018)
Underground rhizome tip, lateral rhizome bud, new shoot tip, root initiated from rhizome, and leaf	PacBio Iso-Seq, RNA-Seq, and PAS-Seq	21,254 AS events/6311 genes with APA	Wang et al. (2017)

(continued)

Table 7.1 (continued)

Types of tissue	Sequencig technology	Results	References
GA treatment seedling	Paired-end reads (Illumina HiSeq™ ^{2500})	22,196 AS events	Zhang et al. (2018)
Shoots just breaking out from ground, shoots reaching a height of ~40 cm, young leaves, shoots, and roots from germinating seeds	Full-length cDNA clone	288 AS events	Peng et al. (2010)
Leaves, roots, and stems of seedlings	LC–MS/MS on an Orbitrap fusion coupled to an EASY-nano-LC system	93,481 unique isoforms could be translated into 163,669 potential amino acid sequences	Yu et al. (2019)
Shoots from eight different heights	Paired-end reads (Illumina HiSeq ™ ^{2500})	18 alternative 5' back-spliced circular RNA isoforms and 26 alternative 3' back-spliced circular RNA isoforms are identified	Wang et al. (2019b)

Table 7.2 List of all abbreviation

Abbreviations	Full name
AS	Alternative splicing
APA	Alternative polyadenylation
PacBio	Pacific biosciences
SMRT	Single-molecule real-time
Iso-Seq	Long-read isoform sequencing
IR	Intron retention
ES	Exon skipping
AltA	Alternative acceptor
AltD	Alternative donor
PAS-Seq	Polyadenylation site sequencing
CesA	Cellulose synthase
Csl	Cellulose synthaselike
PTB	Polypyrimidine tract binding protein
EST	Expressed sequence tags
ONT	Oxford nanopore technologies
ICS	Isochorismate synthase
PAT-Seq	Poly(A) tag sequencing
DPAC	Differential poly(A)-clustering
PRAPI	Post-transcriptional regulation analysis pipeline for Iso-Seq
TAPIS	Transcriptome analysis pipeline for isoform sequencing
ATI	Alternative transcription initiation
NAT	Natural antisense transcripts

(continued)

Table 7.2 (continued)

Abbreviations	Full name
circRNAs	Circular RNAs
ORFs	Open reading frame
PTCs	Premature termination codons
NMD	Nonsense-mediated mRNA decay
miRNAs	MicroRNAs
pri-miRNA	Primary miRNA
pre-miRNAs	miRNA precursors
m^6A	N^6-methyladenosine
m^6A-LAIC-seq	m^6A-level and isoform-characterization sequencing

References

Abdel-Ghany SE, Hamilton M, Jacobi JL et al (2016) A survey of the sorghum transcriptome using single-molecule long reads. Nat Commun 7:11706

Barbazuk WB, Fu Y, McGinnis KM (2008) Genome-wide analyses of alternative splicing in plants: opportunities and challenges. Genome Res 18:1381–1392

Bartel DP (2004) MicroRNAs: genomics, biogenesis, mechanism, and function. Cell 116:281–297

Bhasi A, Pandey RV, Utharasamy SP, Senapathy P (2007) EuSplice: a unified resource for the analysis of splice signals and alternative splicing in eukaryotic genes. Bioinformatics 23:1815–1823

Blaustein M, Quadrana L, Risso G et al (2009) SF2/ASF regulates proteomic diversity by affecting the balance between translation initiation mechanisms. J Cell Biochem 107:826–833

Blekhman R, Marioni JC, Zumbo P et al (2009) Sex-specific and lineage-specific alternative splicing in primates. Genome Res 20:180-189.

Blencowe BJ (2000) Exonic splicing enhancers: mechanism of action, diversity and role in human genetic diseases. Trends Biochem Sci 25:106–110

Brodersen P, Sakvarelidze-Achard L, Bruun-Rasmussen et al (2008) Widespread translational inhibition by plant miRNAs and siRNAs. Science 320:1185–1190

Cartegni L, Chew SL, Krainer AR (2002) Listening to silence and understanding nonsense: exonic mutations that affect splicing. Nat Rev Genet 3:285–298

Chang YF, Imam JS, Wilkinson MF (2007) The nonsense-mediated decay RNA surveillance pathway. Annu Rev Biochem 76:51–74

Chen FC, Wang SS, Chaw SM et al (2007) Plant gene and alternatively spliced variant annotator. a plant genome annotation pipeline for rice gene and alternatively spliced variant identification with cross-species expressed sequence tag conservation from seven plant species. Plant Physiol 143:1086–1095

Chen L, Zhang P, Fan Y et al (2018) Circular RNAs mediated by transposons are associated with transcriptomic and phenotypic variation in maize. New Phytol 217:1292–1306

Chen S, Huang V, Xu X et al (2019) Widespread and functional RNA circularization in localized prostate cancer. Cell 176:831–843

Chico JM, Chini A, Fonseca S et al (2008) JAZ repressors set the rhythm in jasmonate signaling. Curr Opin Plant Biol 11:486–494

Chung HS, Howe GA (2009) A critical role for the TIFY motif in repression of jasmonate signaling by a stabilized splice variant of the JASMONATE ZIM-domain protein JAZ10 in arabidopsis. Plant Cell 21:131–145

de la Grange P, Dutertre M, Correa M et al (2007) A new advance in alternative splicing databases: from catalogue to detailed analysis of regulation of expression and function of human alternative splicing variants. BMC Bioinform 8:180

Filichkin SA, Priest HD, Givan SA et al (2010) Genome-wide mapping of alternative splicing in arabidopsis thaliana. Genome Res 20:45-58

Fu H, Yang D, Su W et al (2016) Genome-wide dynamics of alternative polyadenylation in rice. Genome Res 26:1753–1760

Gao Y, Wang H, Zhang H et al (2018) PRAPI: post-transcriptional regulation analysis pipeline for Iso-Seq. Bioinformatics 34:1580–1582

Gao ZM, Li CL, Peng ZH (2011) Generation and analysis of expressed sequence tags from a normalized cDNA library of young leaf from Ma bamboo (Dendrocalamus latiflorus Munro). Plant Cell Rep 30:2045–2057

Gilbert W (1978) Why genes in pieces? Nature 271:501

Graveley BR (2000) Sorting out the complexity of SR protein functions. RNA 6:1197–1211

Gregory BD, O'Malley RC, Lister R et al (2008) A link between RNA metabolism and silencing affecting arabidopsis development. Dev Cell 14:854–866

Guo H, Ingolia NT, Weissman JS et al (2010) Mammalian microRNAs predominantly act to decrease target mRNA levels. Nature 466:835–840

Guo ZH, Ma PF, Yang GQ et al (2019) Genome sequences provide insights into the reticulate origin and unique traits of woody bamboos. Mol Plant 12:1353–1365

Haas BJ, Delcher AL, Mount SM et al (2003) Improving the arabidopsis genome annotation using maximal transcript alignment assemblies. Nucleic Acids Res 31:5654–5666

Han MH, Goud S, Song L et al (2004) The arabidopsis double-stranded RNA-binding protein HYL1 plays a role in microRNA-mediated gene regulation. Proc Natl Acad Sci U S A 101:1093–1098

Henderson IR, Jacobsen SE (2008) Sequencing sliced ends reveals microRNA targets. Nat Biotechnol 26:881–882

Holste D, Huo G, Tung V et al (2006) HOLLYWOOD: a comparative relational database of alternative splicing. Nucleic Acids Res 34:D56-62

Hong L, Ye C, Lin J et al (2018) Alternative polyadenylation is involved in auxin-based plant growth and development. Plant J 93:246–258

Hsu M-T, Coca-Prados M (1979) Electron microscopic evidence for the circular form of RNA in the cytoplasm of eukaryotic cells. Nature 280:339–340

Hung LH, Heiner M, Hui J et al (2008) Diverse roles of hnRNP L in mammalian mRNA processing: a combined microarray and RNAi analysis. RNA 14:284–296

Iida K, Seki M, Sakurai T et al (2004) Genome-wide analysis of alternative pre-mRNA splicing in arabidopsis thaliana based on full-length cDNA sequences. Nucleic Acids Res 32:5096–5103

Ip JY, Schmidt D, Pan Q et al (2011) Global impact of RNA polymerase II elongation inhibition on alternative splicing regulation. Genome Res 21:390–401

Jeck WR, Sorrentino JA, Wang K et al (2013) Circular RNAs are abundant, conserved, and associated with ALU repeats. RNA 19:141–157

Johnson JM, Castle J, Garrett-Engele P et al (2003) Genome-wide survey of human alternative pre-mRNA splicing with exon junction microarrays. Science 302:2141–2144

Jones-Rhoades MW, Bartel DP (2004) Computational identification of plant microRNAs and their targets, including a stress-induced miRNA. Mol Cell 14:787–799

Ke S, Alemu EA, Mertens C et al (2015) A majority of m6A residues are in the last exons, allowing the potential for 3' UTR regulation. Genes Dev 29:2037–2053

Kim N, Alekseyenko AV, Roy M et al (2007) The ASAP II database: analysis and comparative genomics of alternative splicing in 15 animal species. Nucleic Acids Res 35:D93-98

Kim P, Kim N, Lee Y et al (2005) ECgene: genome annotation for alternative splicing. Nucleic Acids Res 33:D75-79

Kim S, Yang JY, Xu J et al (2008) Two cap-binding proteins CBP20 and CBP80 are involved in processing primary MicroRNAs. Plant Cell Physiol 49:1634–1644

Kitagawa N, Washio T, Kosugi S et al (2005) Computational analysis suggests that alternative first exons are involved in tissue-specific transcription in rice (Oryza sativa). Bioinformatics 21:1758–1763

Kriventseva EV, Koch I, Apweiler R et al (2003) Increase of functional diversity by alternative splicing. Trends Genet 19:124–128

Kwan T, Benovoy D, Dias C et al (2008) Genome-wide analysis of transcript isoform variation in humans. Nat Genet 40:225–231

Lam BJ, Hertel KJ (2002) A general role for splicing enhancers in exon definition. RNA 8:1233–1241

Laubinger S, Sachsenberg T, Zeller G et al (2008) Dual roles of the nuclear cap-binding complex and SERRATE in pre-mRNA splicing and microRNA processing in arabidopsis thaliana. Proc Natl Acad Sci USA 105:8795–8800

Li L, Shi Q, Hou D et al (2018) Transcriptome analysis of alternative splicing in different moso bamboo tissues. Acta Physiol Plant 40:89

Licatalosi DD, Mele A, Fak JJ et al (2008) HITS-CLIP yields genome-wide insights into brain alternative RNA processing. Nature 456:464–469

Lin XC, Chow TY, Chen HH et al (2010) Understanding bamboo flowering based on large-scale analysis of expressed sequence tags. Genet Mol Res 9:1085–1093

Loraine AE, Helt GA, Cline MS et al (2003) Exploring alternative transcript structure in the human genome using blocks and InterPro. J Bioinform Comput Biol 1:289–306

Lu T, Cui L, Zhou Y et al (2015) Transcriptome-wide investigation of circular RNAs in rice. RNA 21:2076–2087

Matranga C, Zamore PD (2007) Small silencing RNAs. Curr Biol 17:R789-793

Memczak S, Jens M, Elefsinioti A et al (2013) Circular RNAs are a large class of animal RNAs with regulatory potency. Nature 495:333–338

Molinie B, Wang J, Lim KS et al (2016) m(6)A-LAIC-seq reveals the census and complexity of the m(6)A epitranscriptome. Nat Methods 13:692–698

Moore MJ, Silver PA (2008) Global analysis of mRNA splicing. RNA 14:197–203

Nagasaki H, Arita M, Nishizawa T et al (2006) Automated classification of alternative splicing and transcriptional initiation and construction of visual database of classified patterns. Bioinformatics 22:1211–1216

Ner-Gaon H, Halachmi R, Savaldi-Goldstein S et al (2004) Intron retention is a major phenomenon in alternative splicing in arabidopsis. Plant J 39:877–885

Nodine MD, Bartel DP (2010) MicroRNAs prevent precocious gene expression and enable pattern formation during plant embryogenesis. Genes Dev 24:2678–2692

Pan Q, Shai O, Lee LJ et al (2008) Deep surveying of alternative splicing complexity in the human transcriptome by high-throughput sequencing. Nat Genet 40:1413–1415

Pearson JL, Robinson TJ, Munoz MJ et al (2008) Identification of the cellular targets of the transcription factor TCERG1 reveals a prevalent role in mRNA processing. J Biol Chem 283:7949–7961

Peng Z, Lu T, Li L et al (2010) Genome-wide characterization of the biggest grass, bamboo, based on 10,608 putative full-length cDNA sequences. BMC Plant Biol 10:116

Peng Z, Lu Y, Li L et al (2013) The draft genome of the fast-growing non-timber forest species moso bamboo (Phyllostachys heterocycla). Nat Genet 45:456–461

Qiao G, Yang H, Zhang L et al (2014) Enhanced cold stress tolerance of transgenic Dendrocalamus latiflorus Munro (Ma bamboo) plants expressing a bacterial CodA gene. In Vitro Cellular & Developmental Biology-Plant 50:385–391

Raczynska KD, Simpson CG, Ciesiolka A et al (2010) Involvement of the nuclear cap-binding protein complex in alternative splicing in arabidopsis thaliana. Nucleic Acids Res 38:265-278

Rambaldi D, Felice B, Praz V et al (2007) Splicy: a web-based tool for the prediction of possible alternative splicing events from Affymetrix probeset data. BMC Bioinform 8(Suppl 1):S17

Reddy AS (2007) Alternative splicing of pre-messenger RNAs in plants in the genomic era. Annu Rev Plant Biol 58:267–294

Resch A, Xing Y, Modrek B et al (2004) Assessing the impact of alternative splicing on domain interactions in the human proteome. J Proteome Res 3:76–83

Rossbach O, Hung LH, Schreiner S, Grishina I, Heiner M, Hui J, Bindereif A (2009) Auto- and cross-regulation of the hnRNP L proteins by alternative splicing. Mol Cell Biol 29:1442–1451

Routh A (2019) DPAC: a tool for differential poly(A)-cluster usage from poly(A)-targeted RNAseq data. G3: Genes, Genomes, Genetics 9:1825–1830

Ryan MC, Zeeberg BR, Caplen NJ, Cleland JA, Kahn AB, Liu H, Weinstein JN (2008) SpliceCenter: a suite of web-based bioinformatic applications for evaluating the impact of alternative splicing on RT-PCR, RNAi, microarray, and peptide-based studies. BMC Bioinform 9:313

Sanford JR, Wang X, Mort M, et al (2008) Splicing factor SFRS1 recognizes a functionally diverse landscape of RNA transcripts. Genome Res 19:381-394

Schoning JC, Streitner C, Meyer IM, Gao Y, Staiger D (2008) Reciprocal regulation of glycine-rich RNA-binding proteins via an interlocked feedback loop coupling alternative splicing to nonsense-mediated decay in arabidopsis. Nucleic Acids Res 36:6977–6987

Shen H, Green MR (2004) A pathway of sequential arginine-serine-rich domain-splicing signal interactions during mammalian spliceosome assembly. Mol Cell 16:363–373

Shen H, Kan JL, Green MR (2004) Arginine-serine-rich domains bound at splicing enhancers contact the branchpoint to promote prespliceosome assembly. Mol Cell 13:367–376

Shen S, Park JW, Lu ZX, Lin L, Henry MD, Wu YN, Zhou Q, Xing Y (2014) rMATS: robust and flexible detection of differential alternative splicing from replicate RNA-Seq data. Proc Natl Acad Sci USA 111:E5593-5601

Shen S, Warzecha C, Carstens R, Xing Y (2010) MADS +: discovery of differential splicing events from Affymetrix exon junction array data. Bioinformatics 26:268-269

Shen Y, Venu RC, Nobuta K, Wu X, Notibala V, Demirci C, Meyers BC, Wang GL, Ji G, Li QQ (2011) Transcriptome dynamics through alternative polyadenylation in developmental and environmental responses in plants revealed by deep sequencing. Genome Res 21:1478–1486

Spellman R, Llorian M, Smith CW (2007) Crossregulation and functional redundancy between the splicing regulator PTB and its paralogs nPTB and ROD1. Mol Cell 27:420–434

Stamm S, Riethoven JJ, Le Texier V et al (2006) ASD: a bioinformatics resource on alternative splicing. Nucleic Acids Res 34:D46-55

Staswick PE (2008) JAZing up jasmonate signaling. Trends Plant Sci 13:66–71

Sultan M, Schulz MH, Richard H et al (2008) A global view of gene activity and alternative splicing by deep sequencing of the human transcriptome. Science 321:956–960

Szabo L, Salzman J (2016) Detecting circular RNAs: bioinformatic and experimental challenges. Nat Rev Genet 17:679–692

Thierry-Mieg D, Thierry-Mieg J (2006) AceView: a comprehensive cDNA-supported gene and transcripts annotation. Genome Biol 7:1–14

Tian B, Manley JL (2017) Alternative polyadenylation of mRNA precursors. Nat Rev Mol Cell Biol 18:18–30

Todesco M, Rubio-Somoza I, Paz-Ares J et al (2010) A collection of target mimics for comprehensive analysis of microRNA function in arabidopsis thaliana. PLoS Genet 6:e1001031

Tress ML, Martelli PL, Frankish A et al (2007) The implications of alternative splicing in the ENCODE protein complement. Proc Natl Acad Sci USA 104:5495–5500

Vaucheret H (2008) Plant ARGONAUTES. Trends Plant Sci 13:350–358

Vazquez F, Gasciolli V, Crete P et al (2004) The nuclear dsRNA binding protein HYL1 is required for microRNA accumulation and plant development, but not posttranscriptional transgene silencing. Curr Biol 14:346–351

Vo JN, Cieslik M, Zhang Y et al (2019) The landscape of circular RNA in cancer. Cell 176:869–881

Wang BB, Brendel V (2006) Genomewide comparative analysis of alternative splicing in plants. Proc Natl Acad Sci USA 103:7175–7180

Wang BB, O'Toole M, Brendel V et al (2008) Cross-species EST alignments reveal novel and conserved alternative splicing events in legumes. BMC Plant Biol 8:17

Wang ET, Sandberg R, Luo S et al (2008) Alternative isoform regulation in human tissue transcriptomes. Nature 456:470–476

Wang H, Wang H, Zhang H, Liu S et al (2019) The interplay between microRNA and alternative splicing of linear and circular RNAs in eleven plant species. Bioinformatics 35:3119–3126

Wang PL, Bao Y, Yee M-C et al (2014) Circular RNA is expressed across the eukaryotic tree of life. PloS One 9:e90859

Wang T, Wang H, Cai D et al (2017) Comprehensive profiling of rhizome-associated alternative splicing and alternative polyadenylation in moso bamboo (*Phyllostachys edulis*). Plant J 91:684–699

Wang Y, Gao Y, Zhang H et al (2019) Genome-wide profiling of circular RNAs in the rapidly growing shoots of moso bamboo (*Phyllostachys edulis*). Plant Cell Physiol 60:1354–1373

Wu X, Liu M, Downie B et al (2011) Genome-wide landscape of polyadenylation in arabidopsis provides evidence for extensive alternative polyadenylation. Proc Natl Acad Sci USA 108:12533–12538

Wu X, Zhang Y, Li QQ (2016) PlantAPA: a portal for visualization and analysis of alternative polyadenylation in plants. Front Plant Sci 7:889

Xing Y, Stoilov P, Kapur K et al (2008) MADS: a new and improved method for analysis of differential alternative splicing by exon-tiling microarrays. RNA 14:1470–1479

Yang L, Liu Z, Lu F et al (2006) SERRATE is a novel nuclear regulator in primary microRNA processing in arabidopsis. Plant J 47:841–850

Ye CY, Chen L, Liu C et al (2015) Widespread noncoding circular RNAs in plants. New Phytol 208:88–95

Ye S, Cai C, Ren H et al (2017) An efficient plant regeneration and transformation system of Ma Bamboo (Dendrocalamus latiflorus Munro) started from young shoot as explant. Front Plant Sci 8:1298

Ye S, Chen G, Kohnen MV et al (2020) Robust CRISPR/Cas9 mediated genome editing and its application in manipulating plant height in the first generation of hexaploid Ma bamboo (Dendrocalamus latiflorus Munro). Plant Biotechnol. J 18:1501

Yu X, Wang Y, Kohnen MV et al. (2019) Large scale profiling of protein isoforms using label-free quantitative proteomics revealed the regulation of nonsense-mediated decay in moso bamboo (*Phyllostachys edulis*). Cells 8:744

Yuan Y, Chung JD, Fu X et al (2009) Alternative splicing and gene duplication differentially shaped the regulation of isochorismate synthase in Populus and arabidopsis. Proc Natl Acad Sci USA 106:22020-22025

Yue Y, Liu J, Cui X et al (2018) VIRMA mediates preferential m(6)A mRNA methylation in 3'UTR and near stop codon and associates with alternative polyadenylation. Cell Discov 4:10

Zhang G, Guo G, Hu X et al (2010) Deep RNA sequencing at single base-pair resolution reveals high complexity of the rice transcriptome. Genome Res 20:646–654

Zhang H, Wang H, Zhu Q et al (2018) Transcriptome characterization of moso bamboo (*Phyllostachys edulis*) seedlings in response to exogenous gibberellin applications. BMC Plant Biol 18:125

Zhang P, Fan Y, Sun X et al (2019) A large-scale circular RNA profiling reveals universal molecular mechanisms responsive to drought stress in maize and arabidopsis. Plant J 98:697-713

Zhang XN, Mount SM (2009) Two Alternatively spliced Isoforms of the arabidopsis SR45 protein have distinct roles during normal plant development. Plant Physiol 150:1450–1458

Zhang XO, Dong R, Zhang Y et al (2016) Diverse alternative back-splicing and alternative splicing landscape of circular RNAs. Genome Res 26:1277–1287

Zhang Y, Gu L, Hou Y et al (2015) Integrative genome-wide analysis reveals HLP1, a novel RNA-binding protein, regulates plant flowering by targeting alternative polyadenylation. Cell Res 25:864–876

Zhang Z, Lotti F, Dittmar K et al (2008) SMN deficiency causes tissue-specific perturbations in the repertoire of snRNAs and widespread defects in splicing. Cell 133:585–600

Zhao H, Gao Z, Wang L et al (2018) Chromosome-level reference genome and alternative splicing atlas of moso bamboo (*Phyllostachys edulis*). Gigascience 7: giy115

Zhao H, Peng Z, Fei B et al (2014) BambooGDB: a bamboo genome database with functional annotation and an analysis platform. Database 2014: bau006

Zhao L, Zhang H, Kohnen MV et al (2019) Analysis of transcriptome and epitranscriptome in plants using PacBio Iso-Seq and nanopore-based direct RNA sequencing. Front Genet 10:253

Zhao T, Wang L, Li S et al (2017) Characterization of conserved circular RNA in polyploid Gossypium species and their ancestors. FEBS Lett 591:3660–3669

Zuo J, Wang Q, Zhu B et al (2016) Deciphering the roles of circRNAs on chilling injury in tomato. Biochem Biophys Res Commun 479:132–138

Characterizations and Functions of Transcription Factor Gene Families

8

Jian Gao, Yan Xiang, Zhanchao Cheng, Xiangyu Li, Lihua Xie, Ying Zhang, Jun Liu, Dan Hou, Qingsong Bai, and Long Li

J. Gao (✉) · Z. Cheng · X. Li
Key Laboratory of Bamboo and Rattan Science and Technology, International Center for Bamboo and Rattan, State Forestry and Grassland Administration, Beijing 100102, China
e-mail: gaojian@icbr.ac.cn

Y. Xiang
Anhui Agricultural University, No. 130 Changjiangxi Road, Hefei 230036, Anhui, China
e-mail: xiangyan@ahau.edu.cn

L. Xie
School of Chemical and Environmental Engineering, Pingdingshan University, Southern Section of Weilai Road, Xincheng District, Pingdingshan 467000, Henan, China

Y. Zhang
Beijing Academy of Agricultural and Forestry Sciences, No. 11 Middle Road, Dawning Garden, Beijing 100097, Haidian District, China

J. Liu
Academy of Chinese Medicine, Henan University of Chinese Medicine, No. 156, East Jinshui Road, Zhengdong New District, Zhengzhou 450046, Henan, China

D. Hou
State Key Laboratory of Subtropical Silviculture, Zhejiang Agricultural and Forestry University, No. 666, Wu Su street, Hangzhou 311300, Zhejiang, China

Q. Bai
Guangdong Academy of Forestry, No. 233, Guangshan 1st Road, Guangzhou, Guangdong 510520, China

L. Li
Northwest Agriculture and Forestry University, No. 3 Taicheng Road, Yangling 712100, Shanxi, China

Abstract

As proteins that play important regulatory roles in vivo, transcription factors (TFs) have been extensively studied in various fields of molecular biology. TFs, such as MYB, MADS, Dof, and NAC, which have profound effects on plant growth and development, organ development, stress resistance, senescence, and stem cell localization. The achievements of TFs function research in moso bamboo are reviewed in this chapter, which will be helpful for the exploration of the function of those TFs and provide reference for the research of the functions of TFs in other species.

8.1 Resistance Gene Candidates

8.1.1 MYB Transcription Factor

MYB transcription factor (TF) family is the greatest plant gene family, which participates in several biological events. In moso bamboo, altogether 114 R2R3MYB genes have been initially discovered in the full-length non-chimeric (FLNC) reads and the genome (Hou et al. 2018). qRT-PCR and microarray data-based expression analysis have revealed distinct R2R3MYBs expression profiles in tissue growth and upon different abiotic stresses; for example, by forming NAC-MYB network (Hou et al. 2018; Zhang

J. Gao (ed.), *The Moso Bamboo Genome*, Compendium of Plant Genomes,
https://doi.org/10.1007/978-3-030-80836-5_8

et al. 2018; Yang et al. 2019). Moreover, some MYB genes were functional analyzed in moso bamboo stress response (Xiao et al. 2013; Hou et al. 2018) or flower organ development (Cheng et al. 2020). The above results not only shed all-sided lights on MYB family involved in moso bamboo, but also provide the potential MYB genes that can be used to study their essential functions.

In the previous study (Hou et al. 2018), altogether 114 R2R3MYB proteins (including 112 R2R3MYBs, 1 4R-type protein and 1 R1R2R3 MYB protein) were primarily obtained; at last, 88 MYB-associated proteins were acquired based on the FLNC reads database and *Moso bamboo* V1 genome (Peng et al. 2013). These proteins were divided into 17 clades, and the bootstrap support was $\geq 50\%$ in the NJ tree (Hou et al. 2018). Additionally, R2R3MYB proteins within Brachypodium, Arabidopsis, moso bamboo and rice were subjected to phylogenetic analysis, which indicated that R2R3MYB proteins of BEP clade in moso bamboo, rice and Brachypodium were closely related to each other (Peng et al. 2013; Hou et al. 2018). The authors established the unrooted NJ phylogenetic tree based on 132 AtMYBs and 85 *PheMYBs* in the latest study (Yang et al. 2019). Among those, 65 *PheMYBs* belonged to 18 function-annotated clades, including Class I participating in SCW generation through the regulation of hemicellulose, cellulose and lignin production and accumulation; Class II related to abiotic/biotic stress responses through modulating ABA pathway; Class III playing vital parts in organ and morphology genesis, like anther, epidermal cells, roots, growth of stomatal and vegetative cells, and embryogenesis; Class IV related to the regulation of secondary metabolism, like generation of flavanol and anthocyanins. Therefore, the above findings indicate that *PheMYBs* exert various activities, which possibly exert vital parts in moso bamboo development.

Based on transcriptome data obtained from diverse tissues of moso bamboo, 85 MYBs were expressed in one or more tissue, whereas the levels of 52 MYBs were discovered in each tissue, and the transcript abundances were between 0 and 110.12 (Yang et al. 2019). Among those, some MYBs were highly expressed in certain tissues For instance, *PheMYBs* that belonged to S1, S2 and S7 showed great expression quantities within leaf and panicle samples, but low expression within the remaining 4 tissues. In addition, *PheMYBs* that belonged to S5 and S6 were highly expressed in each tissue and *PheMYBs* that belonged to S9 were lowly expressed in each tissue (Yang et al. 2019).

According to the previous description, the expression of many *PheMYBs* increased during moso bamboo flowering, particularly for genes related to GA pathways (Gao et al. 2014). As revealed by the expression profiles for such gene family in diverse flower development periods, many *MYB* genes showed high expression levels (RPKM > 10) during one or more flower development stages, whereas just 12% PheR2R3-MYBs showed low expression (RPKM < 1) during over two stages (Hou et al. 2018). A number of *PheMYBs*, including *MYB103-1* and *PheMYB103*, were highly expressed during the formation of floral buds (F1) and the maturation of gradual organs (F2). Such results indicated that *MYB103-1* and *PheMYB103* played important parts in regulating formation of floral meristem in moso bamboo, in particular, for anther growth. In addition, *PheMYB2R-42* and *PheMYB2R-98* showed high expression levels during full-blossoming period (F3), which are identified as the *OsGAMYB* orthologous genes. As a result, these two genes might be associated with pollen growth in moso bamboo.

External factors and stress can regulate many *MYB* genes. For the screened *PheR2R3MYBs* with the only exception of *PheMYB2R-33*, their expression markedly increased in moso bamboo in one or more time courses (Hou et al. 2018). These genes were classified as diverse groups based on the respective expression patterns, among which, many showed high expression levels at 3 h after cold stress (Fig. 8.1). The expression of most of the above genes was triggered by drought stress, many of which had early responses to water depletion (6 h). Nonetheless, drought stress suppressed *PheMYB30*,

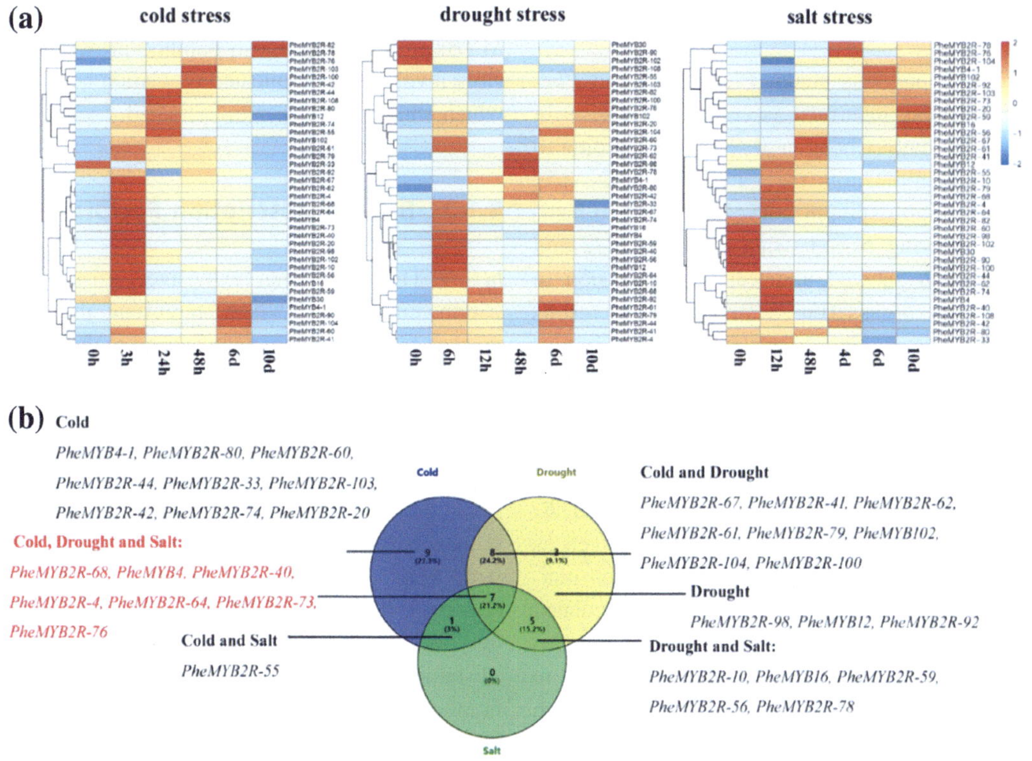

Fig. 8.1 Expression profiles of screened *PheR2R3MYBs* in the presence of different abiotic stresses as well as the moso bamboo response model. **a** Expression profiles of *PheMYBs* following cold treatments (4 °C) for 0/3/24/48 h, 6/10 d; drought treatments (20% PEG) for 0/6/12/48 h, 6/10 d; salt treatments (200 mM NaCl) for 0/12/48 h, 4/6/10 d. Heat maps were drawn to visualize the expression profiles. The colored blocks are indicative of the up-regulated (red) or down-regulated (blue) expression. **b** Venn diagram analysis on the screened *PheMYBs* according to the qRT-PCR data (Hou et al. 2018)

PheMYB2R-90, and *PheMYB2R-102* expressions. By contrast, a greater number of *PheR2R3MYBs* showed decreased expressions upon salt treatment. Apart from the above-mentioned 3 genes, *PheMYB2R-60/82/100/108* exhibited reduced expressions. At last, it was found that salt, cold and drought stresses markedly induced the expression of 7 *PheR2R3MYBs* (foldchange FC > 2), while 14 genes had significant responses to two treatments (FC > 2). As indicated by the above results, *MYB* genes possibly make up the convergence nodes related to diverse stress response pathways.

In the previous study, a MYB-type transcription factor, designated PeMYB2, was cloned from moso bamboo (Xiao et al. 2013). Phylogenetic analysis indicated that PeMYB2 shared the highest homology with 85.98% to OsMYB18 protein from rice. Overexpression of *PeMYB2* caused delayed flower or dwarfism in transgenic Arabidopsis. Under salt and cold stresses, *PeMYB2* overexpressed Arabidopsis had higher survival rate than WT, by up-regulating downstream *NXH1*, *SOS1*, *RD29A*, and *COR15A*. Hou et al. (2018) found that *PheMYB4-1* was a homologous gene of *OsMYB4* in rice, which was related to the responses to several abiotic stresses in moso bamboo. PheMYB4-1 is located in nucleus, which exhibits great transcriptional activity within yeast. Its overexpressed expression is found to mildly promote NaCl and mannitol sensitivity in transgenic Arabidopsis in germination period. Additionally, *PheMYB4-1* OE plants have been reported to have enhanced resistance to freezing and cold, which is associated with the significantly increased expression of CBF as well as the downstream genes upon cold stress.

In the previous study, *PheMYB42* and *PheMYB98* were identified as targets of miR159 in moso bamboo (Cheng et al. 2020). According to in situ hybridization (ISH) analysis, miRNA159 showed high expression level in the moso bamboo flowering period, whereas *PheMYB42* and *PheMYB98* showed undetectable expression in the flower organs (Fig. 8.2). The above results indicate the possible role of miRNA159 in regulating *PheMYB42* and *PheMYB98* during the formation and growth of flower organs. Moreover, 12 *PheMYBs* related to SCW synthesis were also selected for further expression validation (Yang et al. 2019). As a result, 11 *PheMYBs* exhibited markedly increased transcript levels as the bamboo shoot height increased, which indicated that they possibly positively regulated SCW formation. Among those, the expression of nine *PheMYBs* was down-regulated within 6.7 m shoots in comparison with 3.0 m shoots, which suggested the potentially vital parts that the above genes might play in later biochemical and physiological responses in early bamboo shoot development. Particularly, the effects of PheMYBs in regulating moso bamboo growth possibly involved a complicated regulation network, which should be further investigated.

8.1.2 BHLH Transcription Factors

bHLH TFs have been extensively distributed within eukaryotes, which represent the 2nd greatest family in plants only second to the MYB family (Riechmann et al. 2000). In bHLH TFs, their DNA binding domains possess around 60 amino acids, among which 10–15 comprise one basic region, whereas the rest around 40 comprise one helix-loop-helix region in the TF sequence N-terminal that is mainly associated with TFs binding to certain DNA sequences (Ledent and Vervoort 2001). As molecular biology develops rapidly, it is found that the bHLH TF family contains many members; to be specific, the bHLH TF genes in plants are discovered in diverse varieties, like rice and Arabidopsis (Li et al. 2006). The bHLH TF gene family exert vital parts in the development and secondary

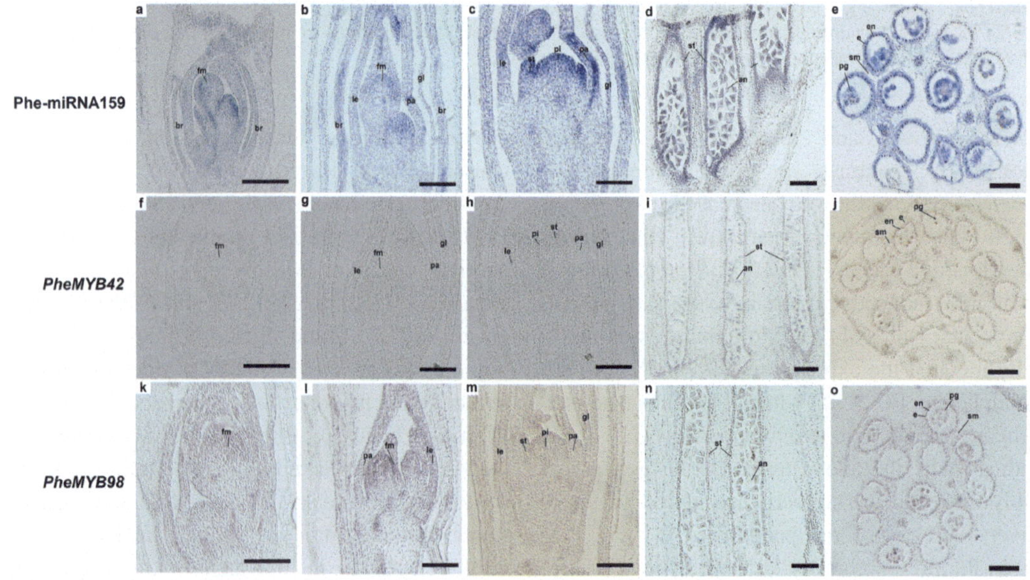

Fig. 8.2 ISH assays on Phe-miRNA159 and *PheMYB42/98*. **a**, **f**, **k** Floral bud; **b**, **g**, **l** inflorescence primordium; **c**, **h**, **m** floral primordium; **d**, **i**, **n** stamen; **e**, **j**, **o** anther. **a–e** Phe-miRNA159 abundances within diverse floral tissues; **f–j** PheMYB42 was undetectable within floral tissues; **k–o** PheMYB98 was undetectable within floral tissues. br, bract; fm, floral meristem; le, lemma; gl, glume; st, stamen; pa, palea; an, anther; pi, pistil; en, endothecium; e, epidermis; sm, septum; pg, pollen grains. Bars ¼100 mm (Cheng et al. 2020)

metabolism of plants; besides, they are involved in various stress responses in plants. In higher plants, their development and morphology genesis involve an extremely intricate process, and it is completed via protein and DNA interactions. To be specific, genes that encode HEC and SPT are found to coordinate flowering female vine (Gremski et al. 2007). With regard to Arabidopsis, bHLH TFs that interact with phytochromes are detected to be responsible for regulating light morphogenesis (Ni et al. 1998). In maize, the R gene-encoded Lc protein is the first bHLH TF gene reported in plants, which can modulate 2 or more structural genes related to the anthocyanin metabolic pathway in maize (Ludwig et al. 1989). Additionally, bHLH TFs participating in regulating anthocyanin production in Arabidopsis are classified as subgroup III, including the MYCl, GL3, EGL3 and TT8 TF families (Bailey et al. 2003). As reported by Zimmermann and colleagues (2004), these TFs interacted with R2R3-MYB TFs of subgroup VI, which included *PAP1* (*AtMYB75*) as well as *PAP2* (*AtMYB90*) (Zimmermann et al. 2004; Su et al. 2018).

Altogether 137 bHLHs were discovered from moso bamboo through evolutionary analysis, and they were classified as 16 subfamilies according to the discovered evolutionary relations of Arabidopsis with rice; at the same time, 51 paralogous gene pairs were discovered as well. Among them, only 3 pairs showed the Ka/Ks ratio of over 1, suggesting the negative selection of many *PhebHLHs*. There were 45 pairs of paralogous genes discovered from rice and 19 from Arabidopsis, along with 15 pairs of orthologous genes discovered between Arabidopsis and rice and 46 between moso bamboo and rice. *PhebHLHs* were found in multi-sequence analysis to possess unique bHLH structure. In addition, according to gene ontology (GO) results, *PhebHLHs* exhibited diverse molecular activities. As suggested by analysis results of promoter cis-elements, many *PhebHLHs* possessed the cis-elements responsible for responding to a variety of events associated with biotic/abiotic stresses. Such *PhebHLH* expression profiles in tissue suggested that the expression of many of these genes was detected in stem, root and leaf samples (Fig. 8.3). According to qRT-PCR, 21 screened *PhebHLHs* were under differential regulation following ABA, drought and MeJA stresses (Figs. 8.4 and 8.5).

8.1.3 bZIP Transcription Factors

Genes encoded by the basic leucine zipper (bZIP) domain are usually related to stress tolerance, which can regulate diverse critical plant biological processes. In addition, they constitute a greatest TF family among eukaryotes, and their names are given based on the conserved bZIP domain that contains one abasic region as well as one leucine zipper sequence (comprising 60–80 aa residues) (Talanian et al. 1990). The bZIP transcription factors recognize that the core sequence is the *cis*-acting element of ACGT. bZIP mainly modulates downstream gene expression levels through binding to gene promoter region containing ABRE, thus participating in ABA signal transduction (Narusaka et al. 2003). So far, bZIP TF genes are discovered by whole-genome analysis in many plant varieties. Further, bZIP TFs are increasingly found to exert vital parts in the regulation of organ and tissue growth and integration, like floral genesis and growth, vascular development, embryogenesis, seed maturation and germination.

Pan et al. identified 154 candidate *PhebZIPs* from the genome of moso bamboo. According to their phylogenetic analysis results, proteins of *PhebZIPs* were classified as 9 groups, whose gene structures suggested a high group-specificity of all *PhebZIPs* (Pan et al. 2019). Additionally, the authors analyzed the evolutionary patterns of non-synonymous (*Ka*) and synonymous (*Ks*) substitution rates as well as respective ratios, and discovered that, in moso bamboo, the paralogous *PhebZIP* pairs experienced extensive genomic duplication at 7–15 MYA (Pan et al. 2019). Based on analysis of promoter sequences, this study screened 18 genes containing abundant *cis*-regulatory elements to analyze the expression levels (Fig. 8.6). The screened genes were widely related to the

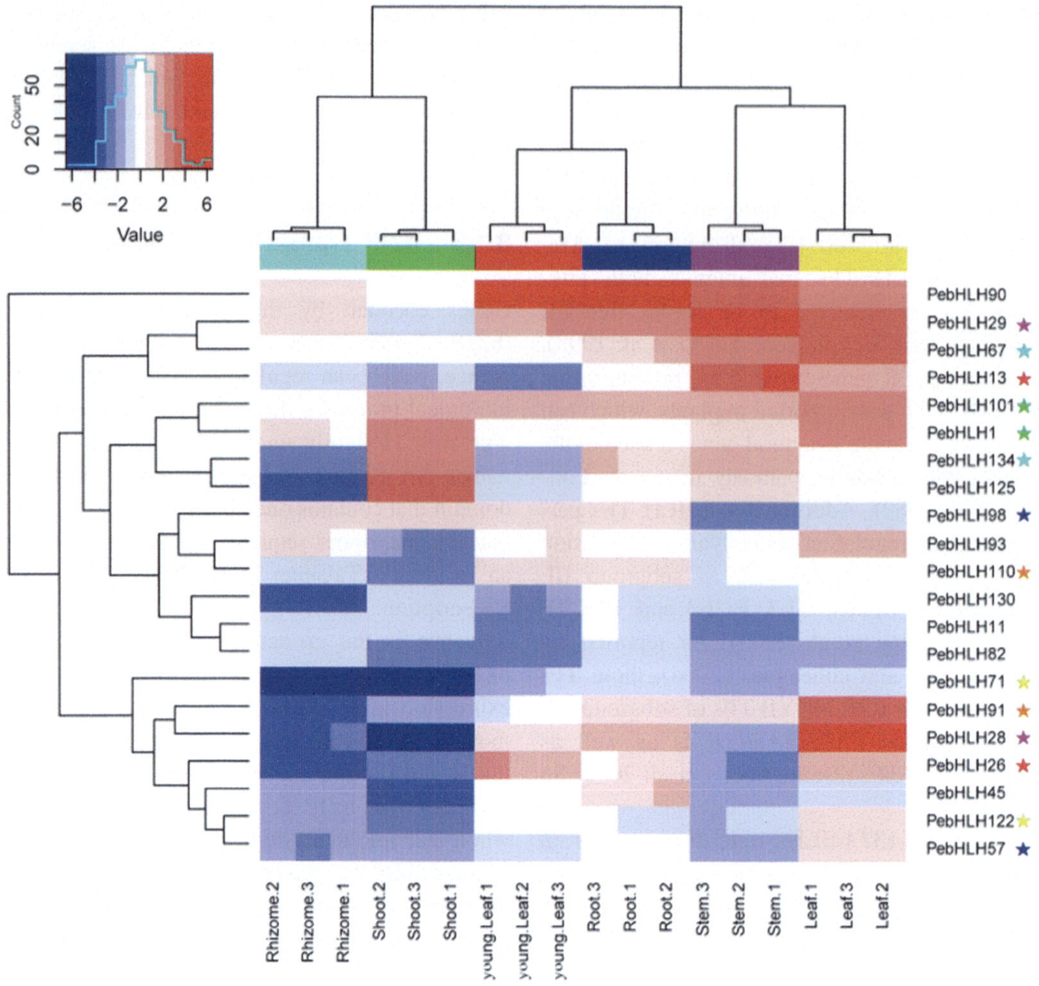

Fig. 8.3 Moso bamboo bHLH gene expression patterns in diverse tissue samples and in diverse development periods measured through qRT-PCR. Heatmap displays that the 21 screened *PhebHLH*s were hierarchically clustered in diverse tissue samples. Altogether 3 biological replicates were set for each experiment (Cheng et al. 2018)

responses to GA, ABA and MeJA stresses (Fig. 8.6a–c), and the underlying mechanisms might be different (Pan et al. 2019). For those screened *PhebZIP*s, their expression patterns within 5 plant organs/tissues and development periods indicated the effects of such genes on moso bamboo organ development, particularly seed growth (Fig. 8.6d) (Pan et al. 2019). *PhebZIP47* is localized in nucleus, which conforms to the prior results that bZIP TFs bind to *cis*-elements within nucleus; as a result, the *PhebZIP47* is typical nuclear localization transcription factor (Fig. 8.7a). Besides, as indicated by transactivation activity, *PhebZIP47* was not transcriptionally activated within yeast (Fig. 8.7b).

8.1.4 GRF Transcription Factor

Growth-regulating factors (GRFs) constitute a diminutive TF family specific to plants, and GRFs play important roles regulating central developmental processes in plants (Omidbakhshfard et al. 2015). The first identified GRF gene was rice *OsGRF1*, which was reported

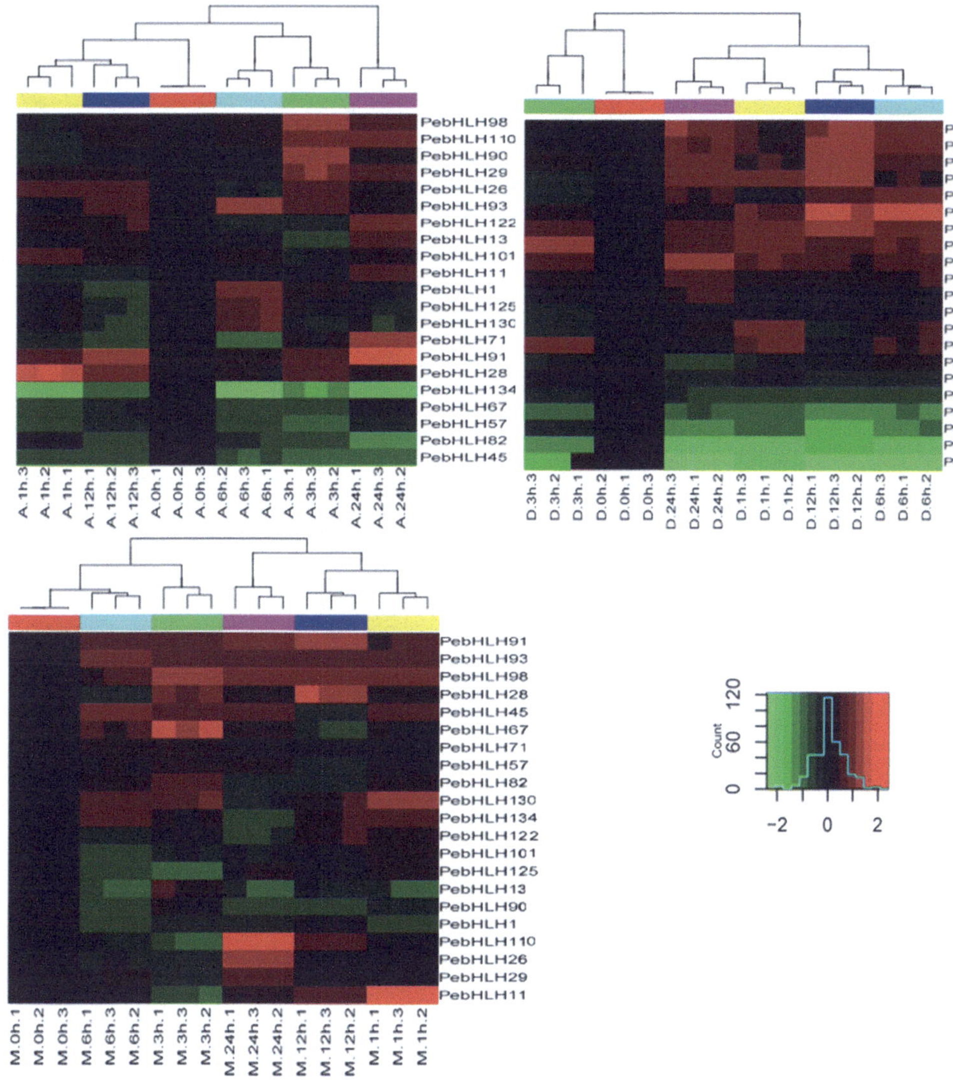

Fig. 8.4 All screened gene expression patterns within moso bamboo upon ABA, drought and MeJA stresses detected through qRT-PCR. Heatmap displays that the 21 *PhebHLHs* were hierarchically clustered in diverse tissue samples. Altogether 3 biological replicates were set for each experiment. A: ABA; D: Drought; M: MeJA (Cheng et al. 2018)

involving in GA-induced stem elongation and leaf development (Knaap et al. 2000). GRF transcription factors, characterized by highly conserved regions in N-terminal, contained two functional domains: WRC (PF08879) and QLQ (PF08880) (Cao et al. 2016). GRFs participate in physiological and biochemical processes in plants, like during stem/root/leaf growth, seed/flower formation, or responses to various biotic as well as abiotic stresses. Studies have shown that the GRF gene is related to many growth and development events, like plant longevity, development of leaves and roots, cell proliferation, stem elongation, meristem development, seed and floral organ formation.

Altogether 18 GRF genes were discovered in the genome of moso bamboo, which were called *PheGRF1-18* genes according to the

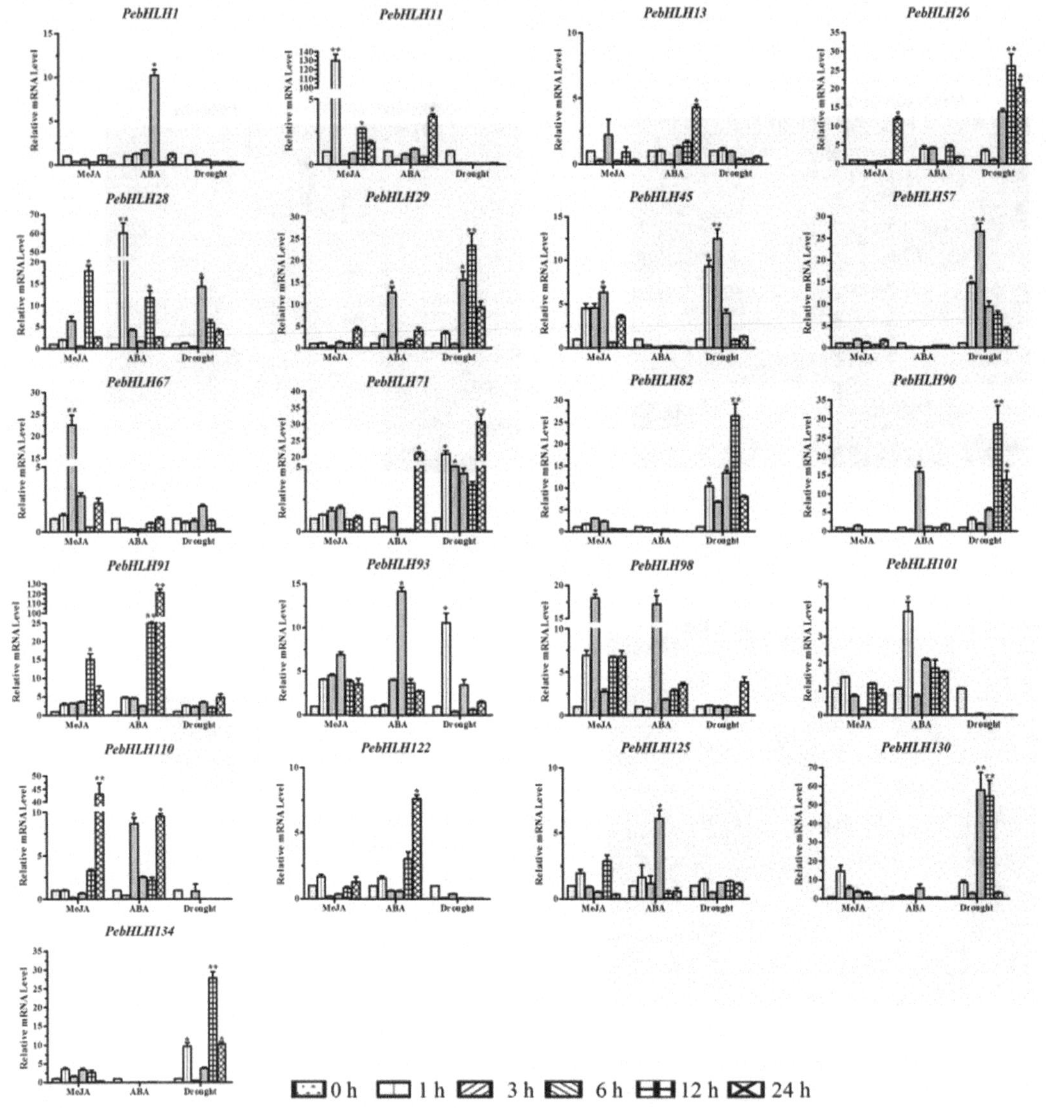

Fig. 8.5 All screened gene expression patterns within moso bamboo upon ABA, drought and MeJA stresses detected through qRT-PCR. X-axis: stress treatment course; Y-axis: relative gene expression; Error bars, 6 ± SE (Cheng et al. 2018)

corresponding scaffold positions (Shi et al. 2019). *PheGRF*s had mildly different CDS lengths, among which, *PheGRF12* was the longest (1656 bp), whereas *PheGRF11* was the shortest (513 bp). Each PheGRF protein possessed the conserved QLQ and WRC domains within respective N-terminal regions, where zinc finger motifs were discovered in the former. Besides, TQL and FFD motifs, which were the 2 smaller amino acid residue stretches, were identified at the C-terminal in PeGRF2, -3, -4, -5, -10 and -14, respectively (Shi et al. 2019). GRFs in moso bamboo showed comparatively even distribution across 3 subfamilies. There were diverse exons (range, 2–8) in the 18 GRF genes, among which, 7 had 3 exons. GRFs exerted vital parts in plant hormone responses and they are the plant metabolites related to plant growth. *PheGRF*s showed high expression levels within the active tissues, which were related to different

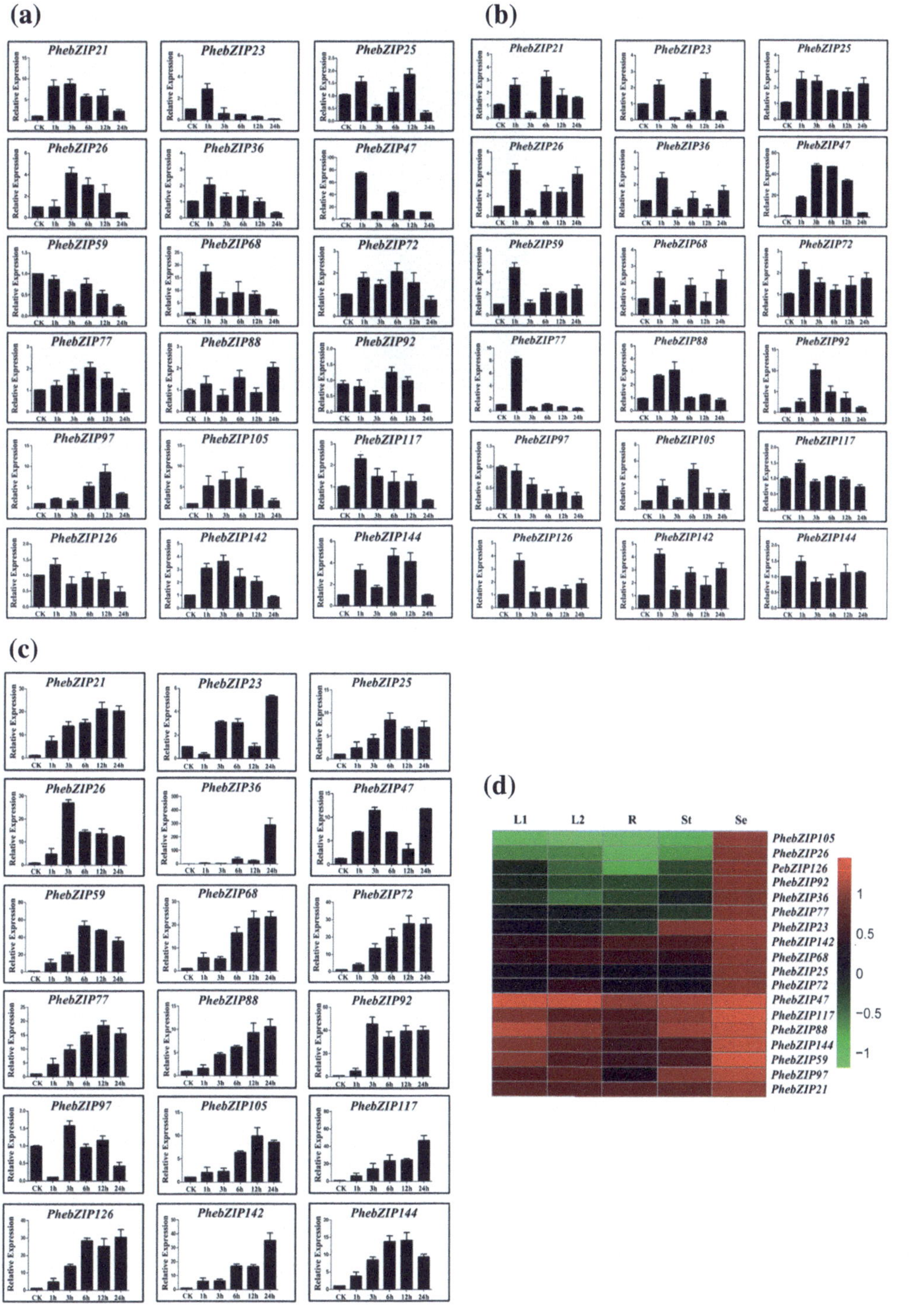

Fig. 8.6 qRT-PCR expression analysis of selected *PhebZIPs* (from Pan et al. 2019). **a** qRT-PCR assays on the screened *PhebZIPs* expression after MeJA exposure; **b** qRT-PCR assays on the screened *PhebZIPs* expression after ABA exposure; **c** qRT-PCR assays to examine screened *PhebZIPs* expression after GA treatment; **d** qRT-PCR assays on the screened *PhebZIPs* expression in diverse tissues and diverse development periods

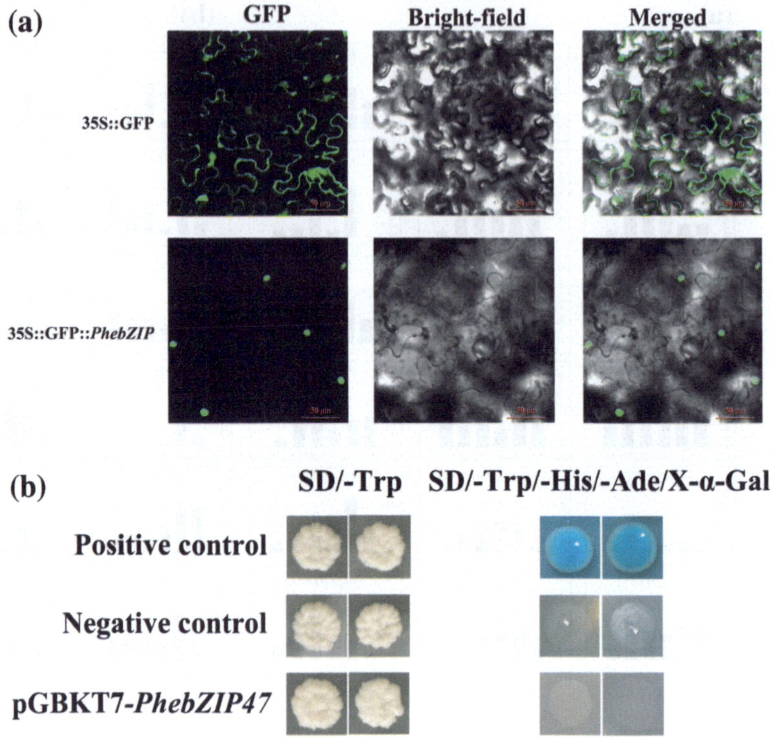

Fig. 8.7 Subcellular localization (**a**) and transactivational analyses (**b**) of *PhebZIP47* (from Pan et al. 2019). **a** GFP (control) and PhebZIP47-GFP fusion protein showed transient expression within *N.tabacum* leaves, as monitored with the fluorescence microscope; **b** Transfection of positive/negative/fusion constructs to the Y2HGold yeast strain, followed by incubation with the SD/-Trp medium within SD-His/-Ade/-Trp plate that contained X-a-GAL

hormone treatment responses, indicating that GRFs in moso bamboo were closely associated with the hormone feedback regulation mechanism (Fig. 8.8).

PheGRF11, which contains WRC domain at its N-terminal, is clearly located in the nucleus (Fig. 8.10a) (Shi et al. 2019). According to BiFC and two-hybridization assays in yeast, PheGIF3 formed a complex with PheGRF11, which was possibly related to the moso bamboo regulation network (Fig. 8.9). However, PheGRF11 did not show transcriptional activity in yeast (Fig. 8.10b) (Shi et al. 2019).

8.1.5 HD-Zip Transcription Factors

Originally detected discovered in *Drosophila*, homeobox gene earned its named because its mutation or ectopic expression could cause homeotype mutation (Garber et al. 1983). The *KNOTTED1* gene in maize (*Zea mays*) was discovered through transposon tagging, which led to the knotted leaves. Therefore, many homeodomain (HD)-encoding plant genes are extracted and examined (Vollbrecht et al. 1991). The genomes of model rice (*Oryza sativa*), plants, and *Arabidopsis thaliana* are completely described, which facilitates to classify these genes as 6 subfamilies according to the HD-encoding sequence, HD location, size, relations with additional gene structures and domains (Ariel et al. 2007). Such subfamilies stand for the plant homeodomain related to finger domain (PHD-finger), homeodomain associated with leucine zipper (HD-Zip), zinc finger related to homeodomain (ZF-HD), BELL (gained the name due to its distinct Bell domain), Knotted-related homeobox (KNOX) and Wuschel-associated homeobox (WOX). HD-Zip genes possess one homeodomain (HD) as well as one leucine zipper (LZ) motif and they can be seen in eukaryotic kingdom TFs, yet they have distinct plant-specific relation with one individual protein (Schena and Davis 1992). HD functions in specific DNA binding, while LZ plays a role of the dimerization motif. The HD-Zip proteins can combine with DNA to be the dimers, while LZ

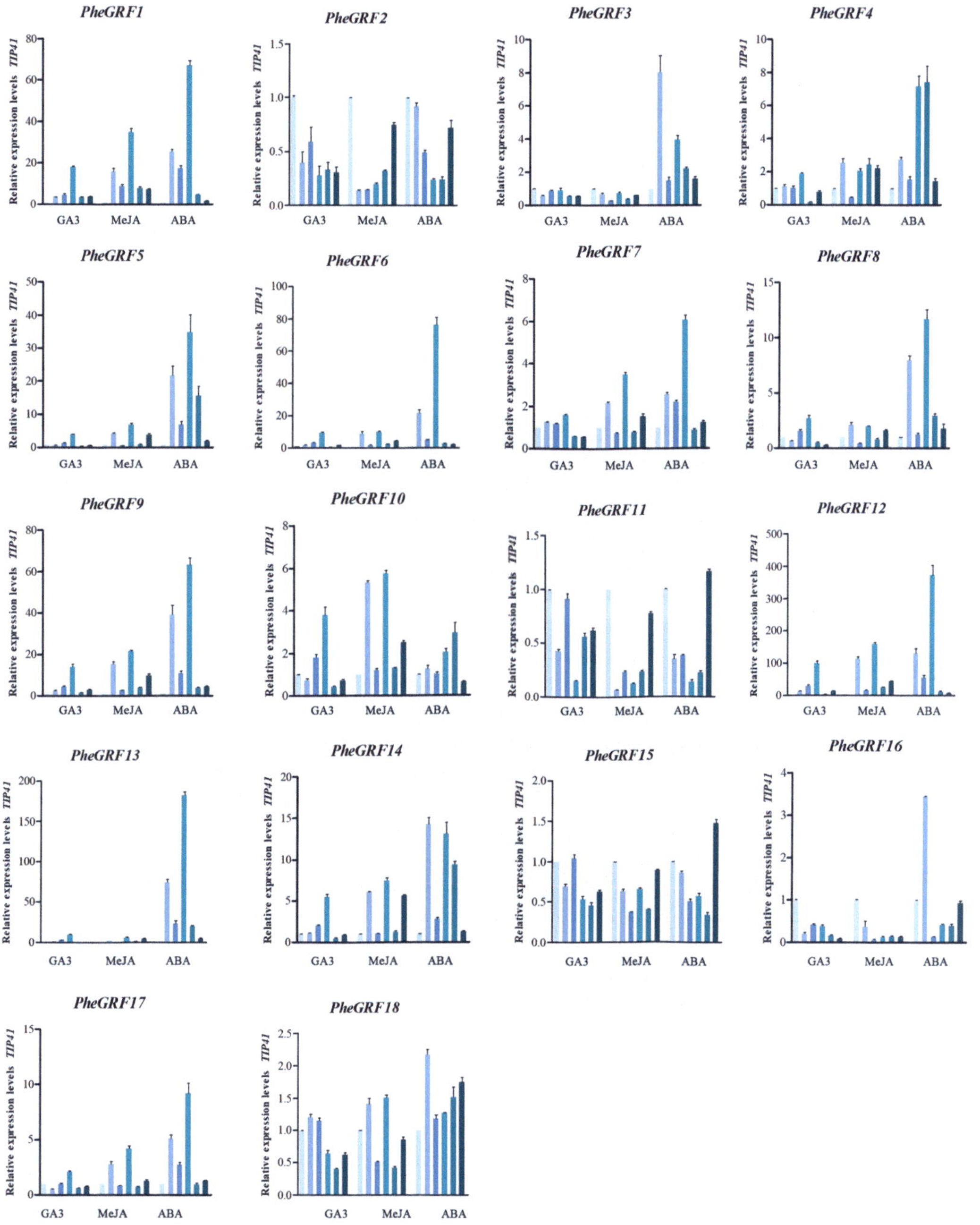

Fig. 8.8 qRT-PCR assays on the responses of moso bamboo *PheGRFs* to ABA, GA3 and MeJA stresses (from Shi et al. 2019)

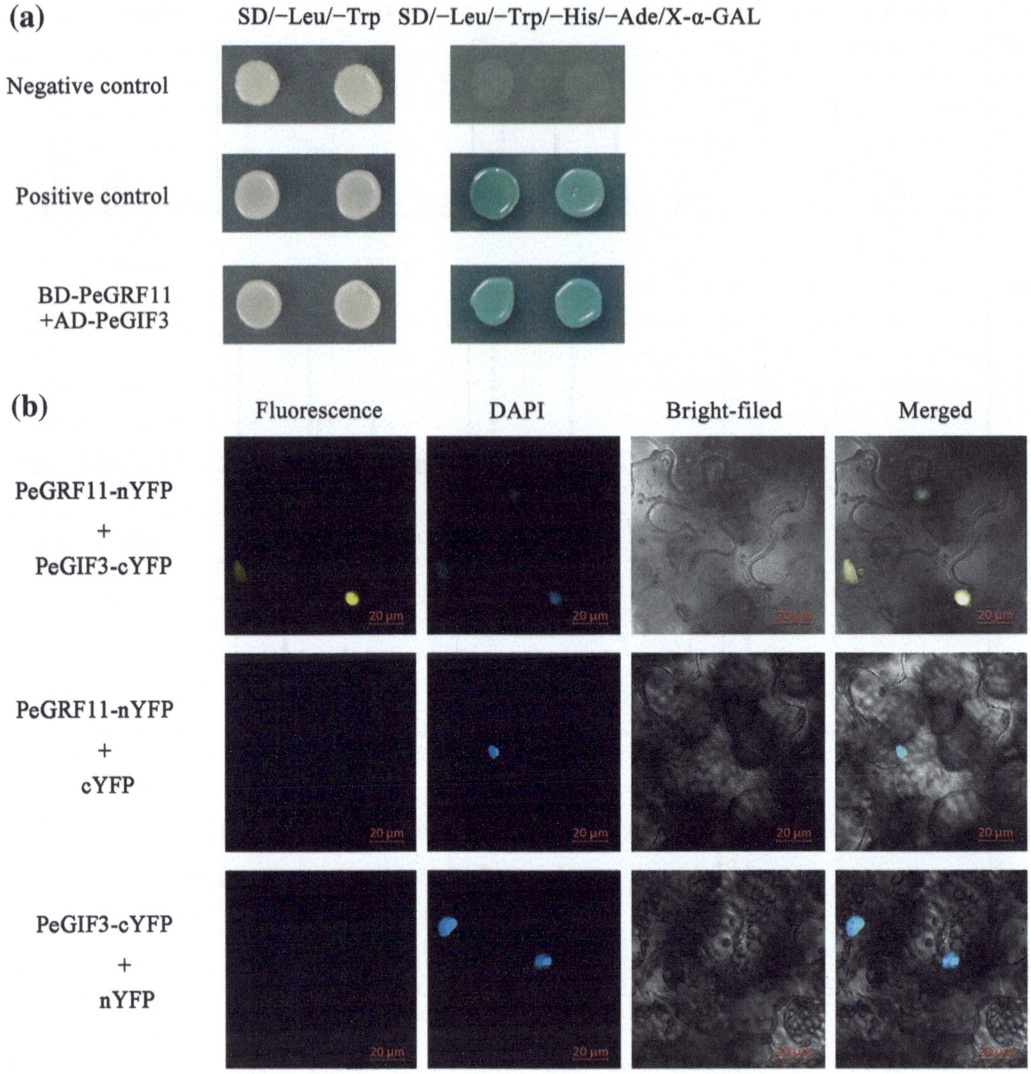

Fig. 8.9 Subcellular localization (**a**) and transactivational analyses (**b**) of PheGRF11 proteins (from Shi et al. 2019)

loss totally deleted their binding capacity, indicating the motif-driven important role of relative monomer orientation in efficiently recognizing DNA (Ariel et al. 2007).

In the genome of moso bamboo, altogether 48 non-redundant HD-Zip genes were found, which were referred to as *PeHDZ1-48*, respectively. As suggested by the phylogenetic tree of moso bamboo, there were 17, 14, 7 and 10 HD-Zip genes in subfamily I-IV, respectively (Fig. 8.11 a). Of them, the proteins encoded had 183–1287 (mean, 466) amino acids in their lengths. For better understanding the diverse HD-Zip gene structures, the exon–intron structural analysis was conducted in moso bamboo (Fig. 8.11b). As a result, those 48 identified HD-Zip genes possessed exons at different amounts (range, 2–26). Besides, in each PeHDZ gene, the LZ and HD domains showed high conservation. Further, motif composition was analyzed, which suggested the presence of START motifs in each gene belonging to HD-Zip III/IV subfamilies, while MEKHLA domain was just detected in proteins belonging to HD-Zip III subfamily

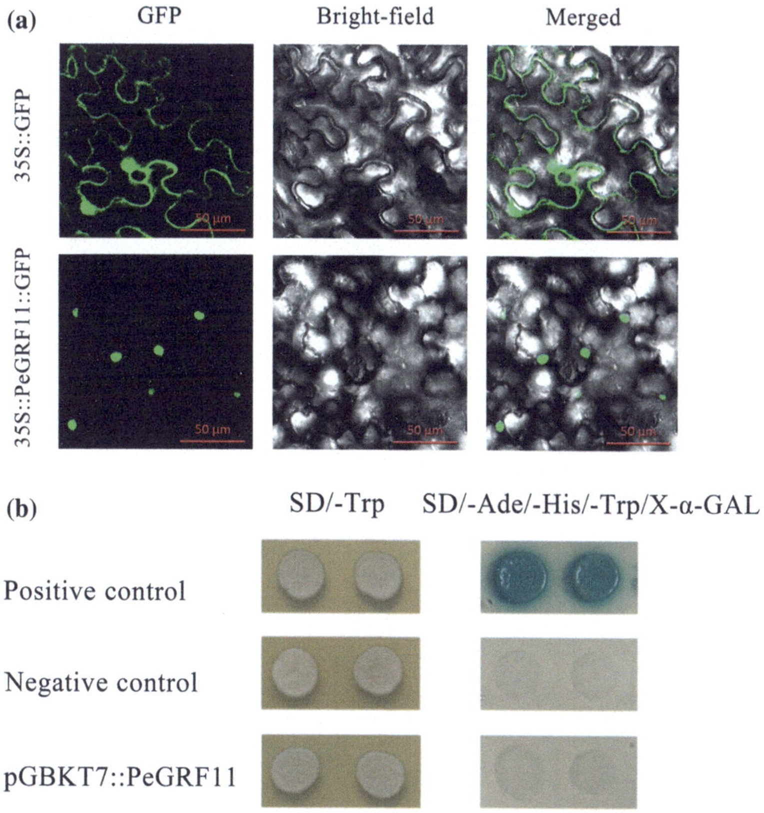

Fig. 8.10 Interactions between PheGRF11 and PheGIF3 within plant and yeast systems (from Shi et al. 2019)

(Fig. 8.11c). For analyzing phylogenetic relations across Poaceae HD-Zip proteins, this study established an integrated phylogenetic tree according to HD-Zip protein full-length sequences in sorghum, maize, rice, and moso bamboo (Fig. 8.11d). Clearly, the smallest one was HD-Zip III subfamily in plant, while HD-Zip I was the greatest. Besides, 27 candidate paralogous pairs (Pe-Pe) suggested that the peaks were detected at 0.2 from the genome of moso bamboo, which indicated the extensive duplication event at around 15 mya. Consistently, for those 64 orthologous pairs between rice and moso bamboo (Pe-Os), the peak was detected at about 0.2–0.3, which implied a 15–23 mya divergence time between the corresponding *HD-Zip* genes. In addition, as for those orthologous pairs between moso bamboo and rice genomes, the Ka/Ks ratio peaked at around 0.1–0.2, while that peak at 0.3 for paralogous pairs from the genome of moso bamboo (Fig. 8.11e). It was important that HD-Zip I genes exerted vital parts in responses to abiotic stresses (Harris et al. 2011). As a result, the present work examined PeHDZ I gene expression by qRT-PCR analysis, which suggested that many PeHDZ I genes had obvious response to high salinity and drought treatments at transcription level (Fig. 8.12).

8.1.6 PHD Transcription Factors

The PHD finger HAT3.1 (Histone acetyltransferases 3.1) was initially discovered from *Arabidopsis thaliana* (Schindler et al. 1993). Since then, an increasing number of PHD-finger proteins are studied within different eukaryotes. For a representative PHD-finger protein, it possesses at least one PHD domain comprising about 40–80 aa and one typical Cys4-His-Cys3 sequence. This is akin to LIM (Cys2-His-Cys5) and RING (Cys3-His-Cys4) finger domains with regard to

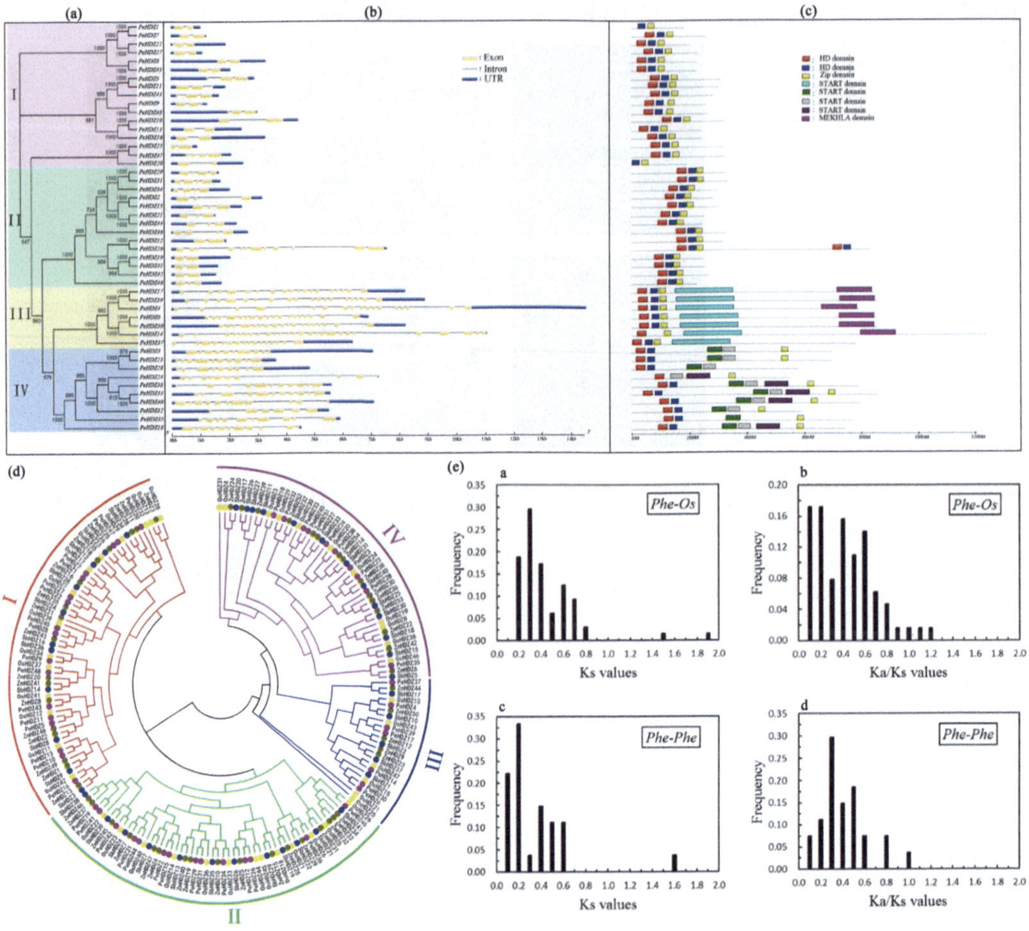

Fig. 8.11 Bioinformatics figures of 48 *PeHDZ*s (from Chen et al. 2017). **a** MEGA6.0 was used to establish a phylogenetic tree in line with full-length aa sequences according to NJ approach under 1000 bootstraps. **b** The yellow boxes stand for exons, blue boxes stand for untranslated regions (UTRs), while black lines stand for introns. **c** Motif composition of amino acids in HD-Zip proteins of moso bamboo. **d** Phylogenetic tree presents the associations among the full-length HD-ZIP proteins in sorghum, maize, rice, and moso bamboo. **e** Ks and Ka/Ks ratio distributions in HD-Zip genes from moso bamboo and rice genomes visualized according to frequency distributions of the above two values

the way that the candidate zinc-binding amino acid region is arranged, but the protein 3D structure is quite different (Capili et al. 2001). Moreover, some PHD-finger proteins contain only one canonical PHD domain, but some harbor several PHD fingers that perform in common or have independent functions. For instance, PHD protein VIL1 in *Arabidopsis* is involved in the vernalization and photoperiod pathways through modulating associated floral repressor levels (Sung and Amasino 2004). *SIZ1* plays a vital role in recognizing H3K4me3 (a histone code) for suppressing *WRKY70* transcription and ATX protein interaction, thus possibly preventing them from binding to histone H3K4 to response to various stress (Miura et al. 2020). In *Arabidopsis*, the ectopically expressed *MtPHD6* enhances drought/osmotic stress resistance through the activation of the stress-responsive WRKY, ZINC FINGER and AP2/EREBP TFs via the ABA-independent and -dependent pathways (Quan et al. 2019). In *Arabidopsis*, *AL5*

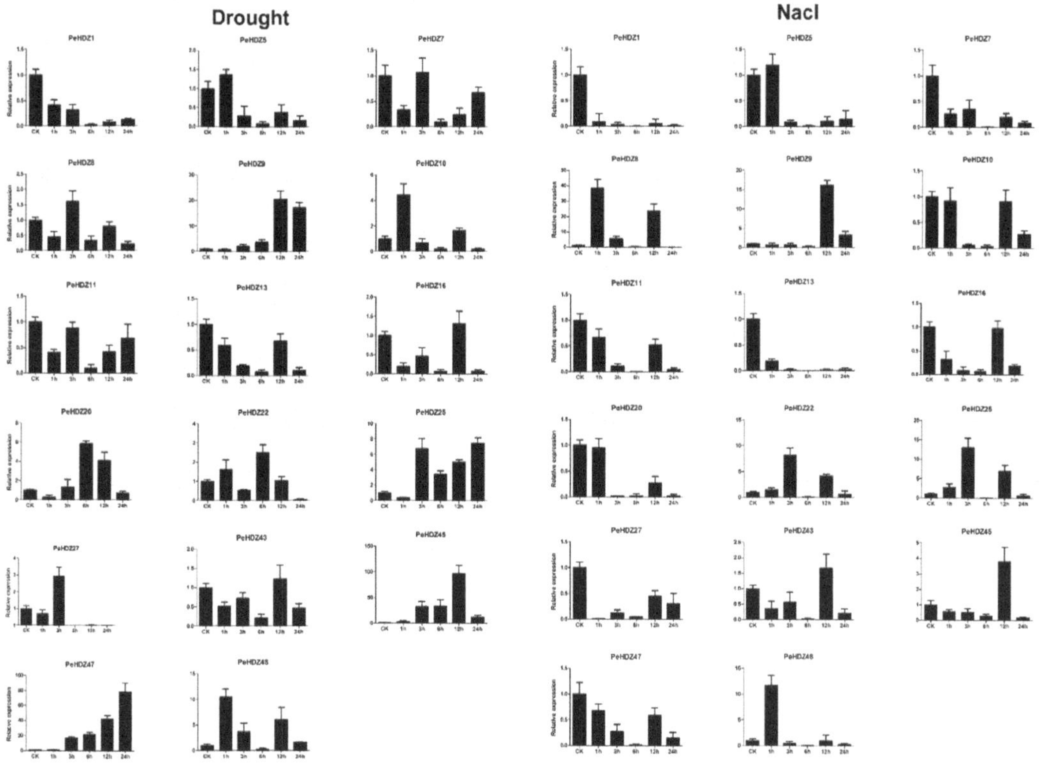

Fig. 8.12 17 PeHDZ I gene expression upon drought and salt stresses detected through qRT-PCR (from Chen et al. 2017). *X*-axis indicates the treatment course for every gene, while the *Y*-axis stands for relative gene level (FC). CK represents the non-treatment control

suppresses several signal transduction pathways to endow plants with resistance to cold, drought and high salt stresses (Wei et al. 2015).

In whole-genome analysis on different species, there are 67 PHD-finger genes in maize (*Zea mays*) (Wang et al. 2015), 73 in poplar (*Populus trichocarpa*) (Wu et al. 2016), 59 in rice (*Oryza sativa*) (Sun et al. 2017), 60 in moso bamboo (*Phyllostachys edulis*) (Gao et al. 2018). We find the numbers in these plant species are very close. 60 *PePHDs* detected from moso bamboo were obviously divided into 11 groups (A–K), and the bootstrap support was $\geq 50\%$ within NJ tree. The introns numbers ranged from 1 (*PePHD31*) to 31 (*PePHD37*), greatly conforming to those in maize and rice. The functional domains, PHD domain, BAH domain, DDT, GANT domain and more were detected in different subfamilies. Moreover, 3 cis-acting elements related to the responses to abiotic/biotic stresses, phytohormones and plant growth were identified for 16 PHD-finger genes (Alfin-like) in their promoter regions (Fig. 8.13 a), which indicated these genes might exert vital parts in adverse stresses and during plant growth. As a result, expression profiles for 16 screened PHD-finger genes within diverse tissues (mature leaves ML, young leaves TL, stems S, roots R, rhizomes Rh, and shoots Sh) and multiple treatments (ABA, drought, cold and salt) were examined using quantitative real-time PCR assays. The results revealed these genes were high expression in moso leaves (Fig. 8.13b) and had significant response to these simulated stresses, especially under ABA and drought stress (Fig. 8.13c). In summary, we identified

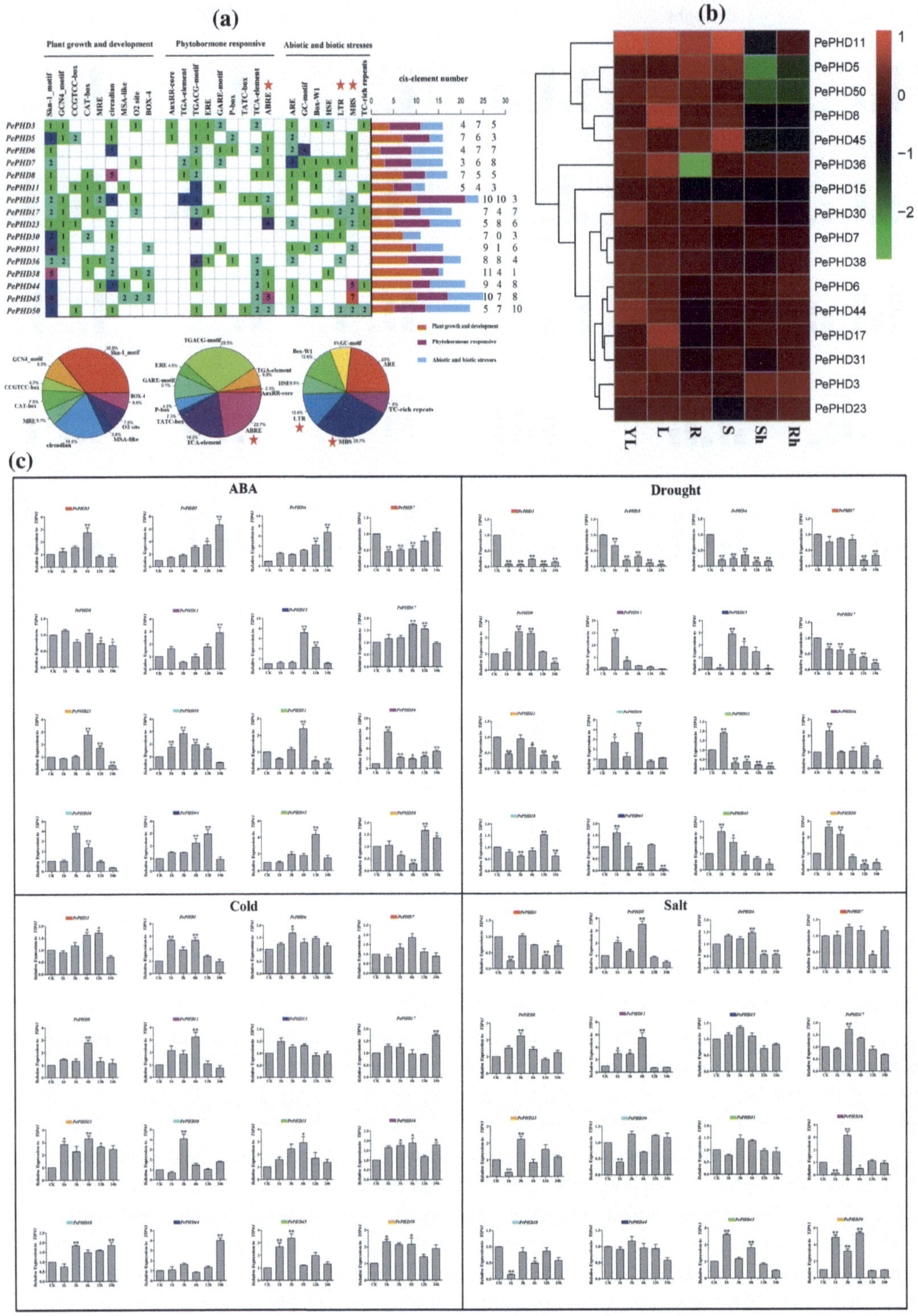
(a)
Plant growth and development
Phytohormone responsive
Abiotic and biotic stresses
cis-element number
(b)
PePHD11
PePHD5
PePHD50
PePHD8
PePHD45
PePHD36
PePHD15
PePHD30
PePHD7
PePHD38
PePHD6
PePHD44
PePHD17
PePHD31
PePHD3
PePHD23
YL
L
R
S
Sh
Rh
(c)
ABA
Drought
Cold
Salt

◀ **Fig. 8.13** The promoter analysis and expression levels of 16 moso bamboo PHD genes (from Gao et al. 2018). **a** In the grid, diverse numbers and colors indicate the amounts of diverse promoter elements within those 16 genes. For every category, the distinct colored histogram displays the overall promoter element amount, whereas pie charts present every promoter element ratio. **b** Expression profiles for the 16 screened *PePHDs* within the 6 tissues: mature leaves (L), young leaves (YL), stems (S), roots (R), rhizomes (Rh) and shoots (Sh). Red suggests increased while green suggests decreased transcript abundances relative to corresponding controls. **c** Expression profiles for the 16 screened *PePHDs* within moso bamboo after treatments with PEG6000, ABA, high salt and 4 C measured through qRT-PCR. The *x*-axis is indicative of treatment course of every stress, while *y*-axis stands for relative level. Mean and SDs were acquired based on 3 technical and 3 biological replicates. Error bars are indicative of SD. Asterisks represent markedly distinct expression compared with control (**P* < 0.05, ***P* < 0.01). Genes denoted in identical color label were the gene pairs

PHD-finger genes within moso bamboo to better analyze their functions in plants, and the exploration of gene function need more experiments to prove and supplement.

8.1.7 SPL Transcription Factors

SQUAMOSA promoter binding protein-like (SPL) proteins have been identified as the TFs specific to plants. Of them, *AmSBP1* and *AmSBP2* are the two genes initially extracted in goldfish grass. Because they can recognize the promoter binding site of flower development gene SQUAMOSA and regulate its expression, they are named Squamosa promoter binding protein (Klein et al. 1996). All SBP transcription factors contain a conserved domain of SBP consisting of about 76 amino acid residues. Many researchers have analyzed the sequence of SBP family genes in different plants and found that this conservative structure is common in the SBP gene family (Yamasaki et al. 2004). *SPL3* is the first SPL gene discovered from arabidopsis, and it can enhance flowering in long-day environment. Besides, *AtSPL3* can combined with the conservative *cis*-element within the *APETALA1* promoter region. *APETALA1* represents the identity gene of floral meristem, which stands for the SQUA analogue, like *AmSBP1* and *AmSBP2* (the two snapdragon genes) (Tsubosa et al. 1997). It has been previously reported in physiological and biochemical research that, TFs in SPL family positively regulate plant growth. The 16 SPL genes are detected from the arabidopsis genome, which exert vital parts in the development of leaves, flowers, shoots, and spores, fertility, alterations in nutrition and reproductive stage, formation of leaf primordia interval, and plant hormone signaling (Guo et al. 2008). Further, SPL genes have been implicated in regulating copper (Cu) homeostasis as well as responses to gibberellic acid (GA).

Altogether 32 candidate *PheSPLs* are detected from the genome of *Moso bamboo*. As revealed by phylogenetic analysis, sequences of *PheSPLs* were classified as 8 groups and the gene structures and conserved motifs suggested a high group-specificity of all *PheSPLs* (Pan et al. 2017). According to our results, *PheSPLs* underwent extensive duplication events around 15 MYA; besides, divergence was observed in *PheSPLs* compared with *OsSPLs* around 34 MYA (Pan et al. 2017) (Fig. 8.14). We examined the expression of *PheSPLs* under diverse drought and PEG (polyethylene glycol-6000; a kind of plant hormone) treatments for simulating abiotic/biotic stresses. As a result, 3 (*PheSPL10/12/17*), 6 (*PheSPL1/10/12/17/20/31*), and 9 (*PheSPL5/8/9/14/15/19/20/31/32) PheSPL* genes still kept stable expression following salicylic acid (SA) (Fig. 8.15), GA (Fig. 8.16), and PEG (Fig. 8.17) treatments, separately (Pan et al. 2017). Besides, this study examined tissue-specific expression levels of moso bamboo SPLs in the growth process, which revealed that the spatiotemporal expression patterns were different, with the increased expression being detected in leaves and flowers (Fig. 8.18) (Pan et al. 2017).

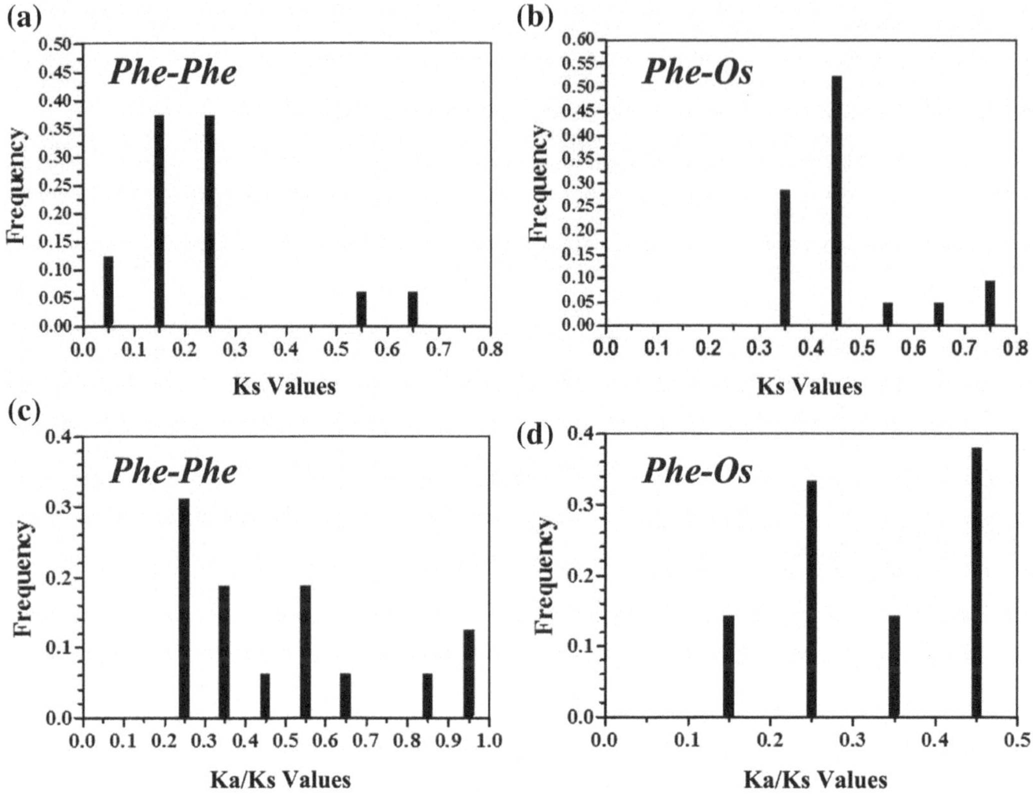

Fig. 8.14 Distributions of Ks and Ka/Ks ratios for SPL genes from paralogous (Phe-Phe) and orthologous (between moso bamboo and rice) gene-pairs in moso bamboo genomes (from Pan et al. 2017). **a**, **c** Distributions of Ks and Ka/Ks acquired from moso bamboo genome paralogous gene-pairs (Phe-Phe); **b**, **d** Orthologous gene-pairs between moso bamboo and rice

8.1.8 TCP Transcription Factors

TCPs (Teosinte Branched 1, Cycloidea, and Proliferating Cell Factors), containing one noncanonical basic helix–loop–helix (TCP domain, PF03634) at N-terminal of amino acid sequences, which exert vital parts in plant growth as well as abiotic stress responses, are plant-specific transcription factors (Martíntrillo and Cubas 2010). These TFs are named on the basis of the initial 3 members identified, which are the maize teosinte branched1 (*TB1*) (Doebley et al. 1997), rice PCF proteins (*PCFs*) (Kosugi and Ohashi 1997) and Antirrhinum cycloidea (*CYC*) (Luo et al. 1999). TCPs play significant regulation roles in regulating diverse plant growth processes, like embryonic growth, seed germination, lateral branching, growth and aging of leaves, development of flowers and gametophytes, hormone pathways, circadian rhythm, as well as mitochondrial biogenesis (Liu et al. 2018). For example, TB1 subfamily members, such as *BRANCHED1* in Arabidopsis thaliana, *ZmTB1* in maize, *OsFC1* in rice, have the functions in the axillary branches development and branch control (Liu et al. 2018). In Arabidopsis, various members including *TCP2*, *TCP3*, *TCP4*, *TCP9*, *TCP10*, *TCP14*, *TCP15*, *TCP20* and *TCP24*, are identified in leaf development regulation. *AtTCP15* regulates internal replication and cytokinin and auxin responses during pistil development. *AtTCP14* and *AtTCP15* work together to regulate internode length and seed germination, and *AtTCP23* participates in flowering time and plant development. In additional, TCPs also functions in environmental signal

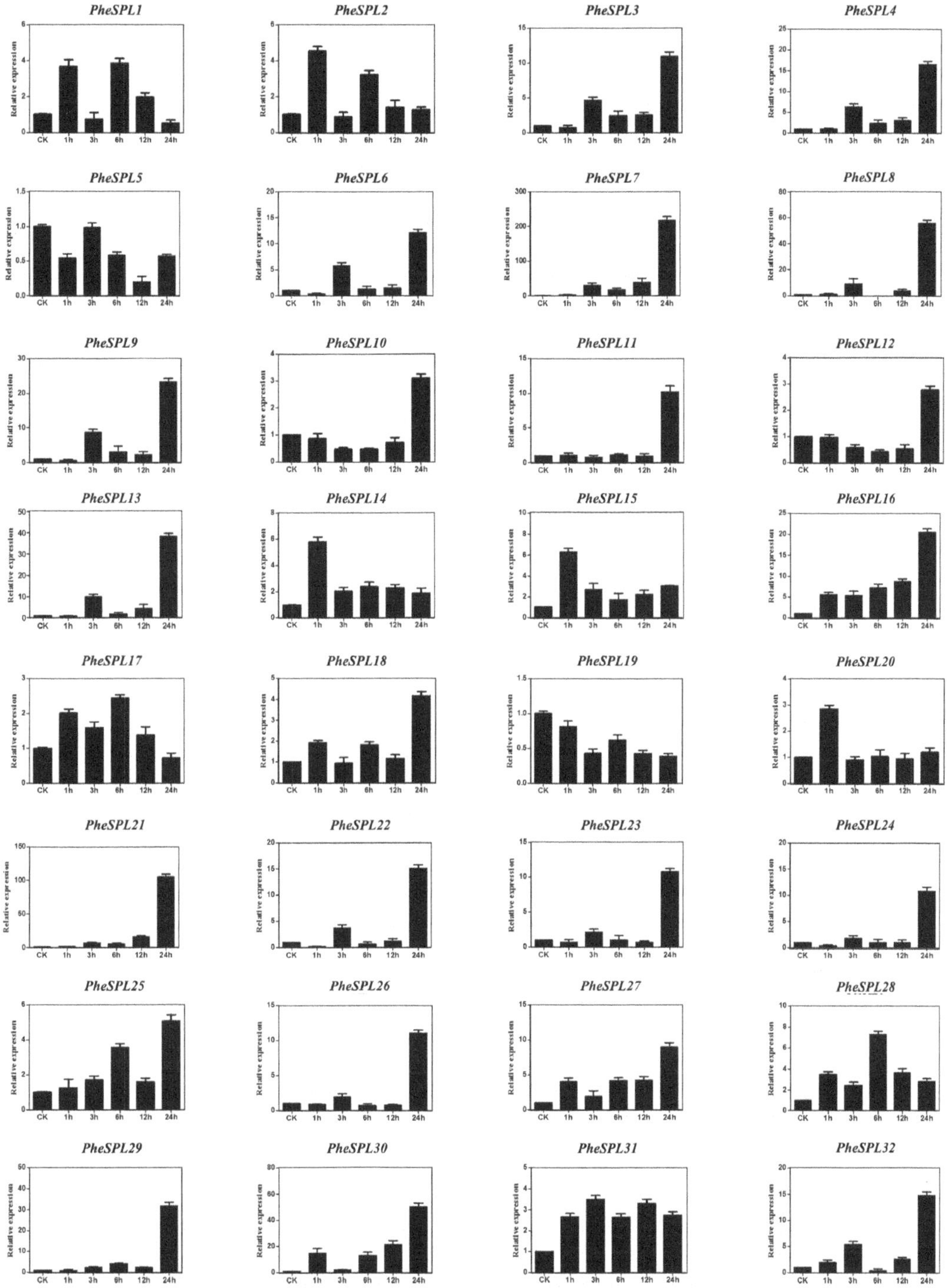

Fig. 8.15 PheSPL gene levels following SA exposure measured through qRT-PCR assay (from Pan et al. 2017). Sampling was completed at 0, 1, 3, 6, 9, 12, and 24 h post-treatment to analyze relative gene expression. The expression level in untreated samples was deemed as 1. *X*-axis stands for time points following SA treatment. *Y*-axis stands for relative expression levels of genes compared with *TIP41* (control gene). Mean and SDs were determined based on 3 technical and 3 biological replicates

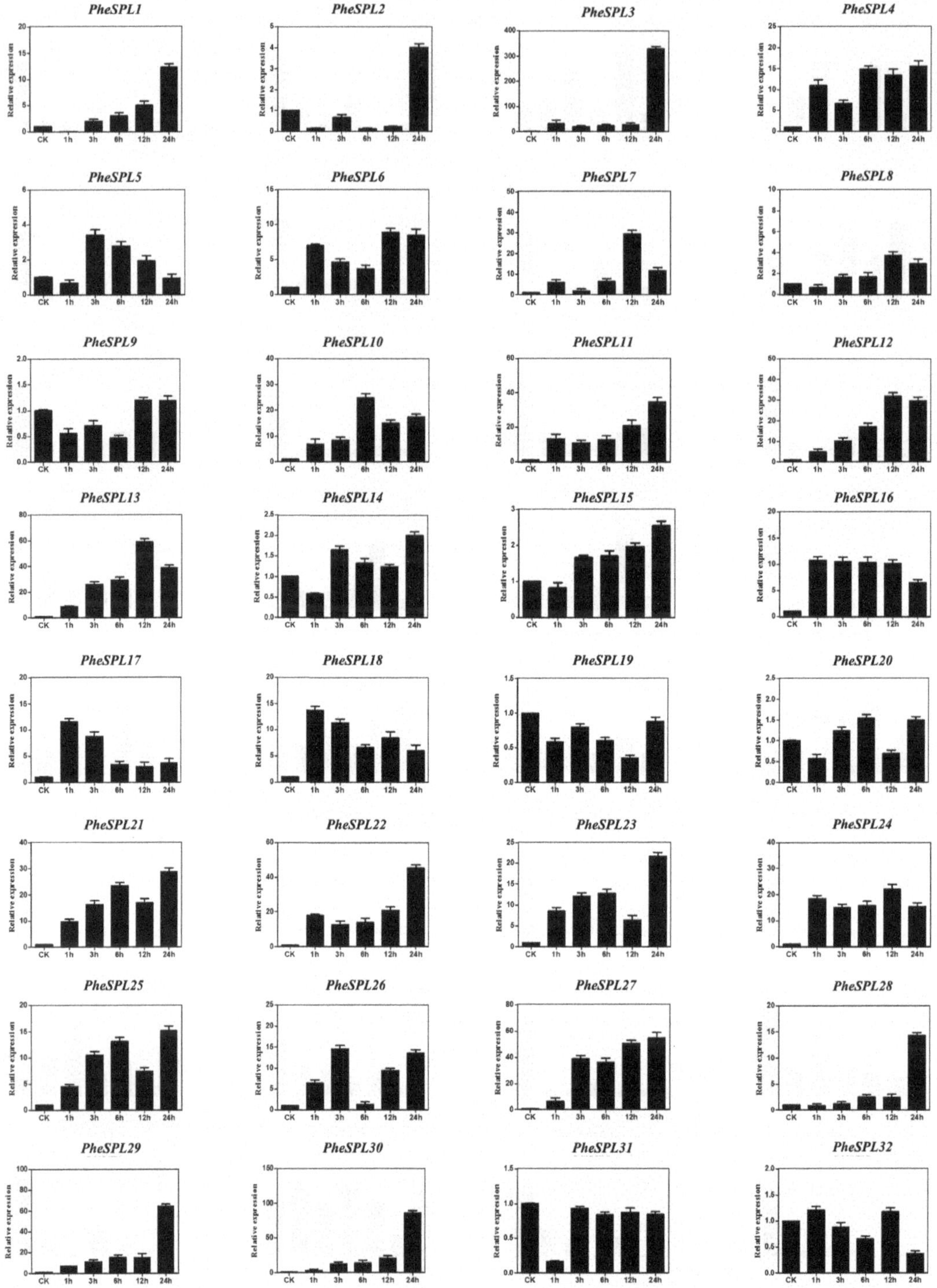

Fig. 8.16 PheSPL gene expression following GA treatment (from Pan et al. 2017). Sampling was conducted at 0, 1, 3, 6, 9, 12, and 24 h post-treatment to analyze relative gene expression. The expression level in untreated samples was deemed as 1. *X*-axis stands for time points following GA exposure. *Y*-axis stands for relative expression levels of genes compared with TIP41 (control gene). Mean and SDs were determined based on 3 technical and 3 biological replicates

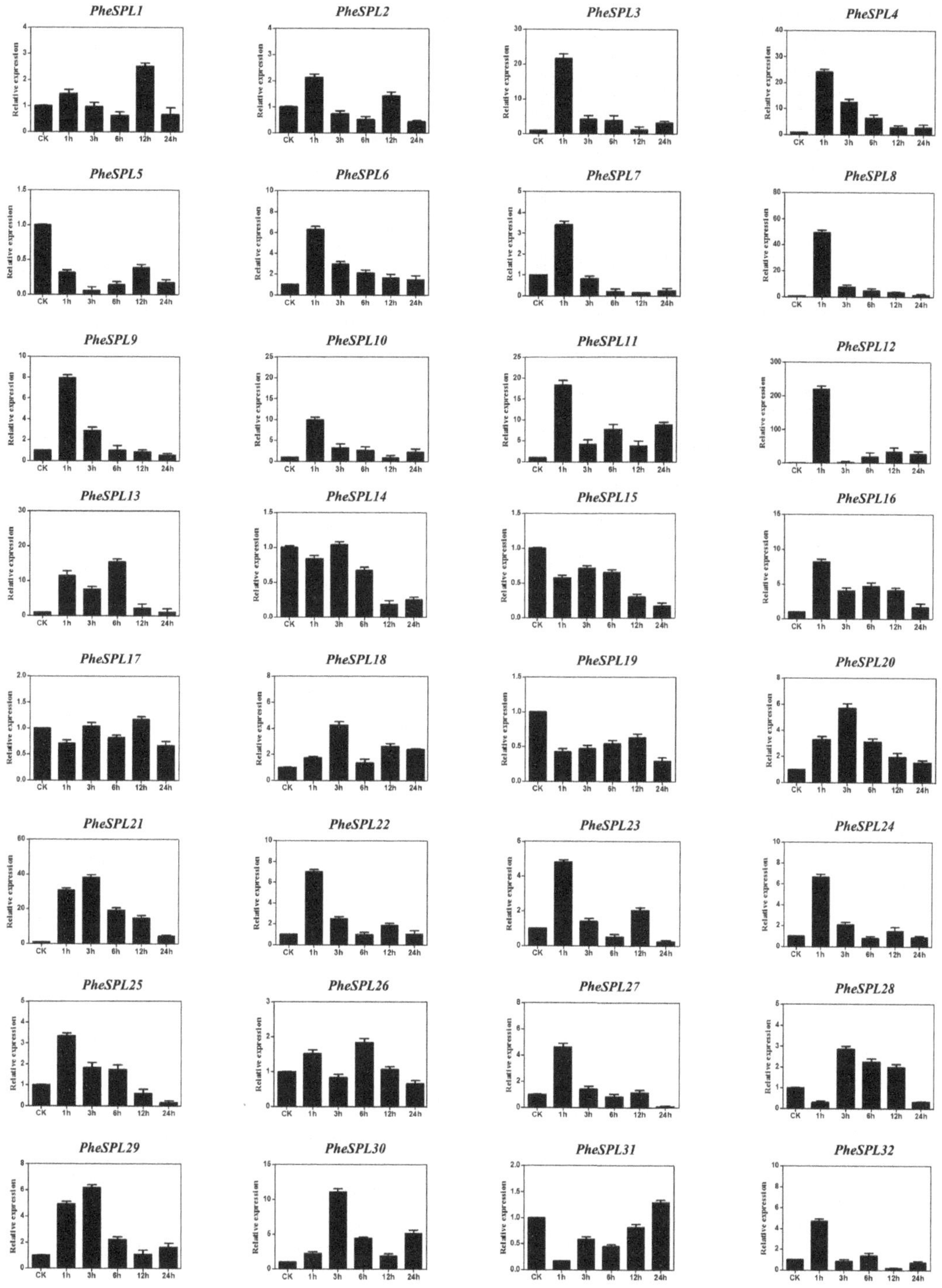

Fig. 8.17 PheSPL gene expression profiles upon drought stress by qRT-PCR (from Pan et al. 2017). Sampling was completed at 0, 1, 3, 6, 9, 12, and 24 h post-treatment to analyze relative gene expression. The expression level in untreated samples was deemed as 1. *X*-axis stands for time points following GA exposure. *Y*-axis stands for relative gene expression compared with *TIP41* (control gene). Mean and SDs were determined based on 3 technical and 3 biological replicates

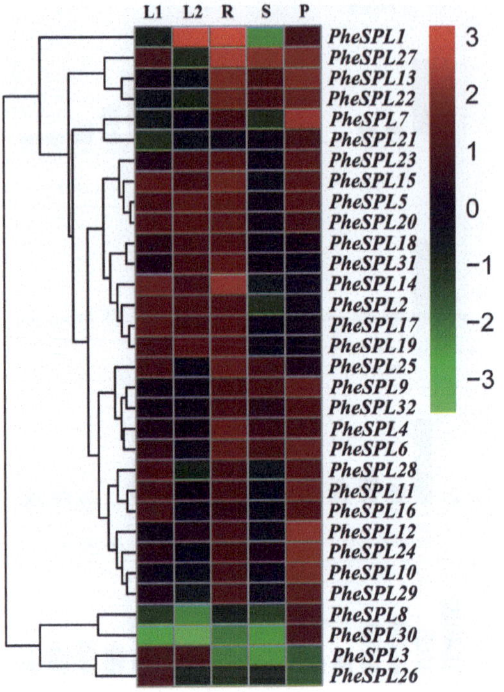

Fig. 8.18 PheSPL gene expression among diverse tissue samples and development periods (from Pan et al. 2017). Sampling was collected from young leaf (L1), mature leaf (L2), root (R), shoot (S) and panicle (P), respectively. The red and green colors stand for increased and decreased transcript abundances in comparison with corresponding control, respectively

response including high light, plant pathogens, osmotic, nitrate, salt, cold and water stress (Danisman 2016).

TCPs are detected in different plant varieties, like Arabidopsis thaliana, maize and rice (Liu et al. 2019) (Fig. 8.19). In moso bamboo (*Phyllostachys edulis*), 16 TCP (PeTCP1-PeTCP16) members are identified (Liu et al. 2018). There are 10 and 6 TCP members in Class I and Class II group, respectively (Fig. 8.19a, b) (Liu et al. 2018). Their CDS lengths were between 303 and 1,221 bp, so the aa sequences were 100–406 aa in length. On the other hand, their molecular weights were 10.40–43.37 kDa, with a theoretical isoelectric point of 4.78–10.76. Alignment results showed that these members contain TCP domain (bHLH structure). Additionally, for TCP members, those in Class II contained 4 aa insertions within the basic regions relative to those in Class I (Fig. 8.19b). Further, an R domain rich in arginine might be detected at the PeTCP13 C terminus (Fig. 8.19b). Gene structure analysis shows that 10 *PeTCPs* lacke an intron, whereas the rest *PeTCP* genes just possess 1–2 introns (Fig. 8.19a). Cis-elements stand for the linear ncDNA nucleotide fragments, which have direct effects on regulating genes related to stress response. Many cis-elements associated with responses to environmental stresses have been discovered from *TCP* promoters, such as elements responsive to ABA, SA and Me-JA (Fig. 8.19c). Additionally, SA, ABA and Me-JA treatments are found to induce most *PeTCPs* expression, as revealed by qRT-PCR aanlysis. Showing that *PeTCPs* may have the regulation functions of environmental stress response during plant growth (Fig. 8.19d). Molecular analysis shows that four TCPs (PeTCP4, PeTCP5, PeTCP10 and PeTCP11) are nucleus localized proteins (Fig. 8.20a) (Liu et al. 2018). And they have self-activation activity in Y2H Gold yeast (Fig. 8.20b) (Liu et al. 2018).

8.1.9 Trihelix Transcription Factors

Genes in the trihelix transcription factor (TTF) family (also called the GT transcription factor family) possess a characteristic DNA-binding domain, which is comprised of 3 helical structures (helix loop-helix-loop-helix). Typically, 5′-G-Pu-(T/A)-A-A-(T/A)-3′ is the core sequence of this gene family, and the genes can bind to DNA photo-responsive element GT (Song et al. 2016; Fang et al. 2010). TTF genes exert vital parts during plant growth (Xi et al. 2012). To take an example, PETAL LOSS (PTL) genes in Arabidopsis (which belong to the GT-2 subfamily) can regulate sepal and petal development as well as sepal fusion (Brewer et al. 2004). As discovered by Fang and colleagues, *GTγ-1* gene in rice promoted salt stress tolerance, whose expression was mildly induced upon drought challenge and ABA treatment; besides, many abiotic stress induced the expression of *GTγ-2* and *GTγ-3* (Fang et al. 2010). In addition, *GmGT-2A* and *GmGT-2B* are

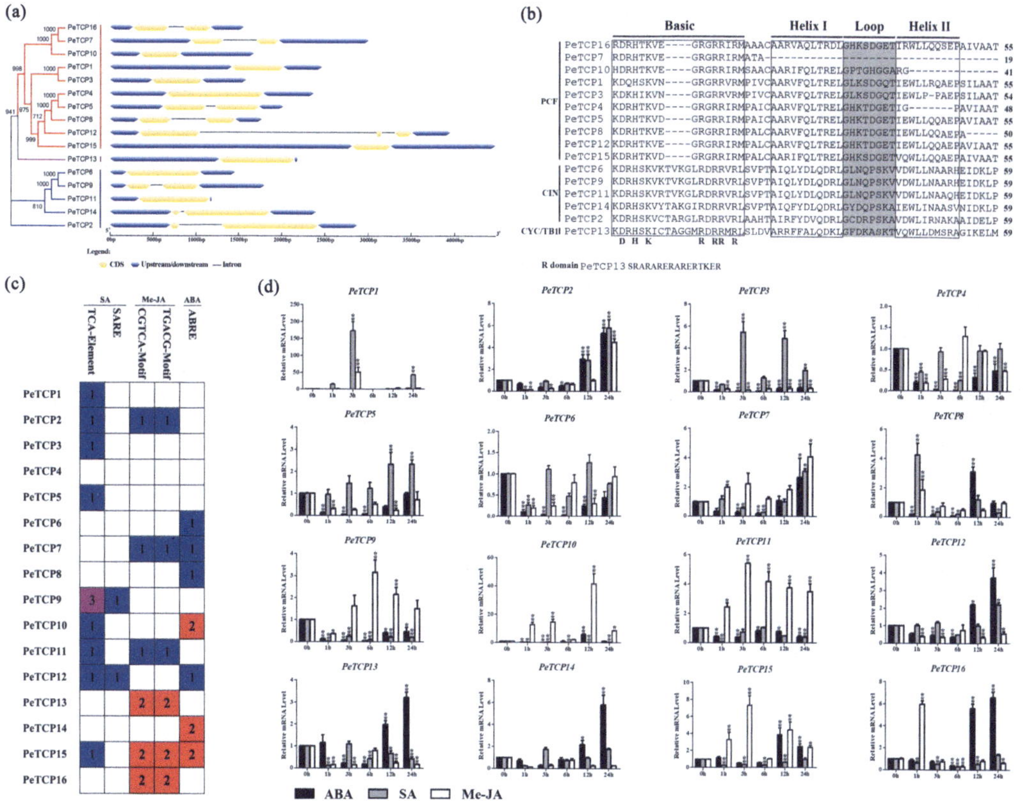

Fig. 8.19 Bioinformatics analysis of PeTCPs. **a** Phylogenetic and gene structures analysis on TCPs from moso bamboo. The GSDS online approach was utilized for gene structure analysis. Yellow rectangles stand for exons, blue rectangles indicate untranslated regions (UTRs), while gray lines represent introns; **b** Multiple sequence analysis on the TCP domain within TCPs from moso bamboo; **c** Cis-elements associated with ABA, SA and Me-JA responses within the *PeTCP*s promoter region. The colored block that had a number on the top stands for the cis-element quantity in *PeTCP*s; **d** *PeTCP*s expression in response to ABA, SA and Me-JA treatments measured through qRT-PCR analysis. X-axis stands for plant hormone treatment course, while Y-axis stands for relative gene expression. Asterisks suggest that the differences in transcript abundance were significant in comparison with control detected through Student's t-test (*$p < 0.05$, **$p < 0.01$) (from Liu et al. 2018)

overexpressed within the transgenic Arabidopsis plants, indicating these two genes promoted resistance to freezing, salt and drought stresses (Xie et al. 2009). All in all, TTFs exert vital parts in regulating plant development.

TTFs from moso bamboo were subjected to evolutionary and functional analyses. Altogether 35 genes were discovered and classified as 5 groups (GT-1, GT-γ, GT-2, SIP1 and SH4) based on the corresponding structural characteristics. As revealed by gene structure analysis, many *PheTTF*s contained few introns. Besides, conserved motif analysis detected a motif (Motif 16) characteristic of GT-γ subfamily. Based on promoter analysis, different cis-elements associated with plant growth, or responses to plant hormone and biotic/abiotic stresses were identified. We utilized data from 35 TTF genes in moso bamboo for generating heat maps, and found that the expression of such gene was detected at diverse development periods and within diverse tissues (Fig. 8.21). Many TTF genes discovered in this study were highly expressed in panicles and leaves, as revealed by expression profile analysis (Fig. 8.22). Further, we conducted qRT-PCR to analyze TTFs expression within young leaves, so

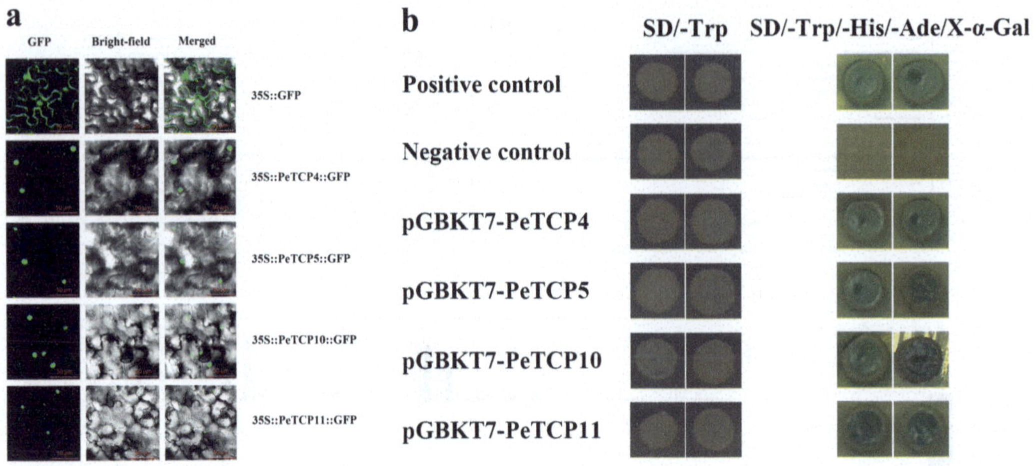

Fig. 8.20 Subcellular localization (**a**) and transactivational analyses (**b**) of PeTCP proteins (Liu et al. 2018)

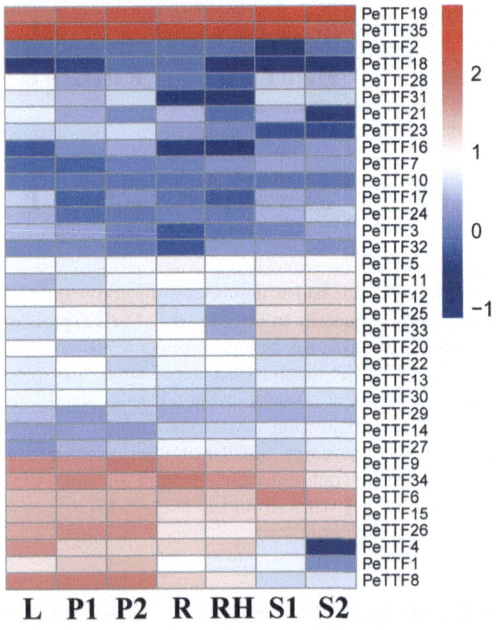

Fig. 8.21 Expression patterns of TTFs at diverse development periods and in diverse tissues of moso bamboo. Heatmap analysis showed that 35 *PheTTF*s were hierarchically clustered across diverse tissues. Leaf (L); early panicle (P1); advanced panicle (P2); root (R); 20 cm shoot (S1); 50 cm shoot (S2); rhizome (RH) (from Cheng et al. 2019)

as to examine the tissue specificity as well as the expression profiles under stress for the sake of functional characterization of individual genes (Fig. 8.23). TTF genes are possibly related to the stress responses in plants. Typically, subcellular localization analysis revealed that PheTTF29 was localized in nuclei, while transcriptional activity analysis suggested that it showed transcriptional activity (Figs. 8.24 and 8.25).

8.1.10 WRKY Family of Transcription Factors (TFs)

WRKY family is among those 10 greatest TF families detected across the higher plants (Ulker and Somssich 2004). To be specific, SPF1 isolated from Pachyrhizus erosus is the first cDNA that encodes the WRKY protein for cloning (Ishiguro and Nakamura 1994). Thereafter, certain proteins in WRKY family are discovered in different plant varieties (Rushton et al. 1995, 2010; Liu and Ekramoddoullah 2009; Zhang and Wang 2005; Jiang et al. 2014), and each of them exerts an important part in plant responses to the changing internal/external stimuli. WRKY TFs gain their names because they possess a WRKY domain comprising about 60 aa. Such TFs exert important parts in regulating the physiological and development processes, like seed dormancy (Pandey et al. 2010), embryonic development (Alexandrova and Conger 2002), growth, metabolism and aging of plant (Chen et al. 2002; Maren et al. 2014;

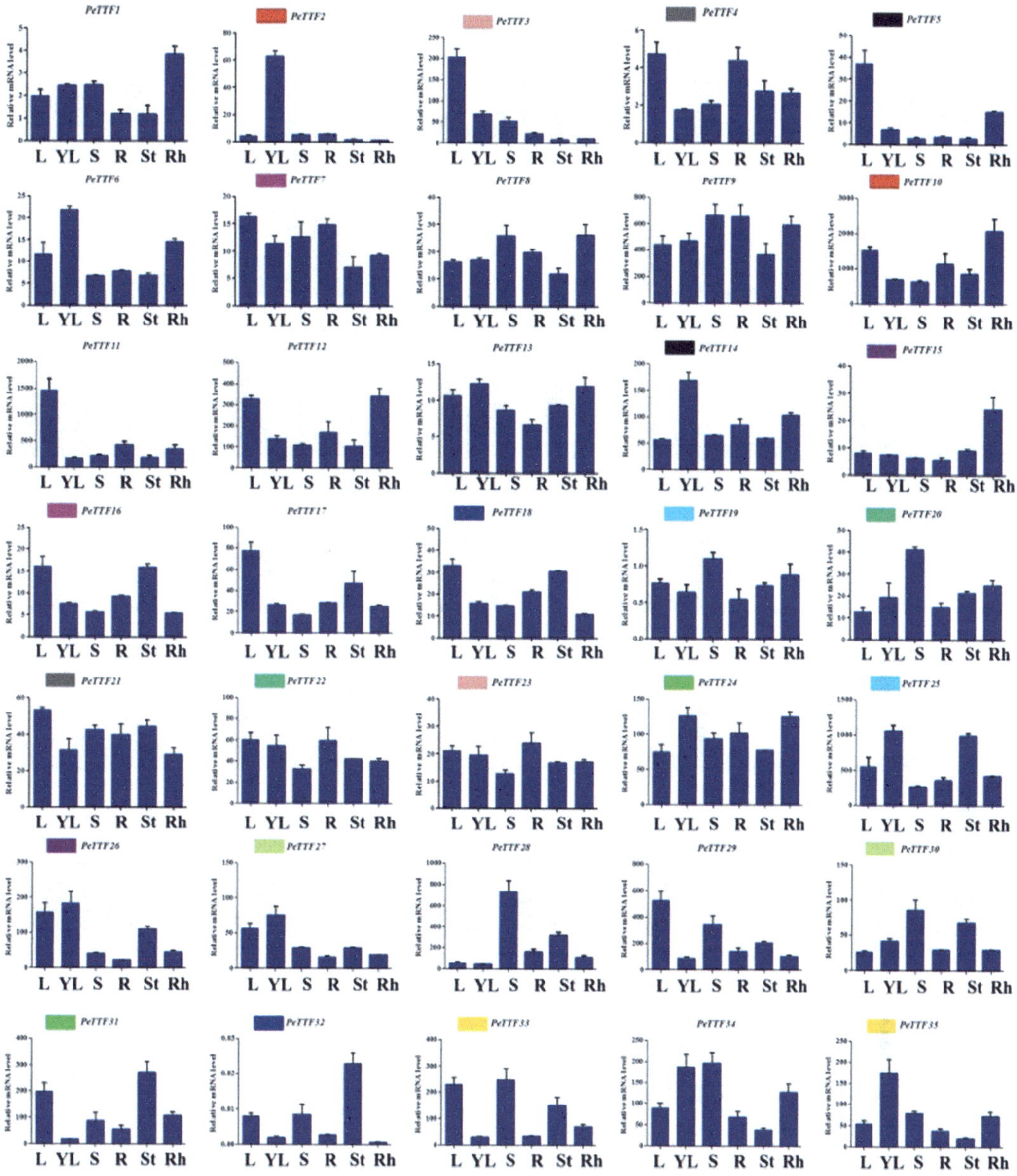

Fig. 8.22 Expression patterns of TTFs at diverse development periods and in diverse tissues of moso bamboo measured through qRT-PCR (Cheng et al. 2019). *X*-axis: stress treatment course; Y-axis: relative gene expression. Error bars = 6 ± SE. Leaf (L); young leaf (YL); root (R); rhizome (RH); shoot (S); stem (St) (from Cheng et al. 2019)

Hinderhofer and Zentgraf 2001; Sun et al. 2003; Johnson et al. 2002). In addition, WRKY TFs participate in diverse biotic/abiotic stress responses, including fungi 15, 21 (Maren et al. 2014), bacteria (Deslandes et al. 2002; Dellagi et al. 2002), nematodes (Higashi et al. 2008), wounding (Eulgem and Somssich 2007), salt, drought, cold and heat (Ciolkowski et al. 2008).

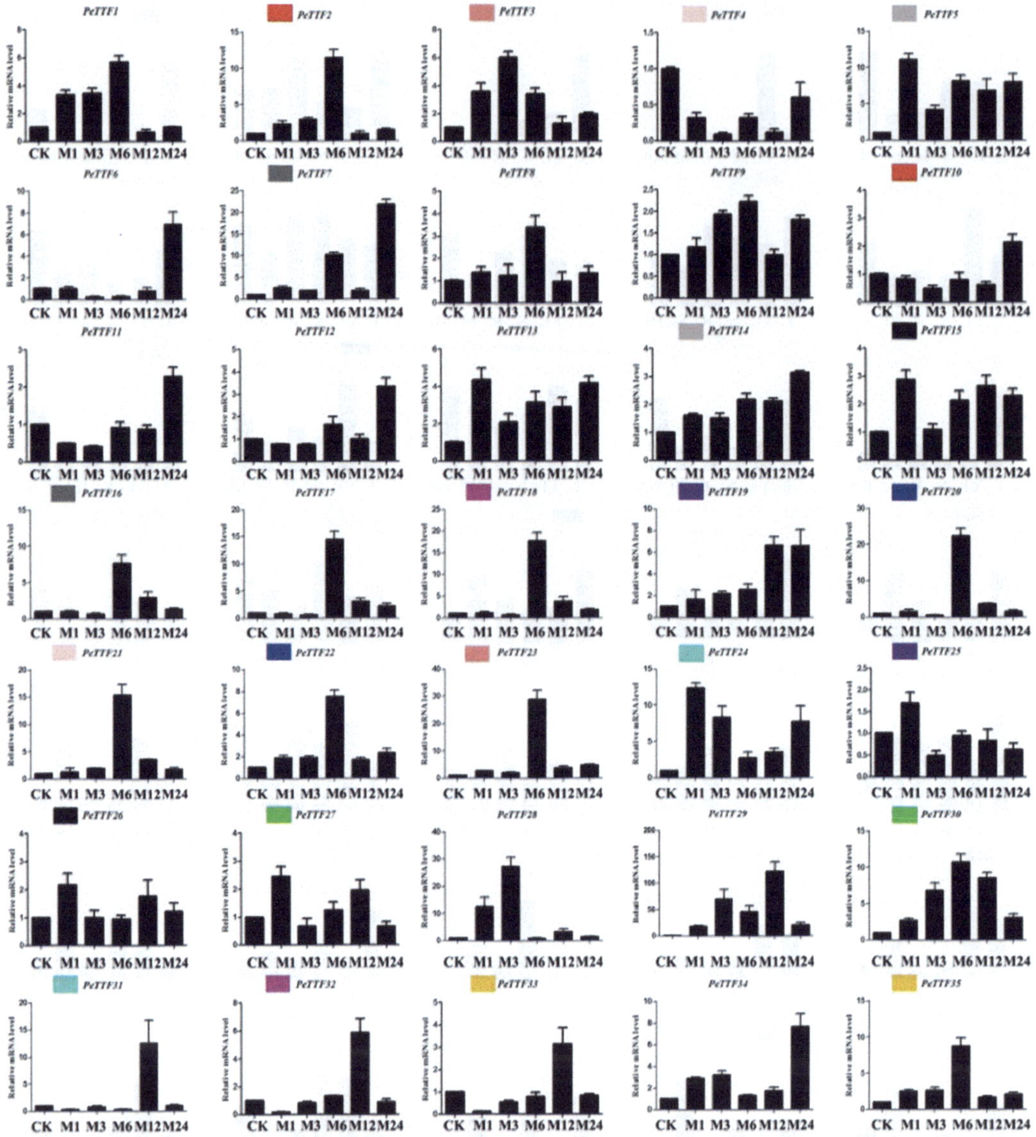

Fig. 8.23 Expression profiles for each chosen gene from moso bamboo exposed to MeJA/PEG/Salt challenge, as measured through qRT-PCR (from Cheng et al. 2019). X-axis: stress treatment course; Y-axis: relative gene expression; Error bars = 6 ± SE

Li and colleagues discovered 121 WRKY TFs from the genomic database of moso bamboo, among which, 5 new TFs were not annotated from the genomic database of moso bamboo before. In addition, the authors measured 81 PheWRKY gene expression within diverse tissues in the 3-month-old seedlings through RT-PCR. As a result, those screened genes were distributed in each subfamily, which contained the up-regulated as well as down-regulated genes related to shoot growth and flower development. Typically, *PheWRKY11-2, 22-2, 45-1, 77-2, 83-1* and *83-3* were highly expressed within roots; *PheWRKY3-2, 13-–3* and *19-1* were highly

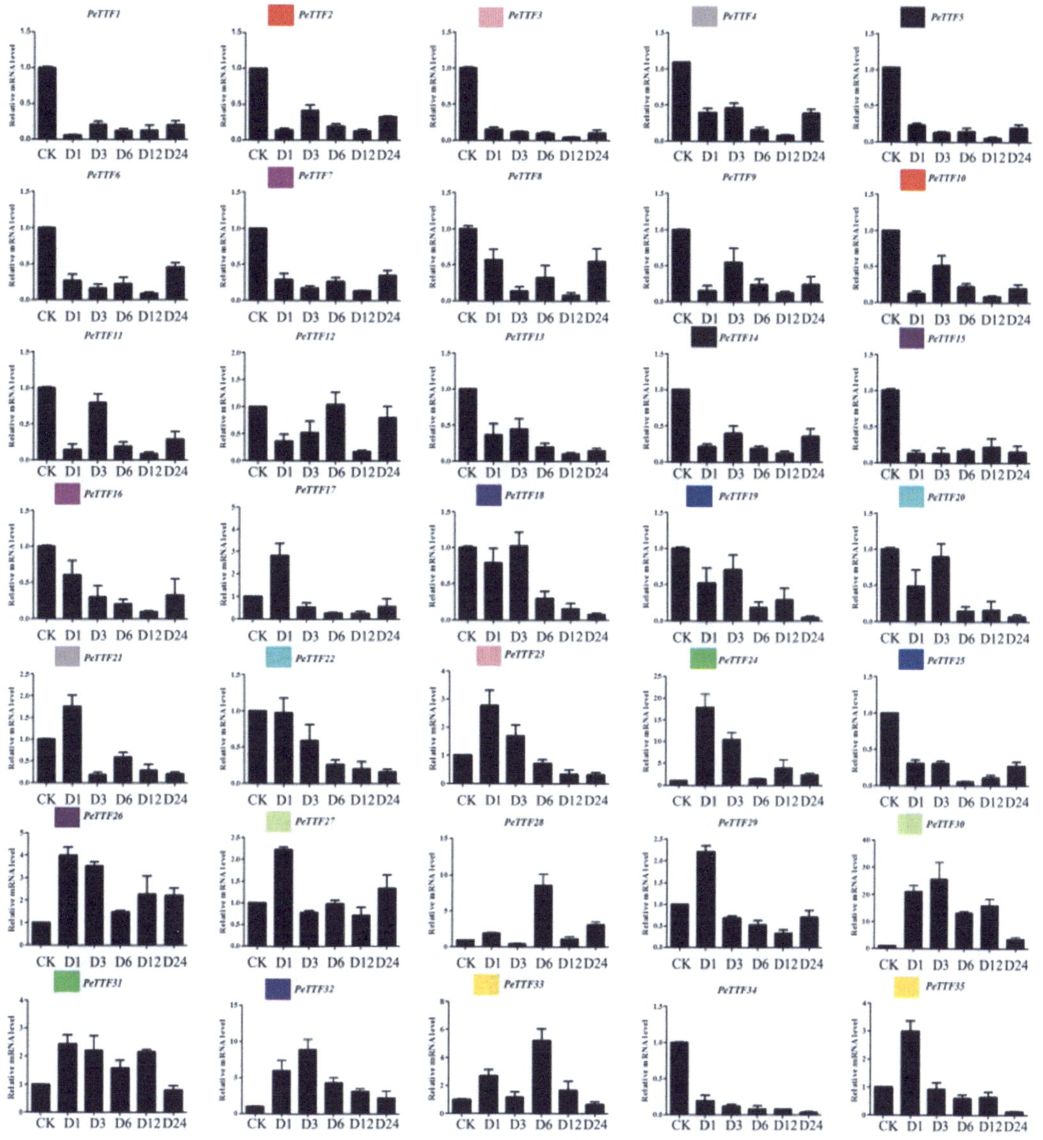

Fig. 8.23 (continued)

expressed within stems; while *PheWRKY26*, *67-1* and *71-2* showed great expression quantities within leaves. We also analyzed *PheWRKY* gene expression to examine their effects on shoot growth and flower development. According to our results, *PheWRKY28*, *29-3*, and *62* showed increased expression levels in shoot growth, indicating their effects on rapid bamboo shoot elongation.

Li and colleagues measured 81 PheWRKY gene expression levels in the presence of cold and drought stresses through qRT-PCR. As a result, 32 genes were up-regulated to diverse levels under one or both stresses, among which, 22 responded to drought stress, while 28 to cold stress (Fig. 8.26). It was interesting that, genes with the highest up-regulation level under drought stress (17/22) always showed increased

Fig. 8.23 (continued)

expression upon cold stress. Besides, *PheWRKY4*, *26-2*, *48-2*, *49*, *62*, *69-1*, *72-3* and *96-1* showed gradual increases in their expression, with the highest expression levels being detected at 12 h after cold stress, while *PheWRKY11-2*, *111*, *34-2*, *45-1*, *114-1*, *69-2*, *70-2* and *76* showed rapid increases in their expression at 1 h but subsequent significant decreases in expression. Under drought stress, *PheWRKY45-1*, *62*, *69-1*, *88-1*, *96-1*, *96-2* had the highest expression at 1 h, whereas *PheWRKY4*, *5-1*, *26-2*, *15-2*, *19-2*, *44*, *49*, *55*, *65-1*, *71-2*, *72-3* had the greatest expression quantities at 24 h. As for *PheWRKY1*, *36*, *69-2*

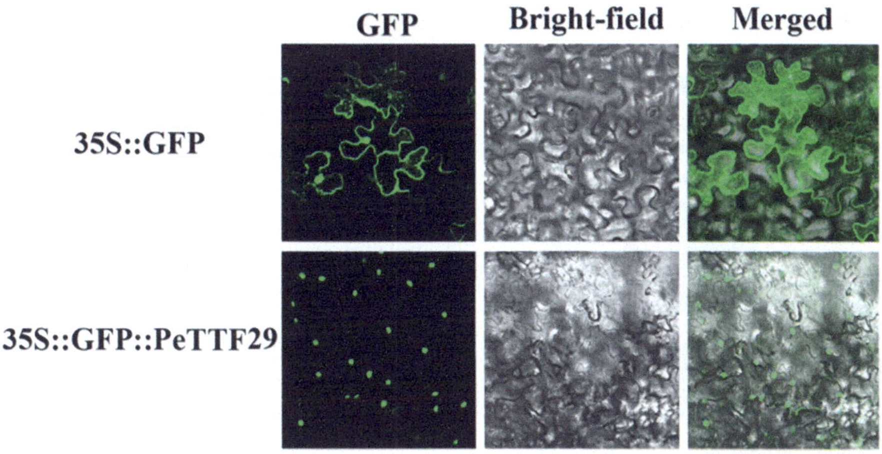

Fig. 8.24 Nuclear localization of PeTTF29 (from Cheng et al. 2019). The 35S::GFP::PheTTF29 construct, as well as the control vector 1305(35S::GFP) was transfected in *Nicotiana tabacum* leaves. Confocal microscopy was performed to observe GFP signals within root cells

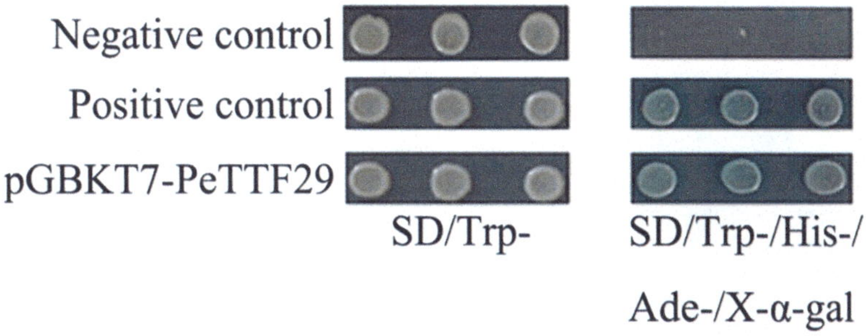

Fig. 8.25 Transcriptional activity analysis of PheTTF29 from (Cheng et al. 2019). pGBKT7-T and pGBKT7–53 were the positive controls, whereas pGBKT7 was the negative control

and *74-2*, their expression upon drought stress showed parabolic trends, which reached the highest levels at 12 h (Fig. 8.26). According to the expression analysis in this study, physiological process (shoot growth and flower development) and cold and/or drought treatment markedly induced the expression of 54 genes. Among them, the expression of 3, 4, 7 and 17 genes was only induced by shoot growth, drought stress, cold stress and flower development, separately.

As revealed by subcellular location analysis, PheWRKY72-2 labeled by green fluorescent protein (GFP) was localized in nucleus, which conformed to its role of a TF. For better investigating the possible function of *PheWRKY72-2* that showed up-regulation upon cold and drought stresses, we transfected *PheWRKY72-2* in Arabidopsis (wild type [WT]). Then, qRT-PCR assay was conducted on *PheWRKY72-2* from both the transgenic and WT lines, which revealed the up-regulation of *PheWRKY72-2* within the transgenic lines whereas deletion within the WT lines (Fig. 8.27). Upon cold stress, differences between transgenic and WT lines were not significant (data not shown). To simulate the drought stress, the surface of transgenic Arabidopsis seeds was disinfected, followed by seed

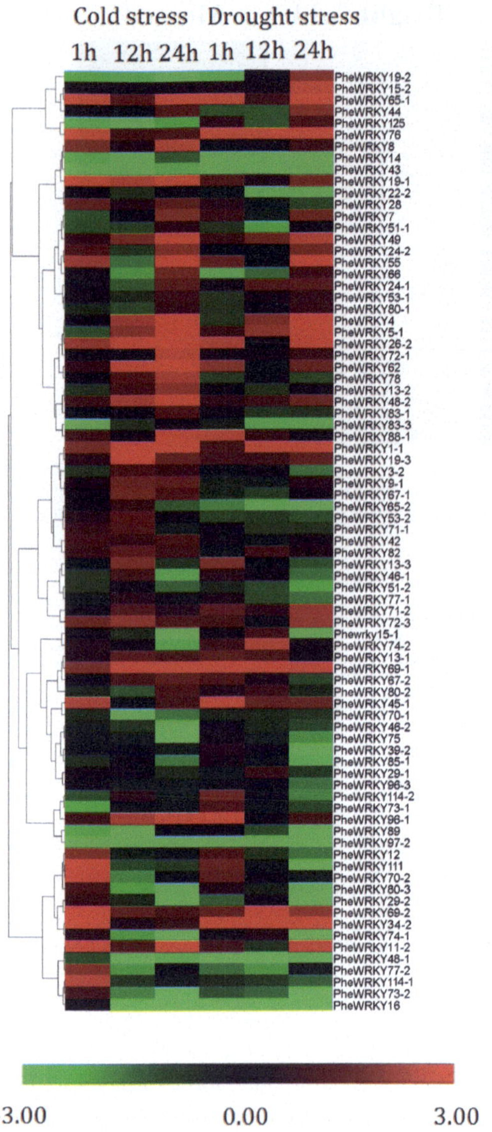

Fig. 8.26 WRKY gene expression patterns upon drought and cold stresses. Expression was measured through qRT-PCR, while the relative PheWRKY gene expression upon different stresses in comparison with controls was adopted to carry out hierarchical clustering using Cluster 3.0. We acquired qRT-PCR results from 3 technical and 3 biological replicates (Li et al. 2017)

germination in the Morishige and Skoog (MS) agar medium that contained 10/20% PEG, respectively (Fig. 8.27b, e). *PheWRKY72-2* was overexpressed (T3generation), as a result, the roots were markedly prolonged relative to WT plants cultured in the medium that contained 10%PEG (Fig. 8.27b, c). Additionally, this study also analyzed root hairs, and there was no significant difference compared with transgenic and WT plants (Fig. 8.27d). Nonetheless, in the presence of 20% PEG, WT plants showed the etiolation symptoms different from transgenic plants (Fig. 8.27e). Thereafter, stomatal apertures were compared upon drought stress, since water loss was mostly determined by stomatal regulation. In this regard, stomatal apertures were compared between WT and transgenic lines upon drought and control stresses (3 h). In addition, we measured stomatal width and length to calculate the length/width ratio, which was later adopted for reflecting the stomatal closure level. In the case of normal development, difference was not significant (Fig. 8.28b). Upon drought treatment, *PheWRKY72-2*ox plants had an increased stomatal length/width ratio relative to WT lines (Fig. 8.28a, b). According to the above findings, stomatal closure showed higher sensitivity to drought treatment in *PheWRKY72-2*ox lines than WT lines. Upon drought stress, OE-PheWRKY72-2 inhibited AtABA2, AtABI2 and AtABI5 expression, but markedly promoted *AtLEA*, *AtABI4*, and *AtAREB* expression (Fig. 8.28c).

8.2 Flowering and Stamen Development Gene Candidates

8.2.1 Dof Transcription Factor

The DNA binding with one finger (Dof) proteins, which form a vital TF family, participate in regulating plant growth, gene transcription, flowering and stress responses. In generally, these proteins possess about 50–52 highly conserved aa and include one C-terminal C2C2-type zinc-finger motif (Yanagisawa 2002). Moso bamboo is the most widely distributed and extensive use among the bamboo species, it is one of the dual-use bamboo species that

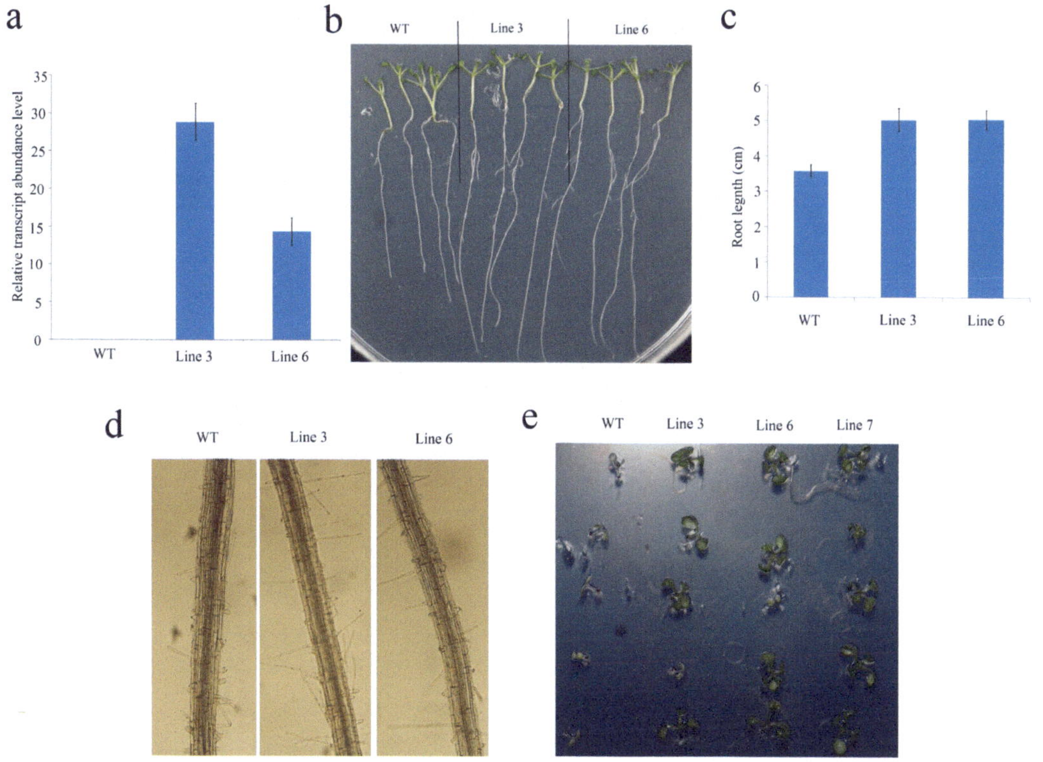

Fig. 8.27 **a** Relative *PheWRKY72* expression within the transgenic plants. β-tubuli (endogenous reference) expression was deemed as 1.0. **b** Response to drought stress (10% PEG in 35S-PheWRKY72-2 transgenic Arabidopsis plants. **c** Primary root lengths in both transgenic and WT seedlings upon drought treatment (10% PEG). **d** Root hairs observed upon drought treatment (10% PEG). **e** Drought stress (20% PEG) response in 35S-PheWRKY72-2 transgenic Arabidopsis plants (Li et al. 2017)

combines economic, ecological and social benefits. However, moso bamboo is a plant that flowers only once in the lifetime, which has a long flowering cycle and uncertain flowering period, and collectively die after flowering, causing serious economic losses and ecological damage.

Many *PheDofs* are found to participate in drought, low temperature and high salt response of moso bamboo (Cheng et al. 2018). *PheDof4-1* is induced in young stems under the treatment of 4°C and 250 mmol/L NaCl, while the transcription level of *PheDof4-1* in roots of 20% PEG8000 decreased significantly (Liu et al. 2017). GUS staining show that *PheDof12-1* is mainly expressed in roots, hypocotyls, leaf vascular, stamens and petals of transgenic Arabidopsis (Liu et al. 2019). Transcriptomic sequencing results at different flower development stages show that 238 genes are associated with moso bamboo flowering. *PheDofs* gene was up-regulated multiple times during flowering and highly expressed in the early stage of flowering (Ge 2015). Drought induced the up-regulation of *PheDofs*, promoted the transcription of *Hd3a* and *MADS14*, and promoted the flowering of moso bamboo (Zhang 2014), which is speculated that the Dof-Hd3a-MADS14 regulation pathway exerts an important part in moso bamboo flowering pathway (Gao et al. 2014). In situ hybridization results show that *PheDof1* expression can be detected within apical meristem, inflorescence axis, glume and bract of moso bamboo. Western blotting show that *PheDof1* is

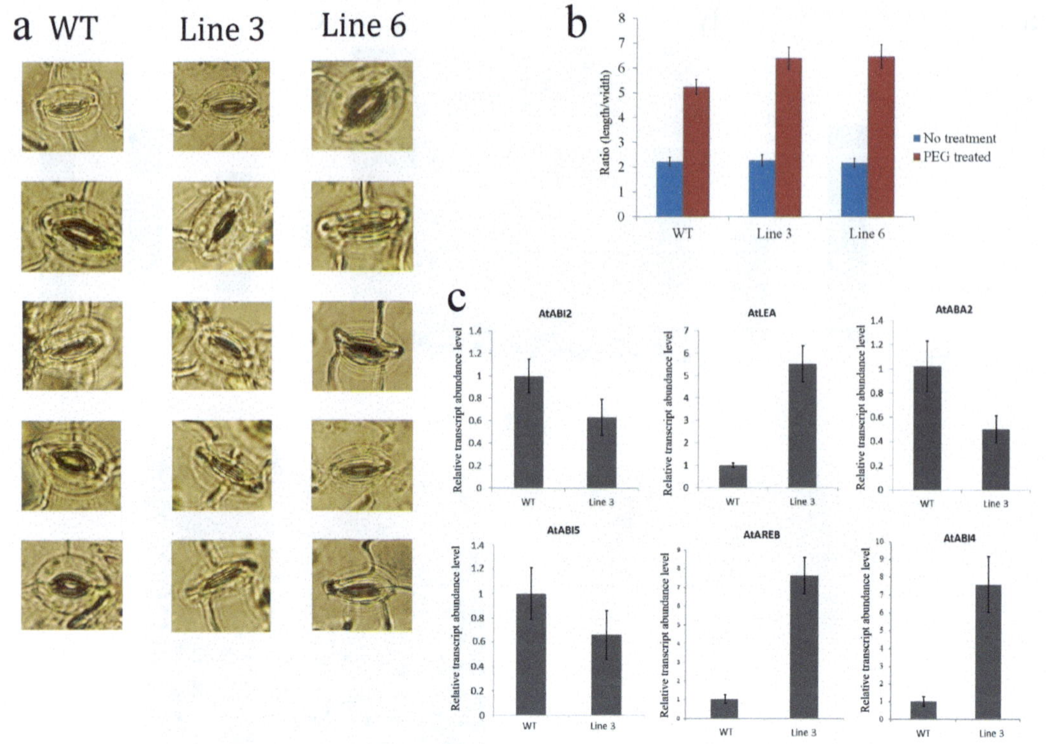

Fig. 8.28 **a** Stomatal apertures compared between WT and transgenic lines upon drought treatment. **b** Stomatal closure rate. The length/width ratio was adopted to measure stomatal closure. **c** ABA-signaling gene expression in 35S-OsWRKY72 transgenic and WT Arabidopsis plants upon 3 h of drought treatment. Gene expression within WT was deemed as 1.0. We acquired qRT-PCR results from 3 technical and 3 biological replicates (Li et al. 2017)

mainly expressed in the bud formation period, inflorescence extension period and flowering period, indicating that *PheDof1* is involved in flowering (Fig. 8.29) (Ge et al. 2017; Cheng et al. 2016). Overexpression of *PheDof12-1* led to early flowering of transgenic Arabidopsis, flowering inducers *FT*, *SOC1* and *AGL24* are up-regulated expression, flowering inhibition of *SVP* and *FLC* are down-regulated expression. *PheDofs* are rhythm genes that are involved in photoperiodic response. Most *PheDofs* genes are highly expressed at the initial stage of light and low at dusk under photoperiodic treatment. Yeast one-hybrid show that *PheDof12-1* may bind to the *PheCOL4* promoter region (Fig. 8.30) (Liu et al. 2019). This indicates that *PheDofs* gene is functionally diverse and conserved, and is involved not only in abiotic stress response, but also in photoperiod regulation of flowering of moso bamboo.

8.2.2 MADS Transcription Factor

As discovered by Peng and colleagues, 30% genes related to flowering included the MADS-box and Dof families (Peng et al. 2013). Additionally, Gao and coworkers reported the significant influence of an activated Dof-Hd3a-MADS-flowering pathway on flowering of moso bamboo (Gao et al. 2014). This showed Dof and MADS-box families might exert vital parts in regulating moso bamboo flowering. According to Cheng and colleagues, *PheMADS14*, *PheMADS15* and

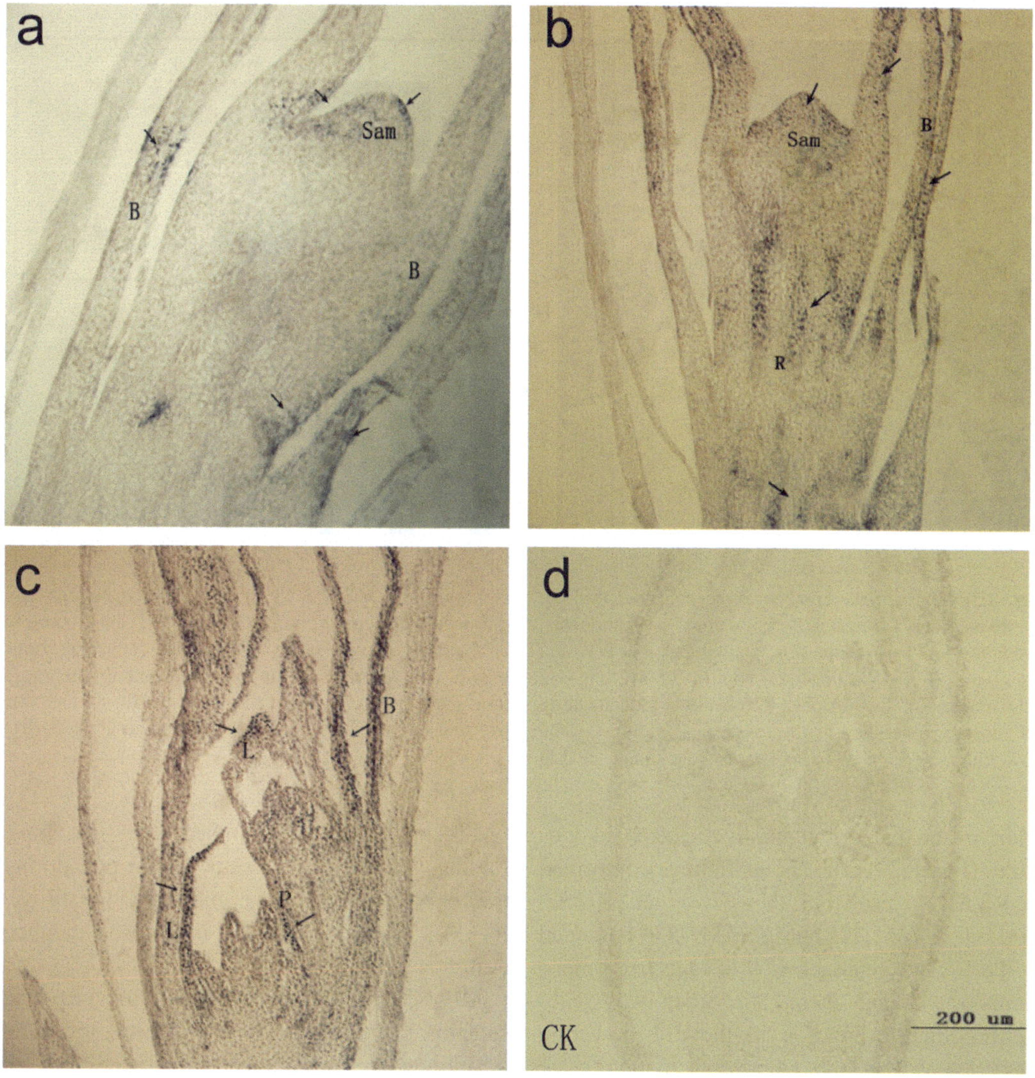

Fig. 8.29 *PheDof1* expression in diverse flowering stages of moso bamboo. B = bract, L = lemma, P = Palea, R = rachis, Sam = shoot apical meristem. **a**, **b** Floral bud formation stage. **c** Bloom stage. **d** Sense probe expression following in situ hybridization (Ge et al. 2017)

PheMADS18-1, the A-class genes, showed up-regulated expression during floral bud formation, whereas *PheMADS29* and *PheMADS30* showed preferential expression at F2–F4 stages (Fig. 8.31) (Cheng et al. 2017). As suggested by in situ hybridization (ISH) results, the expression of *PheMADS15* was detected within early spikelet meristems, reproductive organs and primordia in flower organs (Fig. 8.32). The expression of *PheMADS3* and *PheMADS21* belonging to the C-class and the D-class, respectively, was mostly detected during the formation of pistils and stamen (Fig. 8.31). As indicated by ISH results, the mRNA expression of *PheMADS3* and *PheMADS21* mRNA was up-regulated within lemma, glumes, developing

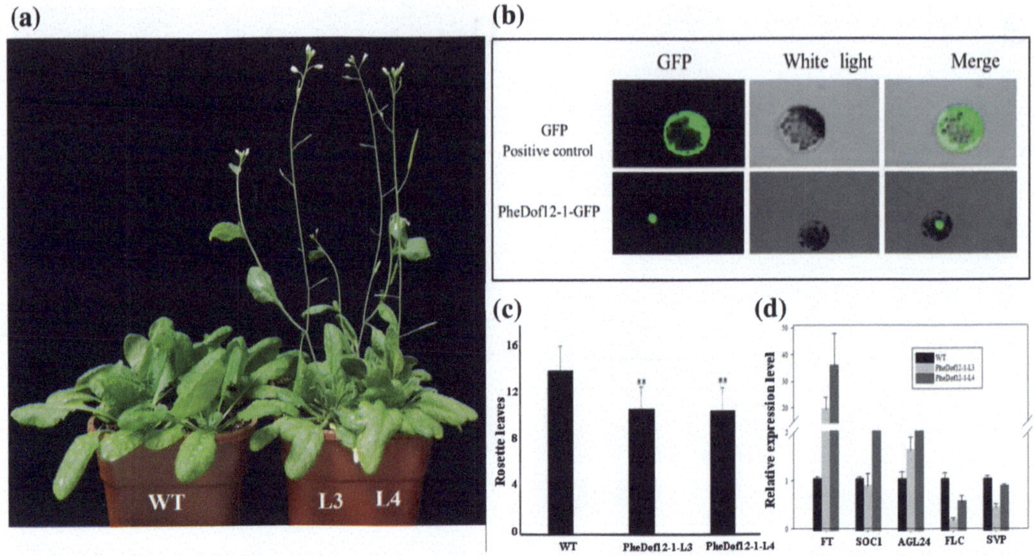

Fig. 8.30 Early flowering phenotype analyzed through overexpressing *PheDof12-1* within Arabidopsis. **a** *PheDof12-1* overexpression phenotypes of WT and transgenic (L3, L4) lines as control upon long-day (LD) condition. **b** PheDof12-1 subcellular localization. **c** Flowering time rated as the rosette leaf number in the flowering stage of WT and transgenic lines in LD condition. **d** *SOC1*, *FT*, *SVP*, *FLC* and *AGL24* transcription levels detected in transgenic and WT lines. *Actin* in Arabidopsis served as an endogenous reference. Error bars = SD. Asterisks stand for significant difference in transgenic plants compared with WT counterparts ($p < 0.01$ upon Student's t-test). (From Liu et al. 2019)

embryo and palea, but hardly detected in lodicules (Fig. 8.32). The above findings conformed to RNA-seq results of these two genes. *PheMADS1*, *PheMADS5* and *PheMADS34* belonged to the E-class genes of *SEP* lineage from moso bamboo. Among them, the first two genes showed high expression in the 3rd flowering stage as well as the embryo formation stage (Fig. 8.31). As for *PheMADS5*, its spatiotemporal expression in moso bamboo was measured between early floral bud and mature floral organ through ISH analysis. Nonetheless, the expression of *PheMADS34* was mainly detected during the formation of floral bud but decreased in floral developmental stage. As suggested by ISH analysis, *PheMADS14* expression was significantly up-regulated in stamen primordia and pistil differentiation, namely, stamen and pistil cell proliferation (Ge et al. 2017) (Fig. 8.33). *PheMADS14* showed high expression within stamens, but it was deleted within endosperms, pistils and embryos. As indicated by the above findings, *PheMADS14* exerts a vital part during the formation of stamen primordia and pistil, yet it does not participate in moso bamboo embryogenesis.

At present, it is very difficult to verify the function of genes related to flowering and floral development, mainly because of the lack of genetic transformation system in moso bamboo. In recent years researchers have reported that the functions of flowering genes were identified in *Arabidopsis*. Gao et al. transferred *PeMADS1* into *Arabidopsis thaliana* that showed different phenotype such as early flowering, curled rosette or stunt, which indicated that *PeMADS1* might participate in regulating the flower development of moso bamboo (Gao et al. 2010). Qi et al. analyzed the expression of *PheTFL1* gene in different tissues and organs of flowering bamboo, non-flowering bamboo, and seedlings. The results showed *PheTFL1* was abundantly

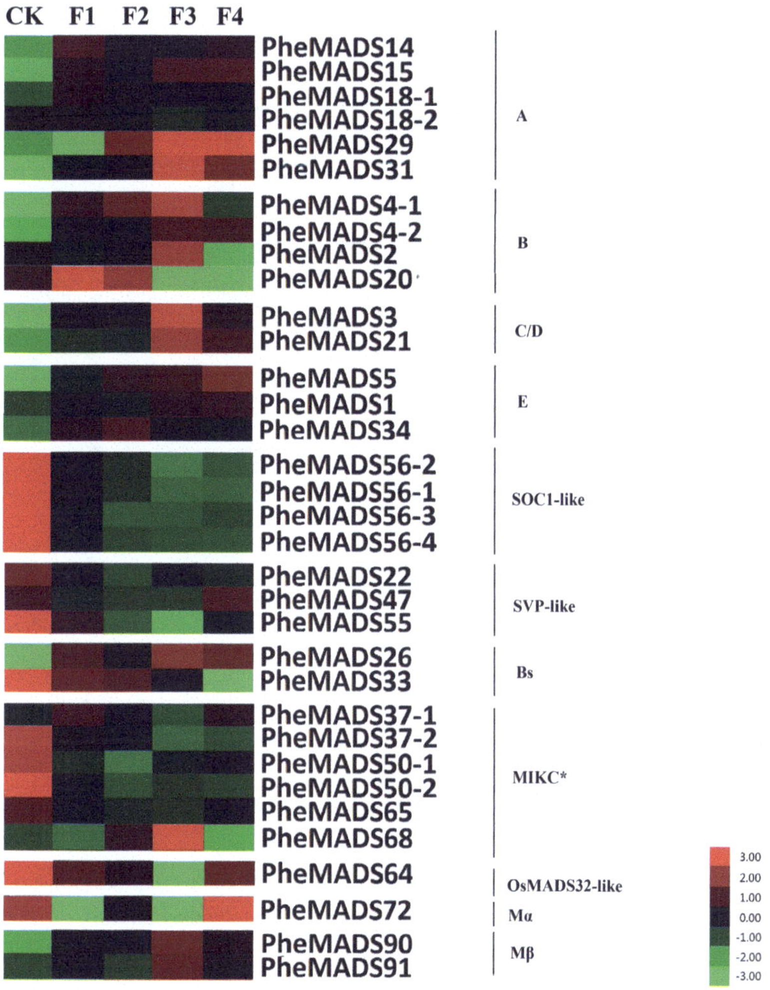

Fig. 8.31 MADS-box gene expression within moso bamboo. Hierarchical clustering analysis on the 34 MADS-box gene expression patterns revealed diverse expression levels in the development of floral organ (CK, F1, F2, F3, and F4). Bottom color bar stands for RPKM level, with green, black and red indicating low, moderate and high levels, separately (Cheng et al. 2017)

expressed in young tissues which had meristematic ability, which was possibly related to regulating meristem and flower development (Qi et al. 2013). Cheng et al. further investigated how *PheMADS15* regulated flowering time of *Arabidopsis* at transcriptional level. Those overexpressed plants showed an early flowering phenotype (Fig. 8.34) (Cheng et al. 2017).

8.2.3 CONSTANS-Like Gene Family

CONSTANS (*CO*) mediates floral integrators and circadian clock to exert an important part in the photoperiodresponse pathway (Suárez-López et al. 2001; Searle and Coupland 2004). The CONSTANS-like (COL) gene family, made up of zinc-finger TFs characterized by 2 conserved domains, is conserved in higher plants. 1–2 B-box domains that are localized around the amino-terminal (Torok and Elkin 2001) and a CCT domain comprising 43 aa adjacent to C-terminal regulate the interactions between proteins (Putterill et al. 1995; Borden 1998; Kurup et al. 2000; Strayer et al. 2000; Robson et al. 2001; Yamashino et al. 2013). It has been previously suggested that diverse *COL* genes possess various activities, particularly in terms of the light response-mediated regulation of flowering and growth (Robson et al. 2001; Griffiths et al.

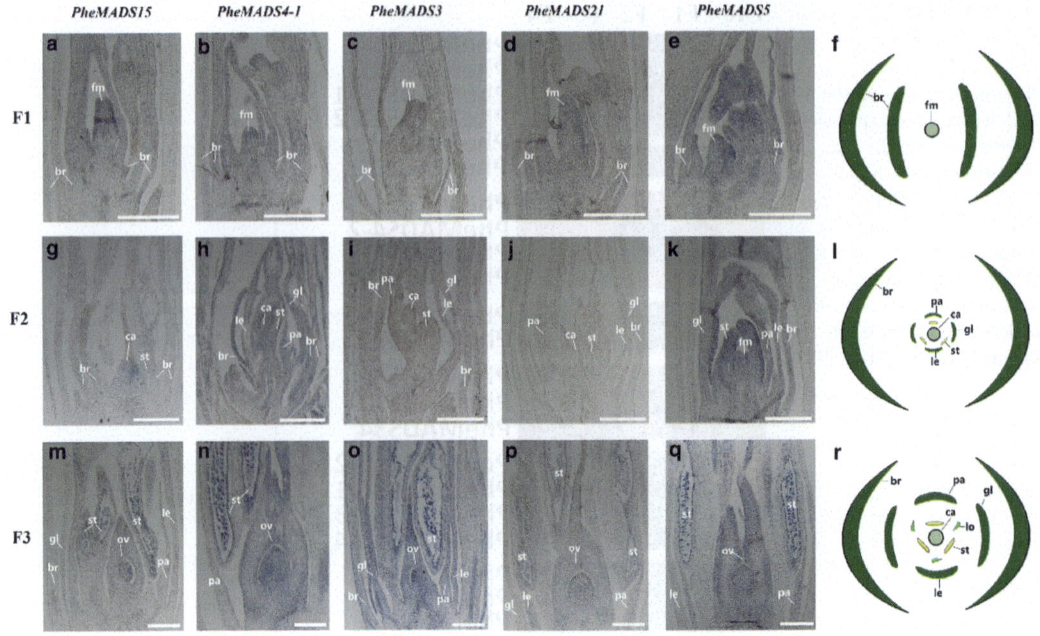

Fig. 8.32 *ISH* results for *PheMADS15*, *PheMADS4-1*, *PheMADS3*, *PheMADS21* and *PheMADS5*. F1: Floral bud formation period. F2: Inflorescence growing period. F3: Flowering period where stamens and pistils emerged from glumes in the bloom period. **a–s** In situ localization for the transcripts of *PheMADS15* (**a**, **g**, **m**), *PheMADS4-1* (**b**, **h**, **n**), *PheMADS3* (**c**, **i**, **o**), *PheMADS21* (**d**, **j**, **p**) and PheMADS5 (**e**, **k**, **q**) between F1 and F3; floral diagrams for F1 (**f**), F2 (**l**), and F3 (**r**). br, bract; fm, floral meristem; le, lemma; gl, glume; st, stamen; pa, palea; ov, ovule; ca, capel. Bars = 100 mm (Cheng et al. 2017)

2003; Bohlenius et al. 2016; Chia et al. 2008; Almada et al. 2009).

Altogether 14 *PheCOL* sequences, designated as *PheCOL1* to *PheCOL14*, were found from the genome of moso bamboo. There were less *PheCOLs* in the genome of moso bamboo than the numbers reported in Arabidopsis and rice (Table 8.1). Bidirectional best-hit and phylogeny-based approaches have been the frequently used methods in comparative genomics to identify the potential orthologous or paralogous genes. As reported by Peng and colleagues, *COL* genes experienced gene evolution following 3 progenitor divergence. Typically, the peak *Ka/Ks* values from moso bamboo genome were around 1.2, while those *Ka/Ks* values between rice and moso bamboo genomes were about 4.6. This suggested that the *COL* genes underwent purifying selection between Brachypodium and moso bamboo, and paralogs from the genome of moso bamboo.

For better detecting the *PheCOLs* expression and possible functions at diverse growth periods, we examined transcriptome data obtained from various growing bamboo shoots (Fig. 8.35). In the early moso bamboo shoot development, four *PheCOL* genes (*PheCOL2*, *PheCOL4*, *PheCOL8*, and *PheCOL14*) were highly expressed. *PheCOL5* accumulated highly in the late growth period. Previous studies had found that *PheCOL* genes were expressed in different flowering developmental stages (Peng et al. 2013; Gao et al. 2014). Most of the *PheCOLs* showed increased transcript abundance within leaves of nonflowering plants than other stages. As flowering develops, the level of expression gradually decreases. Of such genes, *PheCOL5*/*PheCOL6*/*PheCOL11* were down-regulated by over tenfold

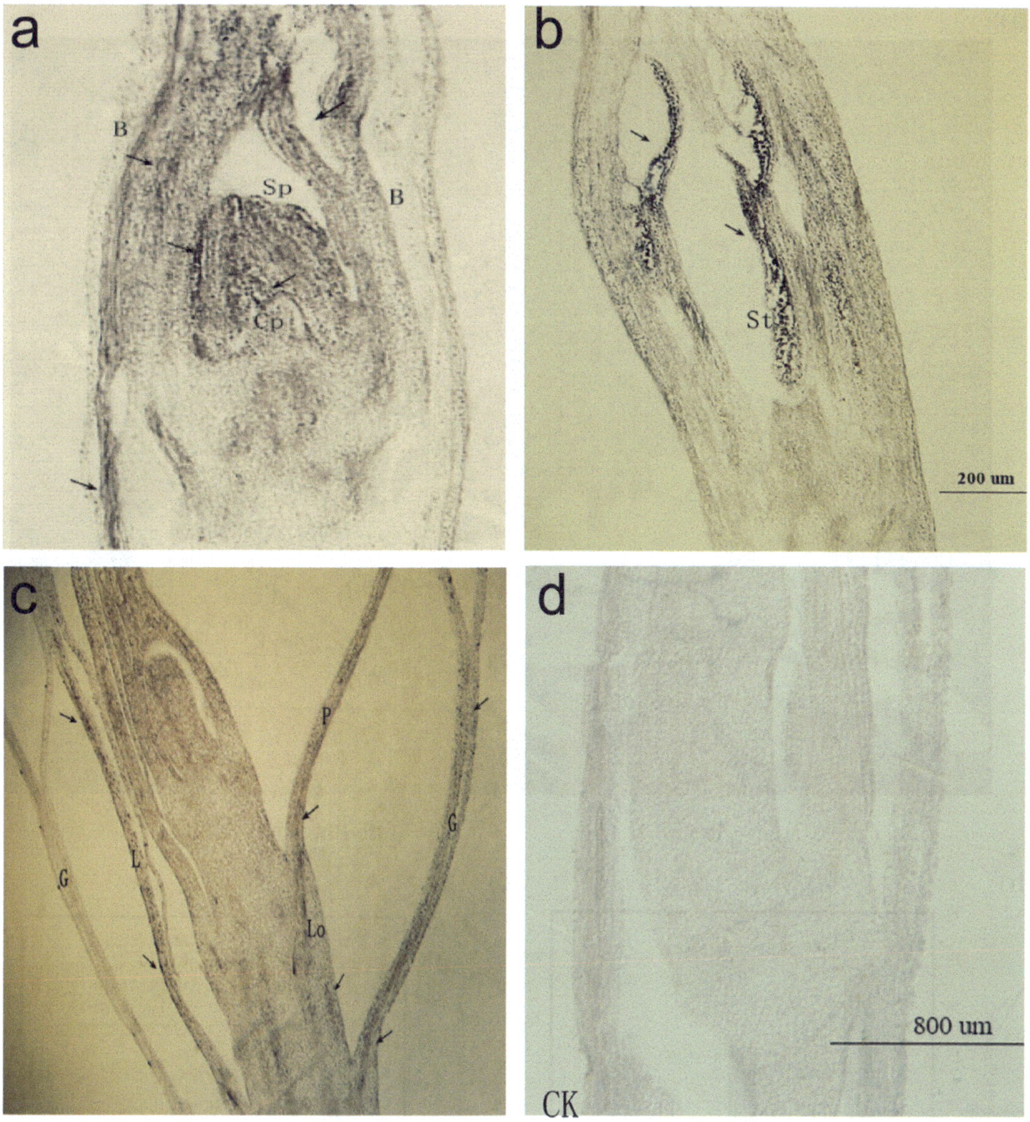

Fig. 8.33 *PheMADS14* expression during the flowering stage of moso bamboo. L: lemma, G: glume, Lo: lodicule, P: Palea, Cp: calyx primordia, B: bract, St: stamen and Sp: stamen primordia. **a** Inflorescence development period. **b** Bloom period. **c** Embryogenesis period. **d** Sense probe expression following ISH (Ge et al. 2017)

in F4 period, indicating the potentially diverse roles of *PheCOLs* in flowering development of moso bamboo.

The typical daily *CO* mRNA level in Arabidopsis was associated with its ability to measure photoperiod. In the SD environment, the highest *CO* level was detected at night, while the lowest level was measured in daytime. In the case of LD, the *CO* level elevated in the afternoon, while such elevation was further strengthened through mitigated suppression mediated by *FKF1*, the blue-light photoreceptor (Imaizumi et al. 2003). *PheCOLs* transcription in moso bamboo showed distinct daily changes under long- or short-day condition. The expression of many *PheCOL* genes was suppressed in light environment and promoted in dark environment.

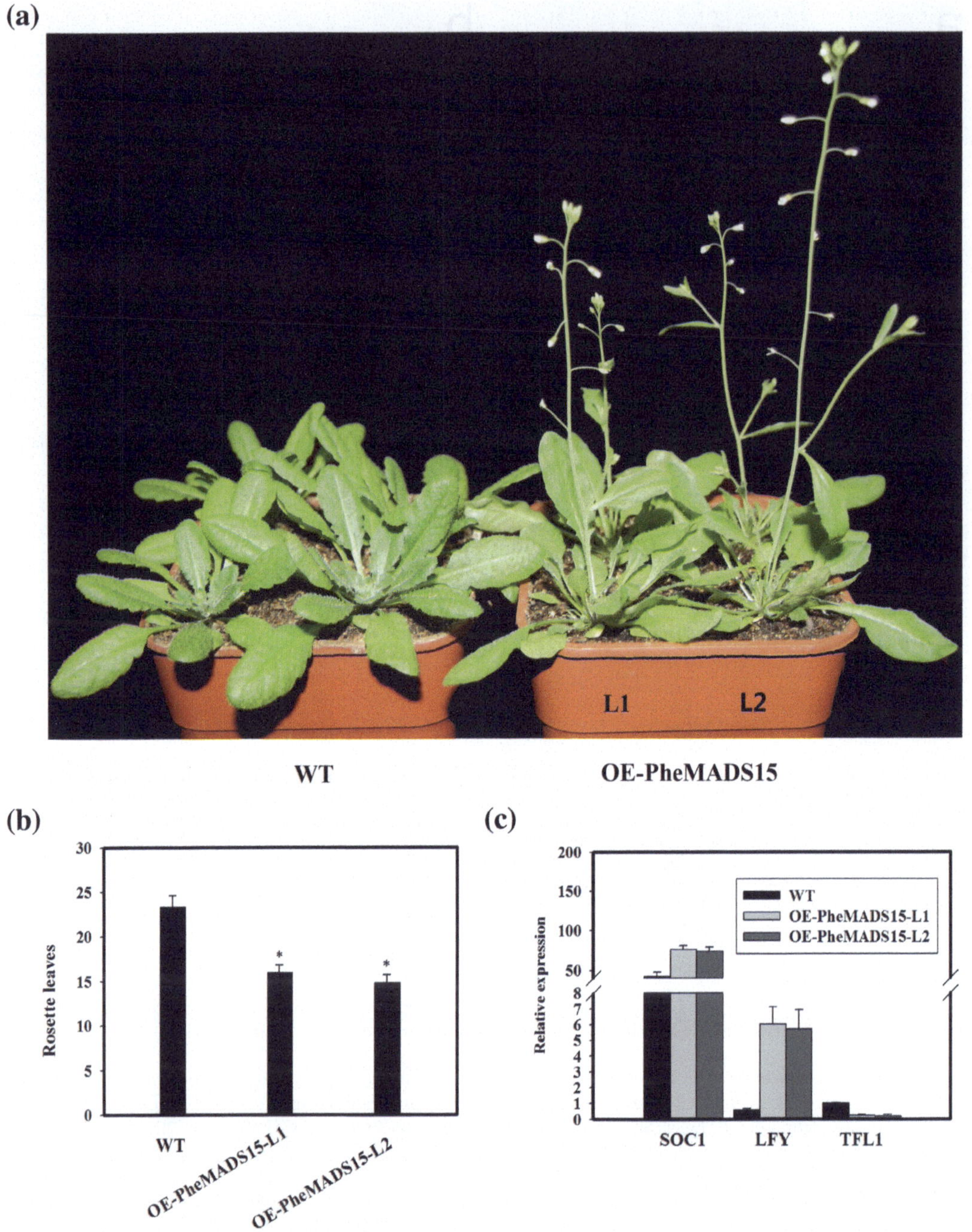

Fig. 8.34 Analysis of a delayed flowering phenotype through over-expressing *PheMADS15* within Arabidopsis. **a** Flowering phenotypes of WT, OEPheMADS15-L1 as well as OEPheMADS15-L2 plants under 4 weeks of growth under 23 °C and LD conditions. **b** Flowering time was rated as the rosette leaf quantity in the flowering period of transgenic and WT lines under 23 °C and LD conditions. **c** SOC1, TFL1 and LFY transcript abundances within transgenic and WT lines (L1 and L2) measured through qPCR analysis. Expression of β-tubulin in Arabidopsis served as the control. Error bars = SDs. Asterisks suggested that the differences were of statistical significance between transgenic and WT lines ($P < 0.05$ upon student's *t*-test) (Cheng et al. 2017)

Table 8.1 Detailed information about the members of the PheCOL gene family (Liu et al. 2019)

Name	Gene ID	Location	CDS length (bp)	Size (aa)	MW (kDa)	pI	Subcellular location
PheCOL1	PH01000142G0230	150,679–153,125 (−strand)	1182	393	43,019.72	8.49	Nucleus
PheCOL2	PH01000459G0310	248,973–250,383 (+strand)	978	326	34,492.53	5.25	Nucleus
PheCOL3	PH01000481G0620	404,909–406,425 (−strand)	1068	356	38,029.77	5.93	Nucleus
PheCOL4	PH01000780G0510	354,908–356,389 (−strand)	996	332	34,866.85	5.25	Nucleus
PheCOL5	PH01001011G0420	283,004–286,459 (+strand)	1152	383	42,280.72	8.96	Nucleus
PheCOL6	PH01001120G0010	1075–8854 (−strand)	1127	408	44,266.14	5.23	Nucleus
PheCOL7	PH01001246G0150	113,837–116,934 (+strand)	1350	449	48,851.82	5.37	Nucleus
PheCOL8	PH01001462G0020	6139–7875 (−strand)	906	301	31,472.47	6.25	Nucleus
PheCOL9	PH01001639G0300	180,211–184,620 (−strand)	1077	358	38,363.60	5.15	Nucleus
PheCOL10	PH01001710G0270	166,735–169,785 (−strand)	1185	394	42,762.45	5.38	Nucleus
PheCOL11	PH01002146G0170	136,505–138,723 (+strand)	1194	397	43,481.03	5.94	Nucleus
PheCOL12	PH01002727G0080	86,255–87,854 (+strand)	888	296	31,670.66	5.46	Nucleus
PheCOL13	PH01003421G0140	73,839–76,559 (−strand)	1194	398	43,198.11	5.64	Nucleus
PheCOL14	PH01005551G0030	11,275–17,886 (−strand)	2094	698	76,847.81	8.24	Nucleus

Network analyses revealed that 11 out of 14 (78.6%) of the moso bamboo *COL* genes analyzed in this study were coexpressed with diverse degrees of connectivity with other *PheCOLs*, and *PheCOL1* had the most interactions with other *PheCOLs* (Fig. 8.36). To investigate possible interactions, *PheCOL14* was identified from moso bamboo for the first time and was localized in the cell membrane (Fig. 8.37b) and PheCOL14 could activate the expression of *PheCOL3* in Y1H system, which was consistent with the coexpression network prediction (Fig. 8.37a). According to the rice orthologous gene interaction network constructed by STRING (Fig. 8.37c), PheCOL14 interacted with those 10 screened proteins. Of these, three MADS proteins (PheMADS2, PheMADS56-1, and PheMADS15), two PEBP proteins (PheFT and PheFTL1), MYB transcription factor (PH01000061G1490), and circadian rhythm and flowering time regulation gene (PH01002142G0290) were found, indicating the possible role of *PheCOL14* in photoperiodic response and flowering time.

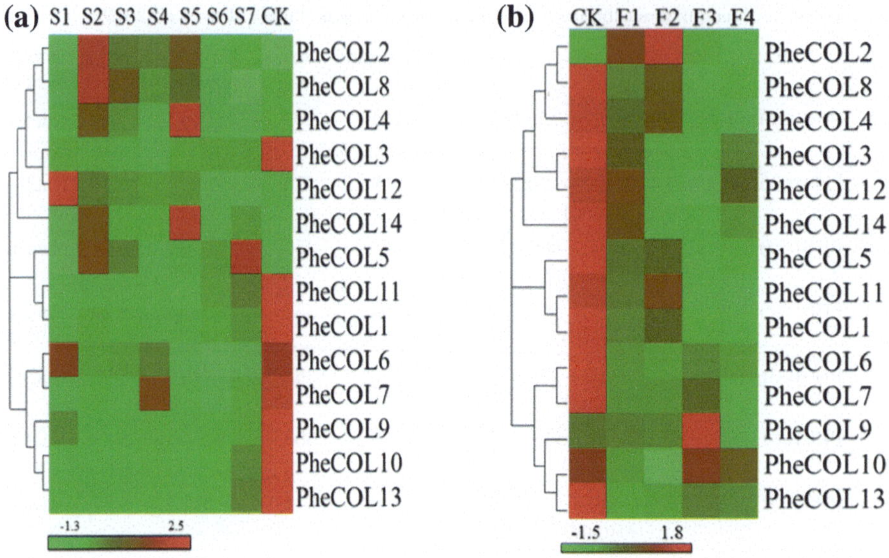

Fig. 8.35 Heat map of the RPKM values of *PheCOL* genes. **a** *PheCOLs* expression profiles within diverse high shoots (0.1, 0.5, 1, 3, 6, 9, 12, and 15 m) of moso bamboo. **b** Heatmap showing the *PheCOLs* expression within moso bamboo in diverse flowering periods (CK, F1, F2, F3, and F4). Heatmap was built based on RPKM. The color scale was shown under the heat map, red and green colors suggest increased and decreased transcript abundances in comparison with controls. Color images are available online. (From Liu et al. 2019)

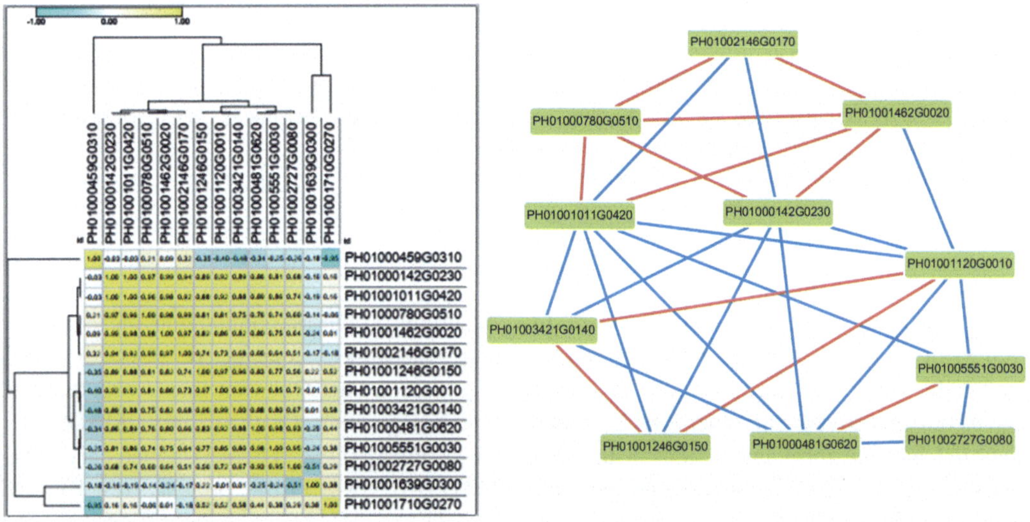

Fig. 8.36 Coexpression network of *PheCOL* genes in moso bamboo. **a** The PCC value of the coexpression network among *PheCOLs*. **b** Coexpression network of *PheCOL* genes. The positive correlation value was shown by blue lines, which represented the PCC value from 0.85 to 0.95, and red lines represented the PCC value > 0.95. PCC, Pearson correlation coefficient. Color images are available online. From Liu et al. (2019)

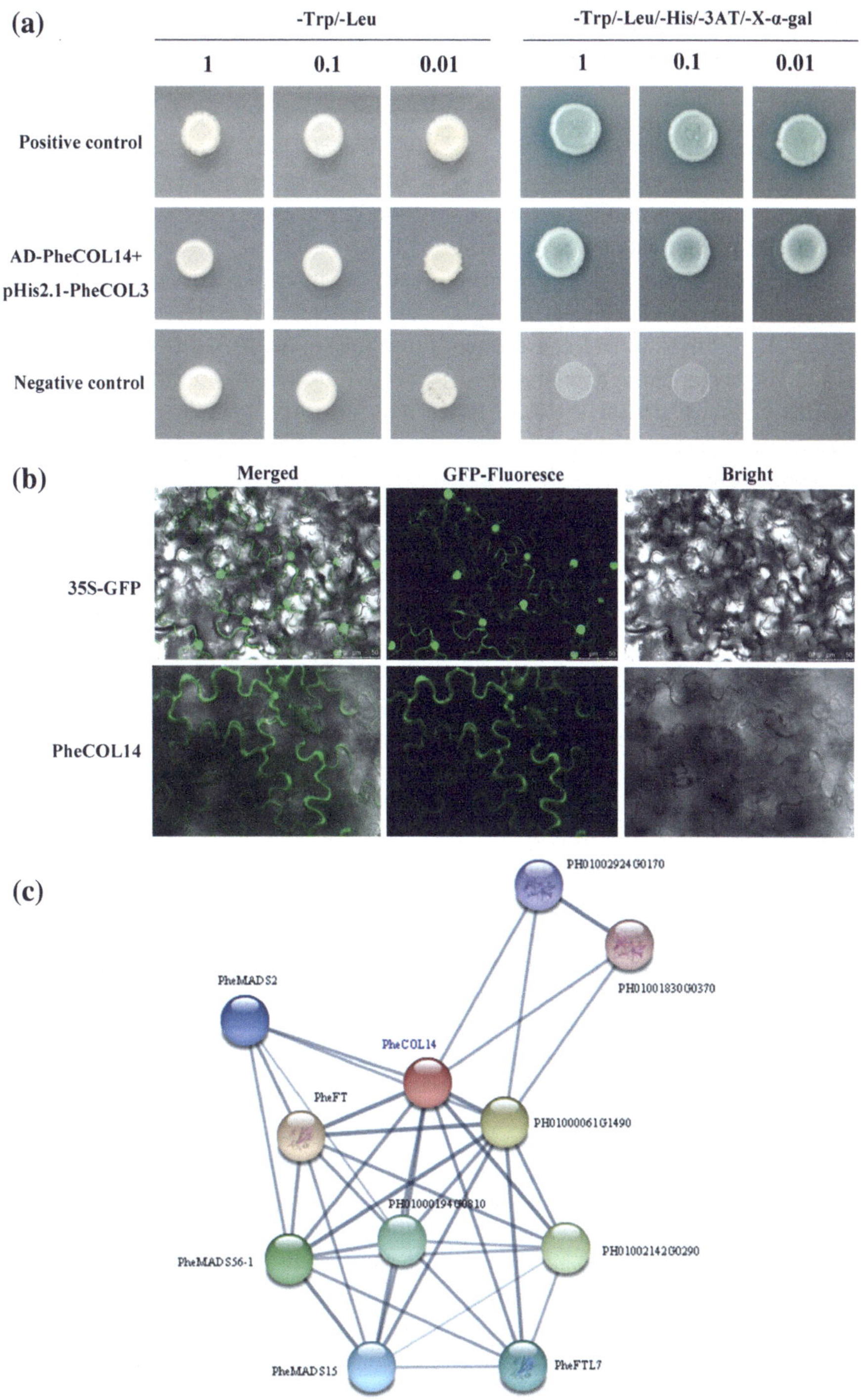

Fig. 8.37 Subcellular localization and Y1H assay of PheCOL14 protein. **a** Y1H assay for the interaction between PheCOL14 and promoter of *PheCOL3*. **b** 35S:: *PheCOL14-GFP* transient expression within the tobacco epidermal cells. **c** PheCOL14 interaction network within moso bamboo. In the constructed network, *colored balls* (protein nodes) served as the visual aid for indicating diverse input proteins and predicting interactors. Protein nodes which are enlarged suggest the available information of 3D protein structure. Gray lines connect proteins, which are associated by recurring text-mining evidence. Color images are available online. From Liu et al. (2019)

8.3 Endogenous Hormones Regulation in Shoot Fast Growth

8.3.1 SAUR Family Members

In appropriate condition, bamboo shoots can grow by 0–20 m in < three months (Peng et al. 2010, 2013). Anatomical structures of bamboo shoot from different developmental stages demonstrated that cell division and cell elongation were both involved throughout the shoot rapid elongation period (Cui et al. 2012). It has been well-known that plant hormone has pleiotropic effects on plant growth, such as cell division/differentiation/elongation, tropic responses, root initiation and apical dominance. The biology function of plant auxin was mainly promoting cell elongation. Several categories of genes were involved in auxin signal pathway. Auxin can mediate such molecular impacts through changing gene expression.

Genome data was usually used to conduct the identification of gene family members. By searching on the bamboo genome database v1.0, a total of 38 *small auxin-up RNA* (*SAUR*) genes (*PheSAUR*) were discovered from the Pfam search for an conserved auxin-inducible signature within the coding or aa sequences (Bai et al. 2017). The aa sequence lengths were 83–1148 (Table 8.2). The molecular weights of PheSAUR proteins were predicted to be 8.88–128.04 kDa, while the theoretical isoelectric point (pl) was 5.47–11.91. The chromosomes in rice is twice than that in moso bamboo while its SAUR number is less than rice (Jain et al. 2006).

Moso bamboo seedlings were used to conduct the tissue specific expression analysis. The results showed that some *PheSAUR* genes have specific expression in root, stem and leaf of moso bamboo. Expression of *PheSAUR2*, *PheSAUR15*, *PheSAUR20*, *PheSAUR28* and *PheSAUR35* was detected in higher levels in stems. In addition, the expression of *PheSAUR4*, *PheSAUR6*, *PheSAUR9*, *PheSAUR17*, *PheSAUR26*, *PheSAUR37* and *PheSAUR44* was specifically detected in leaves, while *PheSAUR18* and *PheSAUR36* were specifically expressed in roots. The tissue specific expression suggests the diverse functions of such genes within diverse organs. After IAA treatment, except *PheSAUR9* and *PheSAUR41*, the remaining 36 *PheSAUR* genes showed increased expression at 1 h post-IAA treatment; typically, *PheSAUR5*, *PheSAUR15*, *PheSAUR31* and *PheSAUR36* showed increased expression within leaves (Fig. 8.38). These genes in response to IAA up-regulated expression may have important functions in the IAA signaling pathway.

GA_3 treated moso bamboo seedlings were used to carry out the identification of GA responsive *PheSAUR*. After GA_3 treatment, the leaves were collected and used for the expression analysis of *PheSAURs* using qRT-PCR (Fig. 8.39). As a result, the expression levels of *PheSAUR4*, *PheSAUR7*, *PheSAUR13*, *PheSAUR16*, *PheSAUR17*, *PheSAUR18*, *PheSAUR21*, *PheSAUR23*, *PheSAUR26*, *PheSAUR28*, *PheSAUR29*, *PheSAUR31*, *PheSAUR32*, *PheSAUR35*, *PheSAUR41* and *PheSAUR43* increased significantly at 3, 6, or 10 h after the application of GA_3. These results indicated that *PheSAUR* gene may be involved in GA pathway as well as IAA response pathway.

By constructing 35S:: PheSAUR and 35S:: PheSAUR-GFP overexpression vectors and transgenic to arabidopsis seedlings, then we can obtain the phenotypic changes comparative to Col. The results showed that the roots of *35S:: PheSAUR29*-GFP transgenic seedlings were longer that Col and *35S:: PheSAUR29* transgenic seedlings under GA_3 treatment. Plant SAUR proteins were mostly located in cell membrane and nucleus (Chae et al. 2012; Spartz et al. 2012). Besides, the nucleus located SAUR proteins are related to the transcriptional regulation (Knauss et al. 2003). By PEG-induced transformation of arabidopsis protoplast analysis and detection of GFP fluorescence, the subcellular localization revealed the membrane localization of PheSAUR4 green fluorescence, PheSAUR28 was localized on cytoplasm, PheSAUR29 and PheSAUR34 were localized on cell membrane and nucleus simultaneously (Fig. 8.40). Locating the PheSAUR protein on the cell membrane

Table 8.2 Overview of *PheSAUR* genes in moso bamboo

Name	Length	MW/kDa	PI	Intro number/IMEter v2.0 scores	CELLO localization
PheSAUR1	257	27.08	9.33	1/18.9	Plasma Membrane(1.253)/Chloroplast (1.219)/Mitochondrial(1.026)
PheSAUR2	110	12.55	11.86	0	Nuclear(2.479)
PheSAUR4	178	19.68	5.47	1/9.17	Cytoplasmic(2.045)
PheSAUR5	144	15.66	9.20	0	Mitochondrial(1.799)
PheSAUR6	163	17.86	9.05	0	Nuclear(2.902)
PheSAUR7	141	14.88	8.55	0	Nuclear(1.735)/Chloroplast(1.416)
PheSAUR8	133	14.35	8.54	0	Nuclear(1.315)/Mitochondrial(1.250)/ Chloroplast(1.179)
PheSAUR9	126	13.85	7.89	0	Nuclear(2.049)
PheSAUR11	116	12.01	9.02	0	Mitochondrial(1.356)/Chloroplast (1.075)
PheSAUR12	94	10.02	6.28	0	Cytoplasmic(1.751)
PheSAUR13	116	11.96	9.16	0	Mitochondrial(1.263)
PheSAUR14	83	8.88	8.55	0	Chloroplast(1.875)
PheSAUR15	120	12.98	8.59	0	Nuclear(1.812)
PheSAUR16	121	13.91	8.80	0	Mitochondrial(1.481)/Cytoplasmic (1.388)/Nuclear(1.033)
PheSAUR17	116	13.03	9.26	0	Nuclear(1.382)/Mitochondrial(1.205)/ Cytoplasmic(1.062)
PheSAUR18	160	17.49	8.75	0	Nuclear(1.691)/Cytoplasmic(1.220)
PheSAUR19	109	11.79	9.22	0	Mitochondrial(1.498)/Nuclear(1.036)
PheSAUR20	154	17.28	8.81	0	Mitochondrial(1.312)
PheSAUR21	144	16.04	8.83	0	Nuclear(1.641)
PheSAUR22	154	17.23	11.91	0	Nuclear(3.237)
PheSAUR23	147	16.54	9.96	0	Mitochondrial(1.489)/Nuclear(1.474)
PheSAUR24	179	18.70	9.33	0	Nuclear(1.534)
PheSAUR25	125	13.75	6.10	0	
PheSAUR26	140	15.50	8.92	0	Mitochondrial(1.700)
PheSAUR27	145	16.28	7.64	0	Mitochondrial(1.237)
PheSAUR28	102	11.02	7.77	0	Mitochondrial(2.099)
PheSAUR29	113	12.15	9.62	0	Cytoplasmic(1.762)/Mitochondrial (1.269)
PheSAUR31	155	16.01	8.52	0	Nuclear(1.293)/Mitochondrial(1.280)/ Chloroplast(1.088)
PheSAUR32	129	14.51	5.91	1/21.5	Cytoplasmic(1.772)
PheSAUR34	147	15.67	8.24	0	Chloroplast(2.619)
PheSAUR35	134	14.95	6.81	0	Nuclear(1.331)
PheSAUR36	138	15.10	6.37	0	Mitochondrial(1.560)/Nuclear(1.267)/ Cytoplasmic(1.110)

(continued)

Table 8.2 (continued)

Name	Length	MW/kDa	PI	Intro number/IMEter v2.0 scores	CELLO localization
PheSAUR37	246	27.20	11.05	2/17.12	Mitochondrial(2.541)
PheSAUR38	95	10.68	7.62	0	Nuclear(1.494)/Chloroplast(1.042)/ Cytoplasmic(1.039)
PheSAUR41	109	12.22	7.23	0	Mitochondrial(1.365)/Nuclear(1.195)
PheSAUR42	1148	128.04	6.59	18/13.39	Nuclear(2.750)
PheSAUR43	171	19.44	7.17	1/23.69	Plasma Membrane(1.236)/ Mitochondrial(1.108)
PheSAUR44	112	12.54	6.37	0	Nuclear(1.350)/Cytoplasmic(1.190)/ Mitochondrial(1.114)

From Bai et al. (2017)
Note Length represent the amino acid sequences of PheSAUR proteins
MW (molecular weight) represents the predicted weights of PheSAUR proteins
PI represents the predicted isoelectric point of PheSAUR proteins

indicated that the IAA signal exerted a vital part in cell membrane, while the location on the cell membrane and nucleus of PheSAUR29 and PheSAUR34 are possibly related to the calmodulin related IAA signaling pathway (Park et al. 2007), locating in cytoplasm of PheSAUR28 protein are possibly associated with IAA signaling to the nucleus transfer function (Li et al. 2015). The precise function of these PheSAUR proteins require more further research study.

8.3.2 WUSCHEL-Related Homeobox (WOX) Protein Family

WOX genes are the homeobox (HB) TF family members, and they are responsible for encoding the characteristic plant TFs that share one DNA-binding domain comprising 60–66 residues, namely, the Helix–loop–helix–turn–helix structure (Gehring et al. 1994; Ariel et al. 2007; Van der Graaff et al. 2009). *WOXs* exert vital parts in a variety of events in vascular plants, such as root and shoot apical meristem (SAM) maintenance, adventive organ growth, vascular development, and embryonic genesis (Haecker et al. 2004; Hirakawa et al. 2010; Zhang et al. 2016; Tanaka et al. 2019). *WUS*, *WOX4*, and *WOX5* function in regulating the maintenance of stem cells within cambium, and root/floral/shoot meristems (Brand et al. 2000; Pi et al. 2015; Lu et al. 2015; Tanaka et al. 2015; Ohmori et al. 2013), whereas *WOX3* is involved in the activities of plate and marginal meristems in leaves (Shimizu et al. 2009; Yoshikawa et al. 2016). In *Arabidopsis thaliana*, *WOX8* (*STIMPY-LIKE*, *STPL*) and *WOX9* (*STIMPY*, *STIP*) play important roles in maintaining vegetative SAM and embryo patterning (Haecker et al. 2004; Ueda et al. 2011; Lie et al. 2012). Moreover, *WUSCHEL* (*WUS*) regulates the WUS–CLAVATA3 (WUS–CLV3) signal transduction pathway to keep stem cell niches within SAM (Zhang et al. 2016; Brand et al. 2000; Laux et al. 1996; Somssich et al. 2016). But, the *WUS*, *TILLERS ABSENT1* (*TAB1*) orthologous gene in rice is not detected within SAM, which is not related to the differentiation of stem cells as well; instead, it shows transient expression during early development of axillary meristem and interacts with *FLORAL ORGAN NUMBER2* (*FON2*, the CLAVATA3 orthologous gene), to exert an important part in the development of axillary meristems in rice (Tanaka and Hirano 2020; Tanaka et al. 2015). Rice *WOX4* positively regulates the maintenance of stems cells in axillary meristem and SAM (Ohmori et al. 2013).

Altogether 27 candidates *WOX* genes were identified from the genome of *Phyllostachys edulis* (*PheWOXs*) (v2.0) (Zhao et al. 2018). In

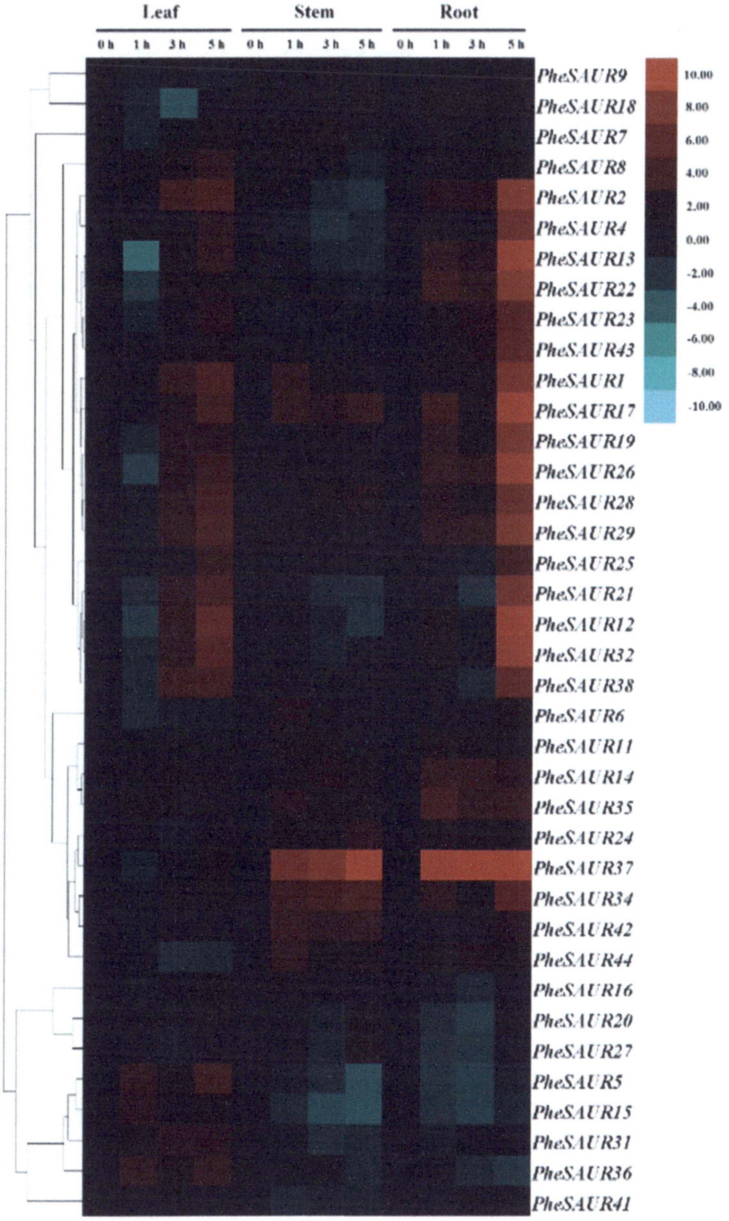

Fig. 8.38 IAA responsive analyses of *PheSAUR* genes using qRT-PCR in moso bamboo seedlings (Bai et al. 2017)

addition, *WOX* genes were also discovered from draft genomes of additional bamboos (*Raddia guianensis*, *Olyra latifolia*, *Bonia amplexicaulis* and *Guadua angustifolia*) (Guo et al. 2019), including 9 from *R. guianensis* (*RguWOXs*), 12 *WOXs* from *O. latifolia* (*OlaWOXs*), 26 from *B. amplexicaulis* (*BamWOXs*) and 18 from *G. angustifolia* (*GanWOXs*). For *Brachypodium distachyon*, *Populus trichocarpa*, *Arabidopsis thaliana*, *Zea mays*, and *Oryza sativa*, WOX proteins were obtained through PlantTFDB v4.0, followed by manual examination. Diverse species had diverse *WOXs* numbers (Table 8.3) (Li et al. 2020). There are numerous polyploidies in bamboo. Typically, polyploidization drives Bambusoideae evolution (Guo et al. 2019). As for the polyploid bamboo varieties, their WOXs absolute copy numbers were higher than those in

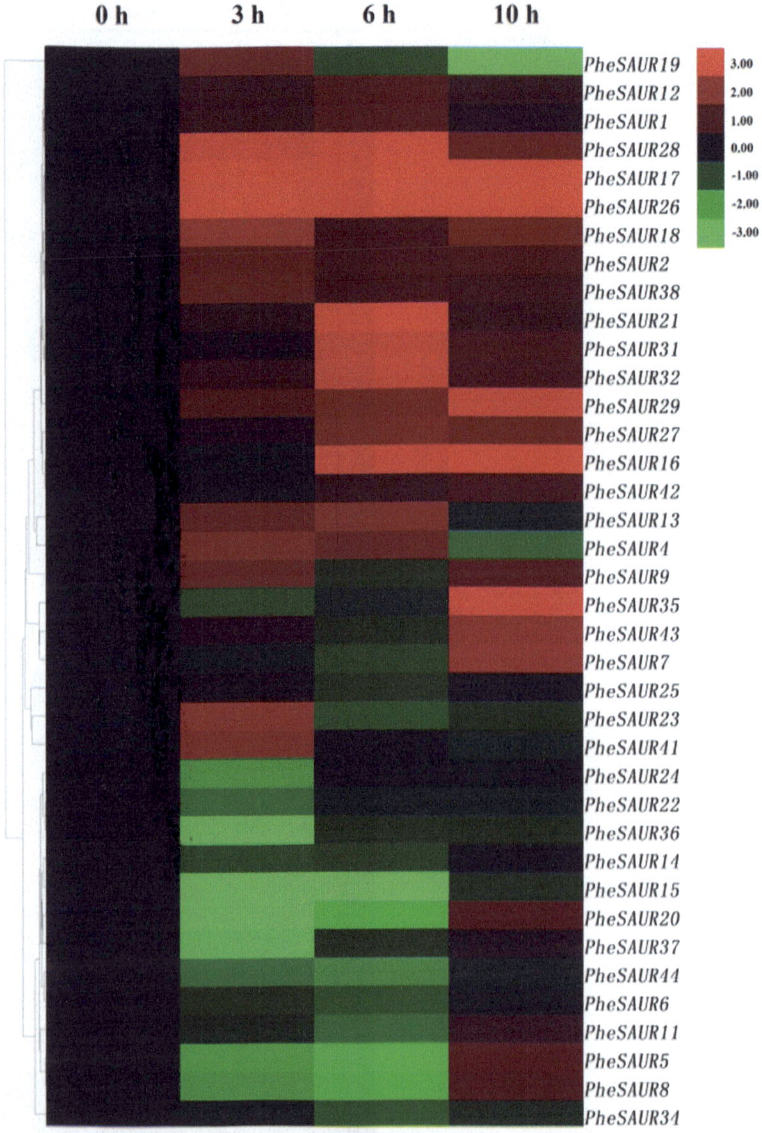

Fig. 8.39 Relative mRNA levels of *PheSAUR* genes after GA_3 treatment (Bai et al. 2017)

the diploid species. We then applied the coding sequences (CDSs) of 91 Bambusoideae *WOXs* in constructing an unrooted phylogenetic tree, and assessed it by the bootstrap value. As revealed by the built phylogenetic tree, Bambusoideae *WOXs* were divided into ancient, intermediate, and WUS clades, and the support value was as high as 80% by Li et al. (2020). We also discovered certain gene structures and motif domains (Fig. 8.41). Among those 91 Bambusoideae *WOXs*, 3 *WOX*-like genes (*PheWUSlike*, *BamWOX8like*, and *BamWOX9like*) encoded the non-HD proteins (Fig. 8.41), while 12 contained part of HDs in motifs 2/3 at the respective N-terminal (Fig. 8.41). Most Bambusoideae WOXs received potent selection pressure for purification during polyploidization. Ks value has been applied in indicating the divergence time for a duplication event (Peng et al. 2013; Guo et al. 2019). As far as the diploid herbaceous species (*R. guianensis* and *O. latifolia*) are concerned, *RguWOX11a-RguWOX11b* and *OlaWOX3a-*

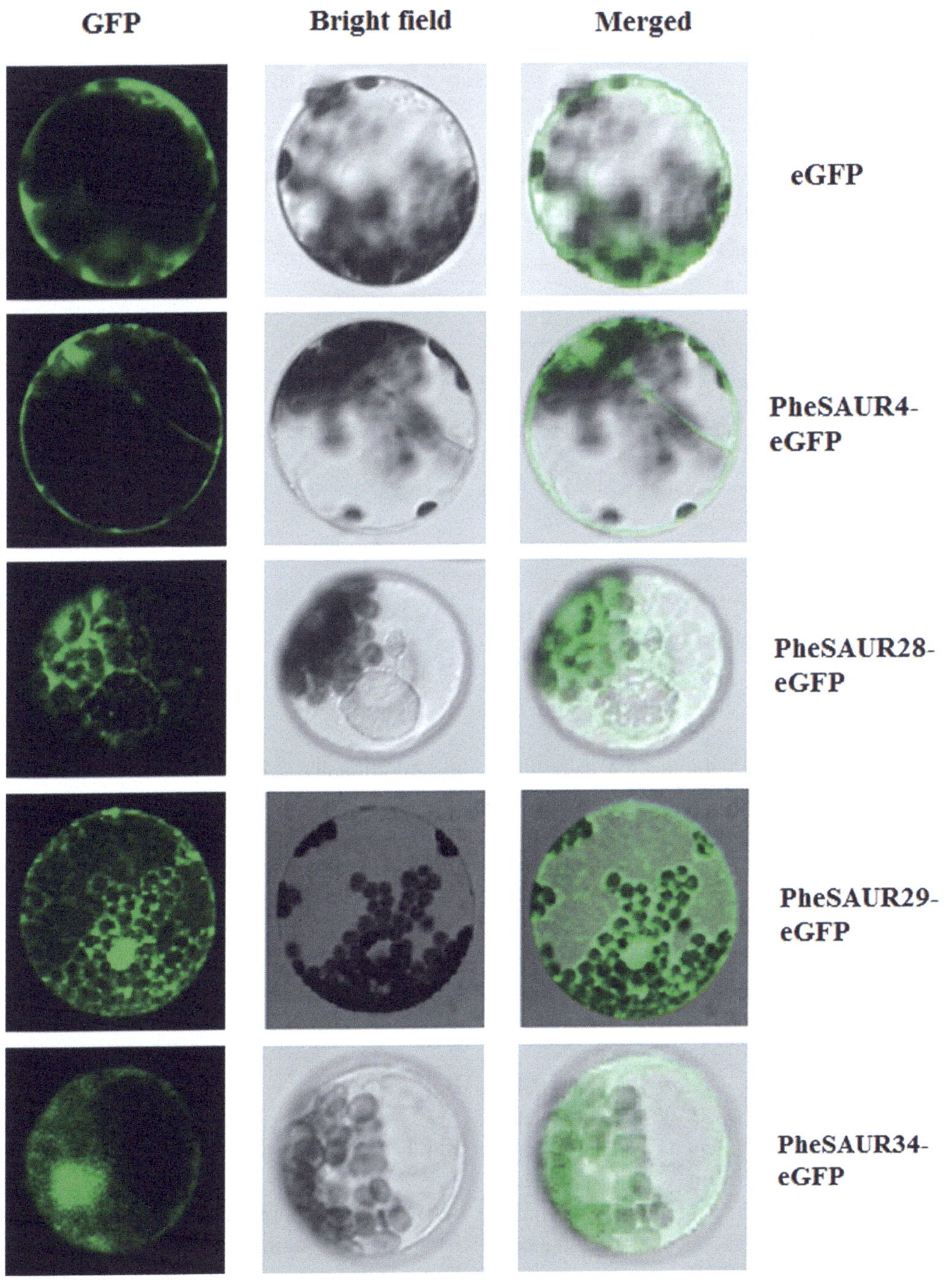

Fig. 8.40 Subcelluar localizations of PheSAUR proteins (Bai2017)

Table 8.3 Comparison of the *WOX*s of Bambusoideae with other species (Li et al. 2020)

	Species	Ancient clade	Intermediate clade	WUSCHEL (WUS) clade	Other	Total
Dicots	*Arabidopsis thaliana*	3	4	8	–	15
	Populus trichocarpa	3	4	11	–	18
Monocots	*Oryza sativa*	1	6	6	–	13
	Brachypodium distachyon	2	5	6	–	13
	Zea mays	3	7	10	–	20
	Olyra latifolia	1	5	6	–	12
	Raddia guianensis	1	3	5	–	9
	Guadua angustifolia	2	9	7	–	18
	Phyllostachys edulis	2	11	14	–	27
	Bonia amplexicaulis	2	6	18	–	26
Chlorophyta	*Micromonas pusilla*	–	–	–	4	4
	Micromonas sp.	–	–	–	1	1
	Ostreococcus tauri,	–	–	–	1	1
	Bathycoccus prasinos	–	–	–	1	1

OlaWOX3b, the 2 WOX duplication pairs, have great Ks value compared with most WOX duplication events detected in the polyploid woody bamboo varieties (*B. amplexicaulis*, *P. edulis* or *G. angustifolia*) (Fig. 8.42). The above-mentioned 2 duplication pairs possibly took place through the segmental duplications prior to the speciation of bamboos. Among the above 3 polyploid woody bamboo varieties, the *WOX3a-WOX3b* pair displayed great Ks value comparable to that of *O. latifolia* (Fig. 8.42), which indicated that *WOX3a-WOX3b* divergence was possibly due to segmental duplications prior to the divergence of woody and herbaceous bamboos. To evaluate the region of the promoters regulating the expression of *WOXs*, the conserved cis-elements of WOXs were processed. In general, MP00302/MP00227 was detected at about –100 bp in these promoters (Fig. 8.43a), rather than in those of AtWOX4 or OsWOX4. MP00083, a B3 protein binding element, was detected in the similar region to PheWOX4b/c and BamWOX4a/c promoter regions (Fig. 8.43 a). Additionally, MP00253/MP00540, the GA-rich element, was conserved within the rice and bamboo WOX3/WUS promoters (Fig. 8.43c). MP00076, the MADS-binding element, was detected within the promoters of many WUS and WOX3, which was predicted as the PH02Gene23951.t1 target, homologous to the rice OsMADS56 (Fig. 8.43c).

8.4 Ageing

8.4.1 NAC Transcription Factor

As a perennial monocotyledonous plant, moso bamboo start aging process after once-in-a-lifetime flowering process. The flowering may result in the decreasing of regeneration ability, the accumulational senescence leaves, and even, the death of stock bamboo and forest aging. Degraded senescence also cause a huge economic and ecological loss. NAC protein stands for a plant-specific TF, which is involved in plant

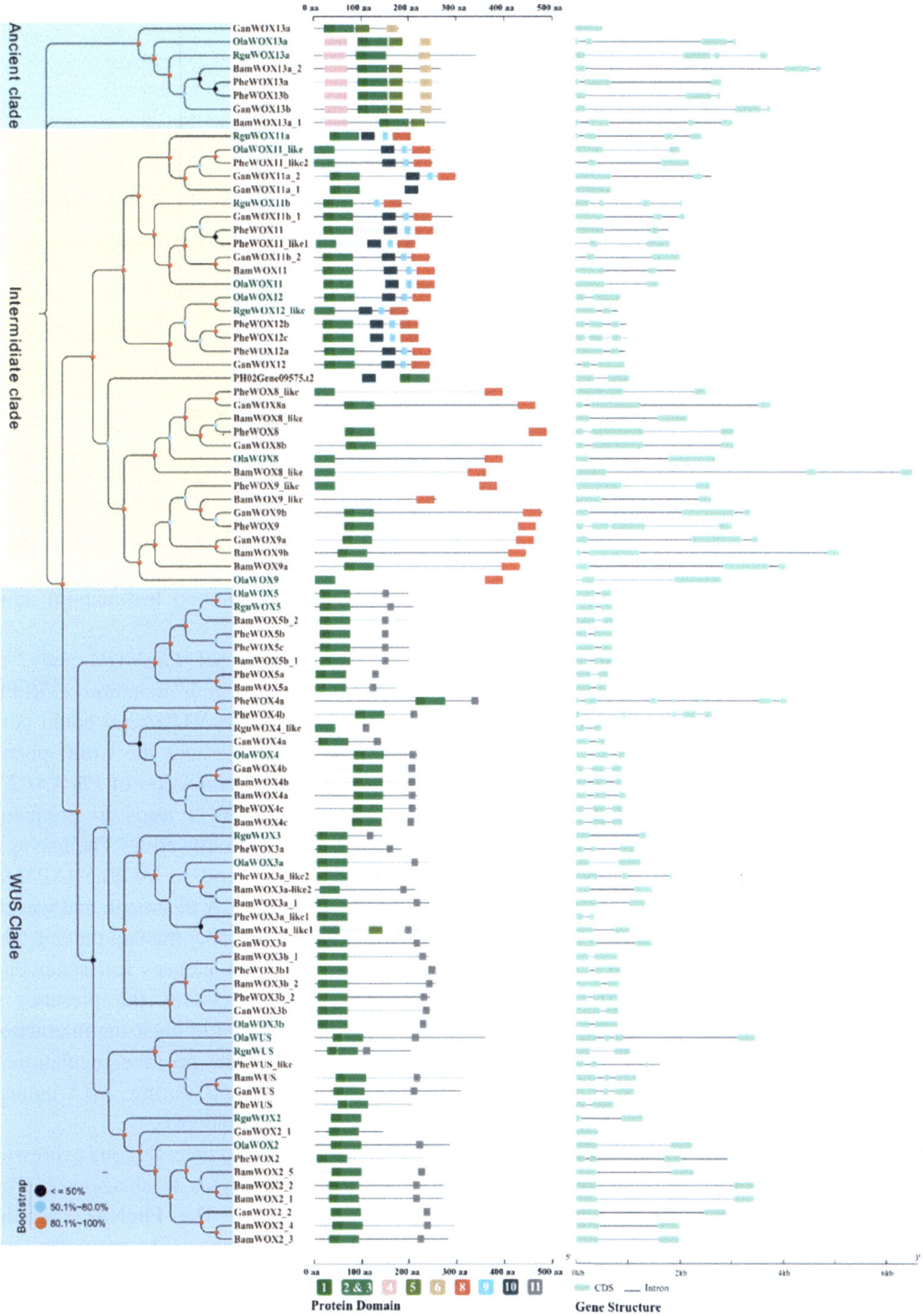

Fig. 8.41 Phylogenetic tree analysis for *WOXs* in Bambusoideae. The full-length CDSs of WOXs in Bambusoideae were used to built a neighbor-joining tree and the bootstrap support value was 1000. One color background indicated one clade. Brown and green leaves in our as-constructed phylogenetic tree indicated the *WOXs* of woody and herbaceous bamboos, respectively. Gene structures and protein domains were plotted after every *WOX* (Li et al. 2020)

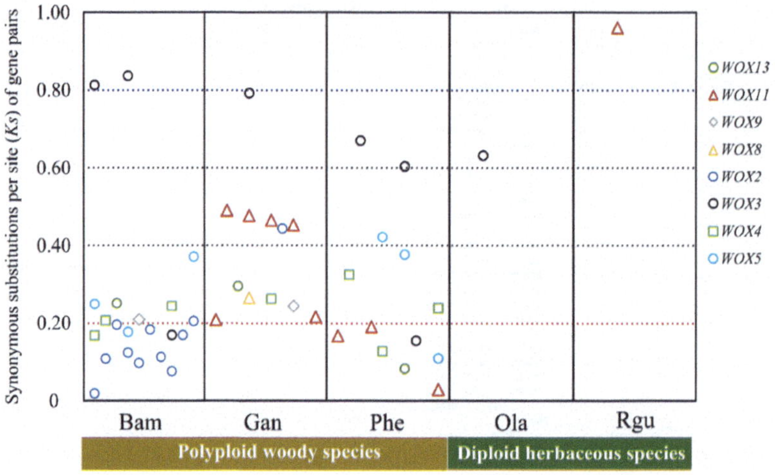

Fig. 8.42 Duplication events of Bambusoideae WOXs in woody and herbaceous bamboos. One color symbol indicated one WOX family member. Every symbol stood for the homologous gene pair displayed in the Ks distribution plot. Y-axis was indicative of Ks value determined for every gene pair (Li et al. 2020)

growth, stress responses, and the regulation of plant aging. Peng and coworkers (2013) predicted that the moso bamboo flowering and drought-responsive genes were related to NAC genes, which was associated with flower senescence.

Xie (2019) identified and analyzed several putative senescence-associated NACs through expression pattern during leaf senescence process and post-harvest shoot senescence process. A total of 7 *PheNACs* genes and one alternative splicing transcript were obtained by cloning, named *PheNAC5-11* and *PheNAC6S*, respectively. Three *PheNACs* (*PheNAC5-7*) were homologous to *OsNAC2* of rice (belongs to ORE1 subfamily), and four *PheNACs* (*PheNAC8-11*) were homologous to *AtNAP* of *Arabidopsis* and *OsNAP* of rice. *PheNAC5-10* were induced significantly in leaves of the flowering moso bamboo and there years old seedling during leaf natural aging (Fig. 8.44and 8.46), phytohormone-induced leaf senescence (incudling ABA, ACC、MeJA and SA) (Fig. 8.45), dark-induced leaf senescence (Fig. 8.45), and post-harvest senescence of shoots (Fig. 8.47). *PheNAC11* was also induced significantly in these different senescence process, except for bamboo leaf natural aging process.

The overexpression of *PheNAP2* and *PheNAP3* in Arabidopsis *nap* mutation (35S:Phe *NAP2/nap* and 35S:Phe *NAP3/nap*) could complement function and restore the Col-0 phenotype (Fig. 8.48). Overexpression of *PheNAP2* in Col-0 (35S:PheNAP2/WT) leads to premature leaf senescence prior to flowering. The leaves of 35S:PheNAP2/*nap* and 35S:PheNAP3/*nap* turned yellow under ABA treatment, and was the same with Col-0, excluding the nap mutant. This demonstrates that ABA induces leaf senescence of Arabidopsis, but requires the presence of *PheNAP2* and *PheNAP3* to elicit the interaction. All of these verified the positive regulation of *PheNAP2* and *PheNAP3* during ABA-induced leaf senescence in Arabidopsis.

Xie et al. (2020) used heterologous expression of in *Arabidopsis* analyses to characterize *PheNAC3* gene function. The PheNAC3 coding sequence, driven by the 35S promoter, was introduced into the wild-type (Col-0) plants to overexpress PheNAC3. After three weeks of growth, the PheNAC3 overexpressors (OE-PheNAC3) displayed the senescence phenotype (the tip of 7th leaf begins to turn yellow), while

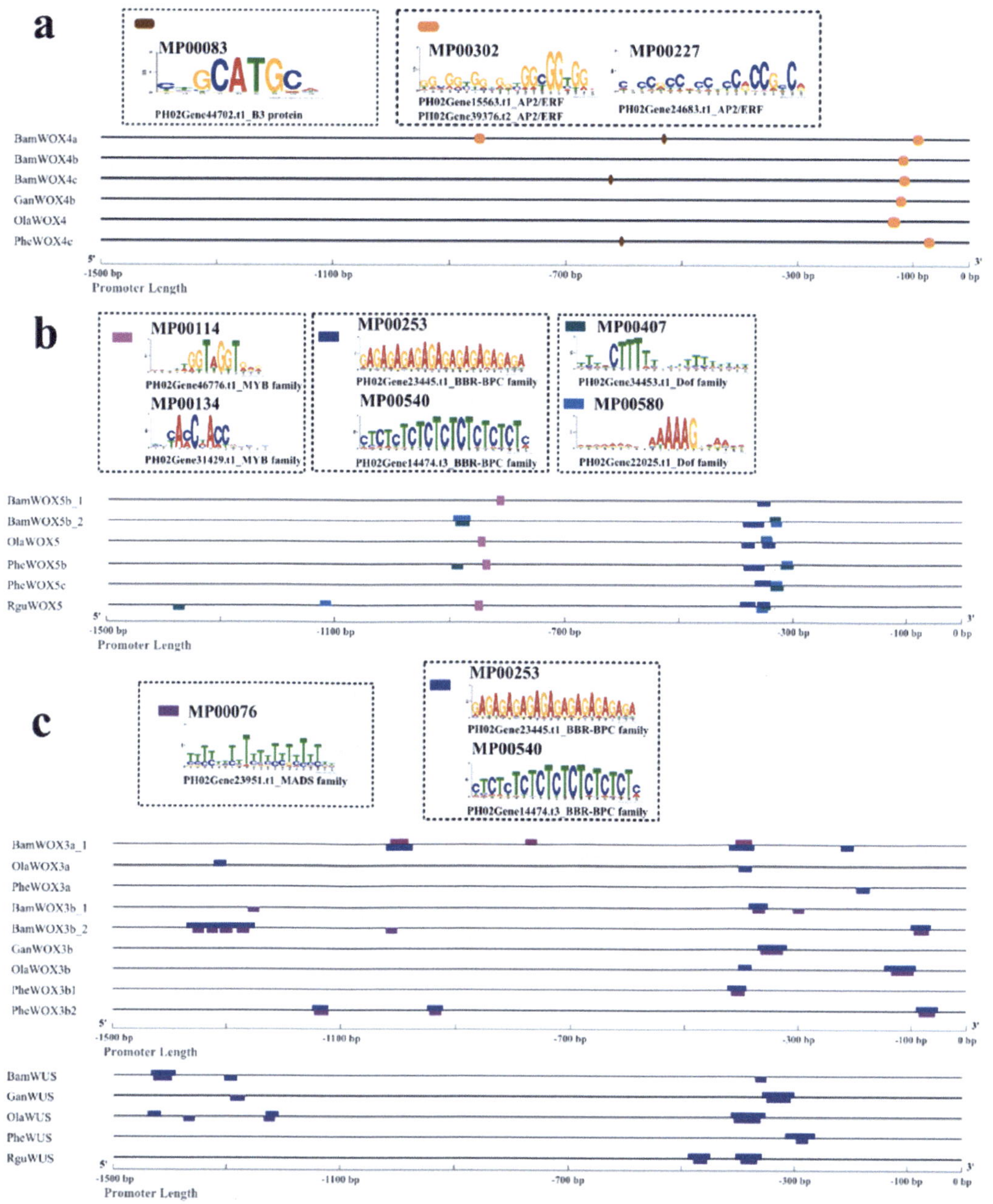

Fig. 8.43 Distributions of conserved cis-elements within Bambusoideae *WOX4* (**a**), *WOX5* (**b**), *WOX3* and *WUS* (**c**) promoters. PlantRegMap was utilized to analyze the promoters that were 1500 bp in length. The cis-element ID conformed to the TF binding motif dataset in PlantTFDB and were labeled in rectangles with diverse colors. Homologous genes were compared with the Arabidopsis TF datasets (http://planttfdb.cbi.pku.edu.cn/index.php?sp=Ath) to predict TFs (Li et al. 2020)

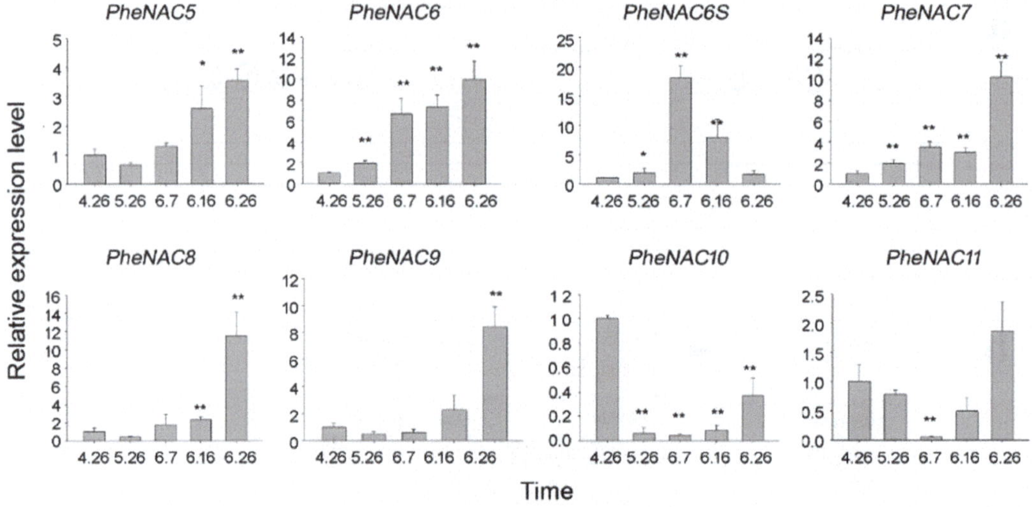

Fig. 8.44 Expression pattrens of eight *PheNACs* gene in leaves of the flowering moso bamboo (Xie et al. 2019)

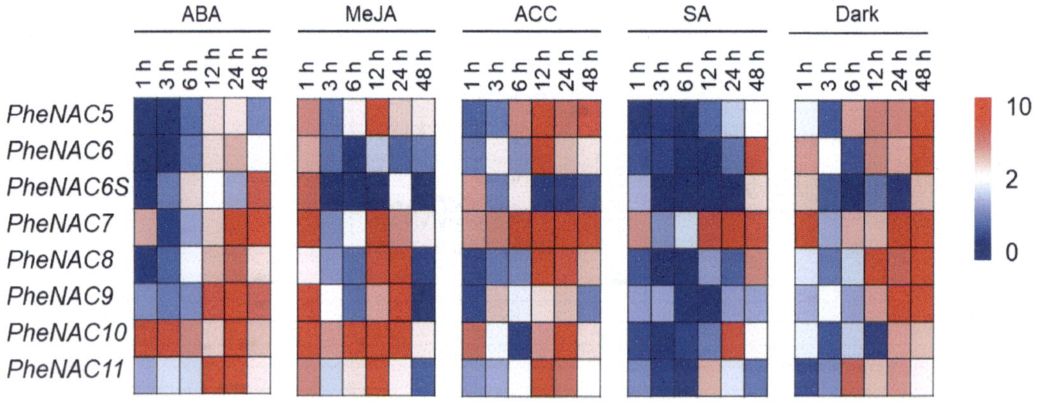

Fig. 8.45 Expression pattrens of eight *PheNACs* genes in detached leaves threated with hormone and dark (Xie et al. 2019)

the non-yellowing phenotype was found in Col-0. After five weeks, both OE-PheNAC3 and Col-0 displayed the senescence phenotype (Fig. 8.49a). The expression level of PheNAC3 in OE-PheNAC3 lines was confirmed by RT-PCR. The detached leaves were divided into three groups (G1 − G3; Fig. 8.49b) according to the senescent condition. G2 and G3 were used to analyze the chlorophyll content and target gene expression levels. Compared with Col-0, the OE-PheNAC3 lines showed much lower chlorophyll content in G2 leaves (Fig. 8.49c). The above results demonstrated that overexpression of PheNAC3 could induce leaf senescence-related physiological changes in *Arabidopsis*. The expression profiles of senescence-specific marker genes AtSAG12 and AtCAB1 were analyzed with qRT-PCR during leaf senescence between OE-PheNAC3 and Col-0 plants (Fig. 8.49e). The results showed that AtSAG12 was up-regulated

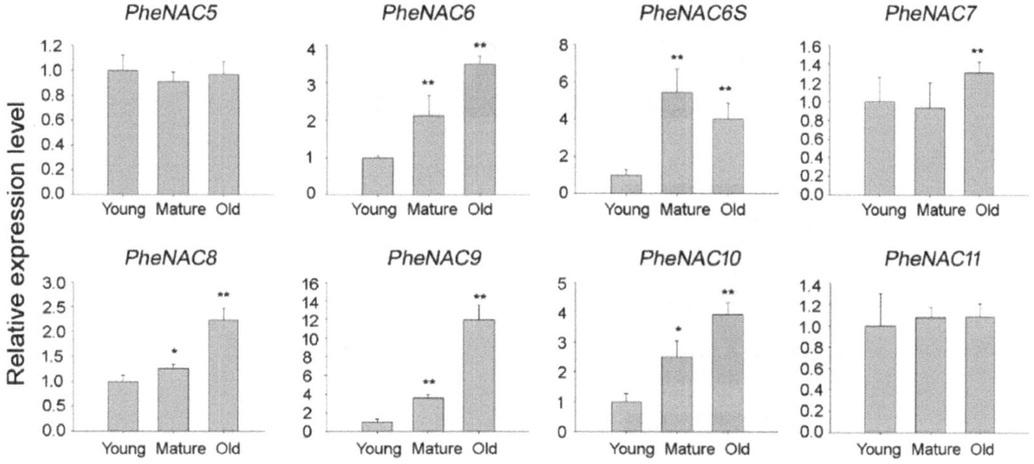

Fig. 8.46 Expression pattrens of eight *PheNACs* genes in leaves of three years old of seedings (Xie et al. 2019)

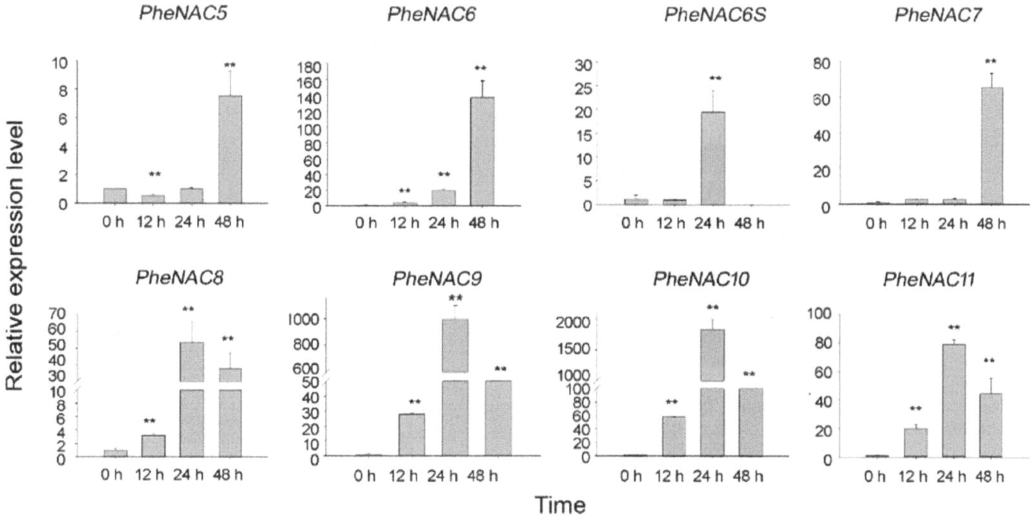

Fig. 8.47 Expression pattrens of eight PheNACs genes in the detached shoot (Xie et al. 2019)

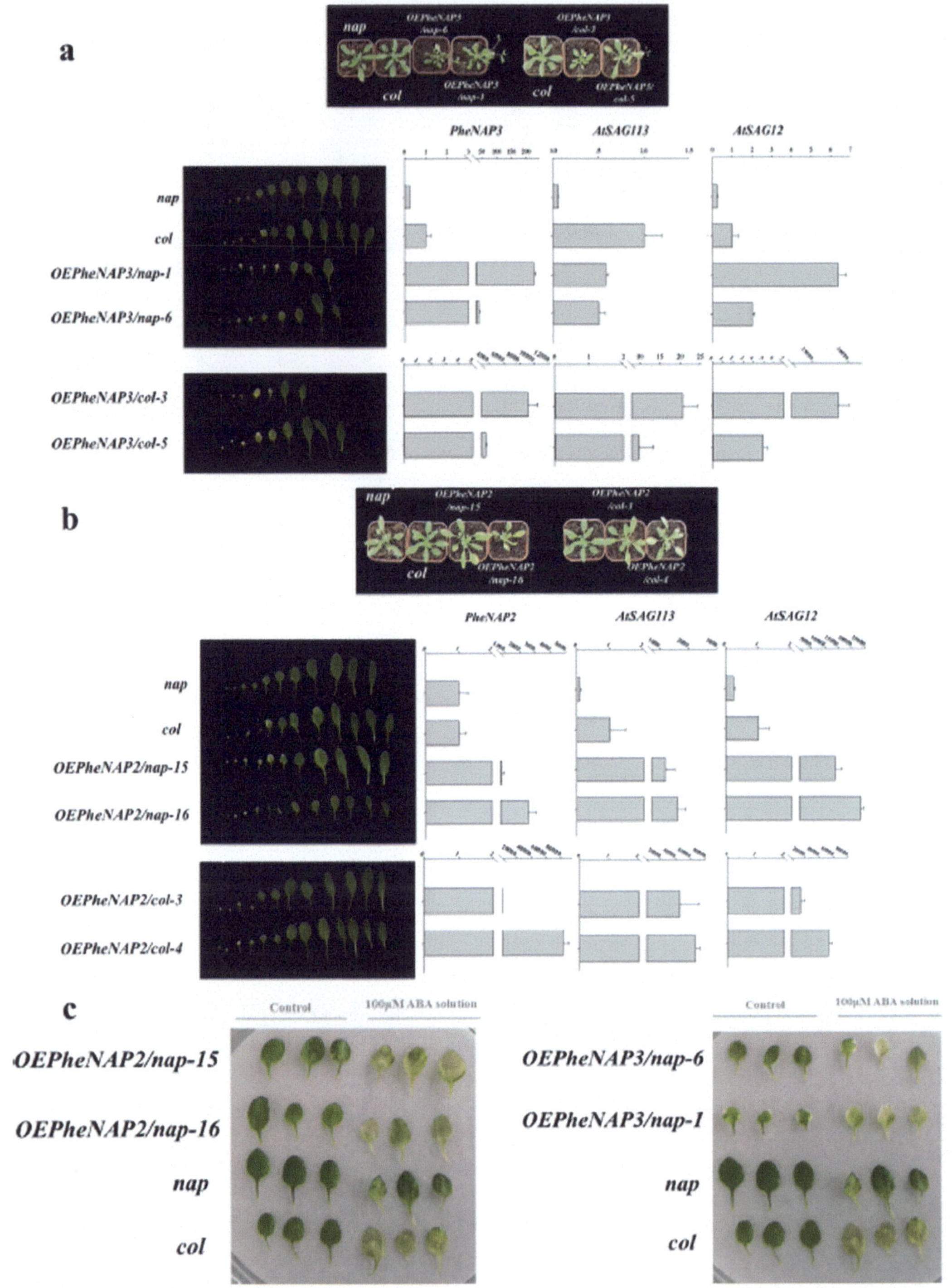

Fig. 8.48 PheNAP2 and PheNAP3 promote leaf senescence of *Arabidopsis*. **a** The phenotype of *Arabidopsis* overexpressing *PheNAP2*. **b** The phenotype of *Arabidopsis* overexpressing *PheNAP3*. **c** Abscisic acid-induced leaf senescence requires a functional NAP gene (Li et al. 2020)

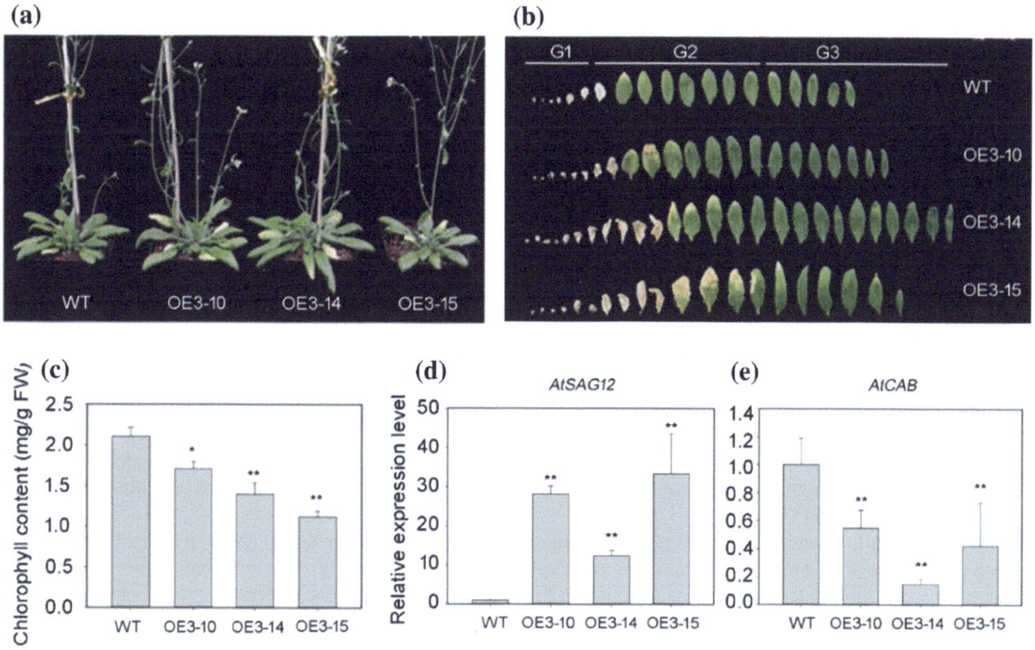

Fig. 8.49 Ectopic expression of PheNAC3 in Arabidopsis induced early leaf senescence. **a** The phenotype of 5-week-old Col and OE-PheNAC3 plants. **b** Leaves detached from plants in (**a**). G1–G3, three groups of leaves divided according to the senescence status and the method of Fan et al. (2015). G1, the first to the sixth leaves which were dry. G2, Leaf yellowing of OE3-15 lines started at an early stage compared with OE3-10 lines, OE3-14 lines and wild-type. Moreover, after five weeks of growth, the 13 or 14 leaves of OE3-15 lines started to turn yellow from tip of leaf. So, the seventh to the 14 leaves which were withered or turned yellow from the tip of leaf were classfied into G2 group. G3, the 15th to the last rosette leaves were G3 group. **c** Chlorophyll content in leaves from 5-week-old plants in (**a**) and (**b**). **d**, **e** Relative expression of two senescence marker genes, including SAG12 and CAB1, in G2 group of 5-week-old WT and OE-PheNAC3 plants. WT, wild-type Arabidopsis; OE3-10, OE3-14 and OE3-15, three independent lines. All experiments were repeated three times, with similar results. Asterisks indicate the significant difference between WT and OE-PheNAC3 (*$P < 0.05$ and ** $P < 0.01$); error bars indicate the ±SE (Xie et al. 2020)

and AtCAB1 was down-regulated compared with Col-0. These results suggested that PheNAC3 acted as a positive regulator of leaf senescence.

References

Alexandrova K, Conger B (2002) Isolation of two somatic embryogenesis-related genes from orchardgrass (Dactylis glomerata). Plant Sci 162:301–307

Almada R, Cabrera N, Casaretto JA et al (2009) *VvCO* and *VvCOL1*, two *CONSTANS* homologous genes, are regulated during flower induction and dormancy in grapevine buds. Plant Cell Rep 28:1193–1203

Ariel F, Manavella P, Dezar C et al (2007) The true story of the HD-Zip family. Trends Plant Sci 12 (9):419–426

Bai Q (2017) Identification, clone and function analysis of *SAUR* and *DELLA* genes in Moso Bamboo. Chinese Academy of Forestry

Bai Q, Hou D, Li L et al (2017) Genome-wide analysis and expression characteristics of small auxin-up RNA (*SAUR*) genes in moso bamboo (*Phyllostachys edulis*). Genome 60(4):325–336

Bailey P, Martin C, Toledo-Ortiz G et al (2003) Update on the basic helix-loop-helix transcription factor gene family in Arabidopsis thaliana. Plant Cell 15 (11):2497–2502

Bohlenius H, Huang T, Charbonnel-Campaa L et al (2016) *CO/FT* regulatory module controls timing of flowering and seasonal growth cessation in trees. Science 312:1040–1043

Brand U, Fletcher J, Hobe M et al (2000) Dependence of stem cell fate in Arabidopsis on a feedback loop regulated by CLV3 activity. Science 289(5479):617–619

Brewer P, Howles P, Dorian K et al (2004) PETAL LOSS, a trihelix transcription factor gene, regulates perianth architecture in the Arabidopsis flower. Development 131(16):4035–4045

Cao Y, Han Y, Jin Q et al (2016) Comparative Genomic Analysis of the GRF Genes in Chinese Pear (*Pyrus bretschneideri Rehd*), Poplar (*Populous*), Grape (*Vitis vinifera*), Arabidopsis and Rice (*Oryza sativa*). Front Plant Sci 7:1750

Capili A, Schultz D, Rauscher F et al (2001) Solution structure of the PHD domain from the KAP-1 corepressor: structural determinants for PHD, RING and LIM zinc-binding domains. EMBO J 20(1):165–177

Chae K, Isaacs C, Reeves P et al (2012) *Arabidopsis SMALL AUXIN UP RNA63* promotes hypocotyl and stamen filament elongation. Plant J 71(4):684–697

Chen C, Chen Z (2002) Potentiation of developmentally regulated plant defense response by AtWRKY18, a pathogen induced Arabidopsis transcription factor. Plant Physiol 129:706–716

Chen W, Provart NJ, Glazebrook J et al (2002) Expression profile matrix of Arabidopsis transcription factor genes suggests their putative functions in response to environmental stresses. Plant Cell 14:559–574

Chen D, Chen Z, Wu M et al (2017) Genome-wide identification and expression analysis of the HD-Zip gene family in Moso Bamboo (*Phyllostachys edulis*). J Plant Growth Regul 36(2):323–337

Cheng Z, Ge W, Li L et al (2017) Analysis of MADS-Box gene family reveals conservation in floral organ ABCDE model of Moso Bamboo (*Phyllostachys edulis*). Front Plant Sci 8:656

Cheng X, Xiong R, Liu H et al (2018) Basic helix-loop-helix gene family: Genome wide identification, phylogeny, and expression in Moso bamboo. Plant Physiol Biochem 132:104–119

Cheng X, Xiong R, Yan H et al (2019) The trihelix family of transcription factors: functional and evolutionary analysis in Moso bamboo (*Phyllostachys edulis*). BMC Plant Biol 19(1):154

Cheng Z, Hou D, Ge W et al (2020) Integrated mRNA, microRNA transcriptome and degradome analyses provide insights into stamen development in moso bamboo. Plant Cell Physiol 61(1):76–87

Chia T, Muller A, Jung C et al (2008) Sugar beet contains a large CONSTANS-LIKE gene family including a *CO* homologue that is independent of the early-bolting (B) gene locus. J Exp Bot 59:2735–2748

Ciolkowski I, Wanke D, Birkenbihl RP et al (2008) Studies on DNA-binding selectivity of WRKY transcription factors lend structural clues into WRKY-domain function. Plant Mol Biol 68:81–92

Cui K, He C, Zhang J et al (2012) Temporal and spatial profiling of internode elongation-associated protein expression in rapidly growing culms of bamboo. J Proteome Res 11(4):2492–2507

Danisman S (2016) TCP Transcription Factors at the Interface between Environmental Challenges and the Plant's Growth Responses. Front Plant Sci 7 (406):1930

Dellagi A, Heilbronn J, Avrova A et al (2002) A potato gene encoding a WRKY-like transcription factor is induced in interactions with Erwinia carotovora subsp atroseptica and Phytophthora infestans and is coregulated with class I endochitinase expression. Mol. Plant-Microbe Interaction 13:1092–1101

Deslandes L, Olivier J, Theulières F et al (2002) Resistance to Ralstonia solanacearum in Arabidopsis thaliana is conferred by the recessive RRS1-R gene, a member of a novel family of resistance genes. Proc Natl Acad Sci USA 99:2404–2409

Ding J, Böhlenius H, Rühl MG et al (2018) *GIGANTEA-like* genes control seasonal growth cessation in Populus. New PhytolOgy 218:1491–1503

Doebley J, Stec A, Hubbard L (1997) The evolution of apical dominance in maize. Nature 386(6624):485–488

Eulgem T, Somssich IE (2007) Networks of WRKY transcription factors in defense signaling. Curr Opin Plant Biol 10:366–371

Fang Y, Xie K, Hou X et al (2010) Systematic analysis of GT factor family of rice reveals a novel subfamily involved in stress responses. Mol Genet Genomics 283(2):157–169

Fornara F, Panigrahi KC, Gissot L et al (2009) Arabidopsis DOF transcription factors act redundantly to reduce *CONSTANS* expression and are essential for a photoperiodic flowering response. Dev Cell 17:75–86

Gao Z, Zheng B, Peng Z (2010) Isolation of *PeMADS1* Gene from *Phyllostachys edulis* and its transformation in *Arabidopsis thaliana*. Scientia Silvae Sinicae 46 (10):37–41

Gao J, Zhang Y, Zhang C et al (2014) Characterization of the floral transcriptome of moso bamboo (*Phyllostachys edulis*) at different flowering developmental stages by transcriptome sequencing and RNA-Seq analysis. PLoS One 9:e98910

Gao Y, Liu H, Wang Y et al (2018) Genome-wide identification of PHD-finger genes and expression pattern analysis under various treatments in moso bamboo (*Phyllostachys edulis*). Plant Physiol Biochem 123:378–391

Garber R, Kuroiwa A, Gehring W (1983) Genomic and cDNA clones of the homeotic locus Antennapedia in Drosophila. EMBO J 2(11):2027–2036

Ge W, Zhang Y, Cheng Z et al (2017) Main regulatory pathways, key genes and microRNAs involved in flower formation and development of moso bamboo (*Phyllostachys edulis*). Plant Biotechnol J 15:82–96

Gehring W, Qian Y, Billeter M et al (1994) Homeodomain-DNA recognition. Cell 78(2):211–223

Gremski K, Ditta G, Yanofsky M (2007) The HECATE genes regulate female reproductive tract development in Arabidopsis thaliana. Development 134(20):3593–3601

Griffifiths S, Dunford R, Coupland G et al (2003) The Evolution of CONSTANS-Like gene families in barley, rice, and Arabidopsis. Plant Physiol 131:1855–1867

Guo A, Zhu Q, Gu X et al (2008) Genome-wide identification and evolutionary analysis of the plant specific SBP-box transcription factor family. Gene 418:1–8

Guo Z, Ma P, Yang G et al (2019) Genome sequences provide insights into the reticulate origin and unique traits of woody bamboos. Mol Plant 12(10):1353–1365

Haecker A, Groß-Hardt R, Geiges B et al (2004) Expression dynamics of WOX genes mark cell fate decisions during early embryonic patterning in Arabidopsis thaliana. Development 131(3):657–668

Harris J, Hrmova M, Lopato S et al (2011) Modulation of plant growth by HD-Zip class I and II transcription factors in response to environmental stimuli. New Phytol 190(4):823–837

Higashi K, Ishiga Y, Inagaki Y et al (2008) Modulation of defense signal transduction by flagellin-induced WRKY41 transcription factor in Arabidopsis thaliana. Mol Gen Genet 279:303–312

Hinderhofer K, Zentgraf U (2001) Identification of a transcription factor specifically expressed at the onset of leaf senescence. Planta 213:469–473

Hirakawa Y, Kondo Y, Fukuda H (2010) TDIF peptide signaling regulates vascular stem cell proliferation via the WOX4 homeobox gene in Arabidopsis. Plant Cell 22(8):2618–2629

Hou D, Cheng Z, Xie L et al (2018) The R2R3MYB gene family in *Phyllostachys edulis*: genome-wide analysis and identification of stress or development-related R2R3MYBs. Front Plant Sci 9:738

Hu X, Xu L (2016) Transcription factors WOX11/12 directly activate WOX5/7 to promote root primordia initiation and organogenesis. Plant Physiol 172 (4):2363–2373

Imaizumi T, Tran H, Swartz T et al (2003) *FKF1* is essential for photoperiodic-specifific light signaling in Arabidopsis. Nature 426:302–306

Ishiguro S, Nakamura K (1994) Characterization of a cDNA encoding a novel DNA-binding protein, SPF1, that recognizes SP8 sequences in the 5′ upstream regions of genes coding for sporamin and β-amylase from sweet potato. Mol Gen Genet 244:563–571

Jain M, Tyagi A, Khurana J (2006) Genome-wide analysis, evolutionary expansion, and expression of early auxin-responsive SAUR gene family in rice (*Oryza sativa*). Genomics 88(3):360–371

Jiang Y, Duan Y, Yin J et al (2014) Genome-wide identification and characterization of the Populus WRKY transcription factor family and analysis of their expression in response to biotic and abiotic stresses. J Exp Bot 65:6629–6644

Jin J, Feng T, De-Chang Y et al (2016) PlantTFDB 4.0: toward a central hub for transcription factors and regulatory interactions in plants. Nucleic Acids Res, pp gkw982

Johnson C, Kolevski B, Smyth D (2002) Transparent Testa GLABRA2, a trichome and seed coat development gene of Arabidopsis, encodes a WRKY transcription factor. Plant Cell 14:1359–1375

Klein J, Saedler H, Huijser P (1996) A new family of DNA binding proteins includes putative transcriptional regulators of the Antirrhinum majus floral meristem identity gene SQUAMOSA. Mol Gen Genet 250:7–16

Knaap E, Kim J, Kende H et al (2000) A novel gibberellin-induced gene from rice and its potential regulatory role in stem growth. Plant Physiol 122:695–704

Knauss S, Rohrmeier T, Lehle L (2003) The auxin-induced maize gene *ZmSAUR2* encodes a short-lived nuclear protein expressed in elongating tissues. J Biol Chem 278(26):23936–23943

Kosugi S, Ohashi Y (1997) PCF1 and PCF2 specifically bind to cis elements in the rice proliferating cell nuclear antigen gene. Plant Cell 9(9):1607–1619

Laux T, Mayer K, Berger J et al (1996) The WUSCHEL gene is required for shoot and floral meristem integrity in Arabidopsis. Development 122(1):87–96

Ledent V, Vervoort M (2001) The basic helix-loop-helix protein family: comparative genomics and phylogenetic analysis. Genome Res 11(5):754–770

Li X, Duan X, Jiang H et al (2006) Genome-wide analysis of basic/helix-loop-helix transcription factor family in rice and Arabidopsis. Plant Physiol 141(4):1167–1184

Li Z, Chen H, Li Q et al (2015) Three SAUR proteins SAUR76, SAUR77 and SAUR78 promote plant growth in *Arabidopsis*. Sci Rep 5:12477

Li L, Mu S, Cheng Z et al (2017) Characterization and expression analysis of the WRKY gene family in moso bamboo. Sci Rep 7:6675

Li X, Li J, Cai M et al (2020) Identification and evolution of the WUSCHEL-related homeobox protein family in Bambusoideae. Biomolecules 10(5):739

Lie C, Kelsom C, Wu X (2012) WOX2 and STIMPY-LIKE/WOX8 promote cotyledon boundary formation in Arabidopsis. Plant J 72(4):674–682

Liu J, Cheng Z, Li X et al (2019) Expression analysis and regulation network identification of the CONSTANS-Like gene family in moso bamboo (*Phyllostachys edulis*) under photoperiod treatments. DNA and Cell Biol 38(7):607–626

Liu J, Cheng Z, Xie L et al (2019) Multifaceted role of PheDof12-1 in the regulation of flowering time and abiotic stress responses in Moso Bamboo (*Phyllostachys edulis*). Int J Mol Sci 20(2):1–13.

Li X, Xie L, Zheng H et al (2019) Transcriptome profiling of postharvest shoots identifies PheNAP2- and PheNAP3-promoted shoot senescence. Tree Physiol 39(12):2027–2044

Liu J, Ekramoddoullah A (2009) Identification and characterization of the WRKY transcription factor family in Pinus monticola. Genome 52:77–88

Liu J, Sheng L, Xu Y et al (2014) WOX11 and 12 are involved in the first-step cell fate transition during de novo root organogenesis in Arabidopsis. Plant Cell 26 (3):1081–1093

Liu J, Huang R, Cheng Z et al (2017). Molecular cloning and expression analysis of the *PheDof4-1* gene in moso bamboo (*Phyllostachys edulis*). Journal of Anhui Agricultural University 44(3): 398-403.

Liu H, Wu M, Li F et al (2018) TCP Transcription factors in Moso Bamboo (*Phyllostachys edulis*): genome-

wide identification and expression analysis. Front Plant Sci 9:1263
Liu M, Wang M, Yang J et al (2019) Evolutionary and comparative expression analyses of tcp transcription factor gene family in land plants. Int J Mol Sci 20:3591
Lu Z, Shao G, Xiong J et al (2015) MONOCULM 3, an ortholog of WUSCHEL in rice, is required for tiller bud formation. J Genet Genomics 42(2):71–78
Ludwig S, Habera L, Dellaporta S, Wessler SR (1989) Lc, a member of the maize R gene family responsible for tissue-specific anthocyanin production, encodes a protein similar to transcriptional activators and contains the myc-homology region. Proc Natl Acad Sci USA 86(18):7092–7096
Luo D, Carpenter R, Copsey L et al (1999) Control of organ asymmetry in flowers of antirrhinum. Cell 99 (4):367–376
Maren P, Silke S, Stefan B et al (2014) Senescence networking: WRKY18 is an upstream regulator, a downstream target gene, and a protein interaction partner of WRKY53. J. Plant Growth Regulation 33:106–118
Martíntrillo M, Cubas P (2010) TCP genes: a family snapshot ten years later. Trends Plant Sci 15(1):31–39
Miura K, Renhu N, Suzaki T (2020) The PHD finger of Arabidopsis SIZ1 recognizes trimethylated histone H3K4 mediating SIZ1 function and abiotic stress response. Commun Biol 3(1):23
Narusaka Y, Nakashima K, Shinwari Z et al (2003) Interaction between two cis-acting elements, ABRE and DRE, in ABA-dependent expression of Arabidopsis rd29A gene in response to dehydration and high-salinity stresses. Plant J 34:137–148
Ni M, Tepperman J, Quail P et al (1998) PIF3, a phytochrome-interacting factor necessary for normal photoinduced signal transduction, is a novel basic helix-loop-helix protein. Cell 95(5):657–667
Ohmori Y, Tanaka W, Kojima M et al (2013) WUSCHEL-RELATED HOMEOBOX4 is involved in meristem maintenance and is negatively regulated by the CLE gene FCP1 in rice. Plant Cell 25(1):229–241
Omidbakhshfard M, Proost S, Fujikura U et al (2015) Growth-regulating factors (GRFs): a small transcription factor family with important functions in plant biology. Mol Plant 8:998–1010
Pan F, Wang Y, Liu H et al (2017) Genome-wide identification and expression analysis of SBP-like transcription factor genes in Moso Bamboo (*Phyllostachys edulis*). BMC Genomics 18(1):486–503
Pan F, Wu M, Hu W et al (2019) Genome-wide identification and expression analyses of the bZIP transcription factor genes in moso bamboo (*Phyllostachys edulis*). Int J Mol Sci 20(9):2203
Pandey SP, Roccaro M, Schön M et al (2010) Transcriptional reprogramming regulated by WRKY18 and WRKY40 facilitates powdery mildew infection of Arabidopsis. Plant J 64:912–923
Park J, Kim Y, Yoon H et al (2007) Functional characterization of a *small auxin-up RNA* gene in apical hook development in *Arabidopsis*. Plant Sci 172(1):150–157
Peng Z, Lu T, Li L et al (2010) Genome-wide characterization of the biggest grass, bamboo, based on 10,608 putative full-length cDNA sequences. BMC Plant Biol 10(116):1–13
Peng Z, Lu Y, Li L et al (2013) The draft genome of the fast-growing non-timber forest species moso bamboo (*Phyllostachys heterocycla*). Nat Genet 45:456–461
Pi L, Aichinger E, van der Graaff E et al (2015) Organizer-derived WOX5 signal maintains root columella stem cells through chromatin-mediated repression of CDF4 expression. Dev Cell 33(5):576–588
Plesch G, Ehrhardt T, Muellerroeber B (2001) Involvement of TAAAG elements suggests a role for Dof transcription factors in guard cell-specific gene expression. Plant J 28(455–464):4
Qi F, Hu T, Peng Z et al (2013) Screening of reference genes used in qRT-PCR and expression analysis of PheTFL1 gene in moso bamboo. Acta Botan Boreali-Occiden Sin 33(1):0048–0052
Quan W, Liu X, Wang L et al (2019) Ectopic expression of Medicago truncatula homeodomain finger protein, MtPHD6, enhances drought tolerance in Arabidopsis. BMC Genomics 20(1):1–16
Riechmann J, Heard J, Martin G et al (2000) Arabidopsis transcription factors: genome-wide comparative analysis among eukaryotes. Science 290(5499):2105–2110
Robson F, Costa M, Hepworth S et al (2001) Functional importance of conserved domains in the flowering-time gene *CONSTANS* demonstrated by analysis of mutant alleles and transgenic plants. Plant J 28:619–631
Rushton PJ, Macdonald H, Huttly AK et al (1995) Members of a new family of DNA-binding proteins bind to a conserved cis-element in the promoters of alpha-Amy2 genes. Plant Mol Biol 29:691–702
Rushton PJ, Somssich IE, Ringler P et al (2010) WRKY transcription factors. Trends Plant Sci 15:247–258
Schena M, Davis R (1992) HD-Zip proteins: members of an Arabidopsis homeodomain protein superfamily. Proc Natl Acad Sci USA 89(9):3894–3898
Schindler U, Beckmann H, Cashmore AR (1993) HAT3.1, a novel Arabidopsis homeodomain protein containing a conserved cysteine-rich region. Plant J 4 (1):137–150
Searle L, Coupland G (2004) Induction of flowering by seasonal changes in photoperiod. Embo J 23:1217–1222
Shi Y, Liu H, Gao Y et al (2019) Genome-wide identification of growth-regulating factors in moso bamboo (*Phyllostachys edulis*): in silico and experimental analyses. Peer J 7:e7510
Shimizu R, Ji J, Kelsey E et al (2009) Tissue specificity and evolution of meristematic WOX3 function. Plant Physiol 149(2):841–850
Somssich M, Je B, Simon R et al (2016) CLAVATA-WUSCHEL signaling in the shoot meristem. Development 143(18):3238–3248

Song A, Wu D, Fan Q et al (2016) Transcriptome-wide identification and expression profiling analysis of chrysanthemum trihelix transcription factors. Int J Mol Sci 17(2)

Spartz A, Lee S, Wenger J et al (2012) The SAUR19 subfamily of *SMALL AUXIN UP RNA* genes promote cell expansion. Plant J 70(6):978–990

Su X, Lu G, Guo H et al (2018) The dynamic transcriptome and metabolomics profiling in Verticillium dahliae inoculated Arabidopsis thaliana. Sci Rep 8(1):15404

Sun C et al (2003) A novel WRKY transcription factor, SUSIBA2, participates in sugar signaling in barley by binding to the sugarresponsive elements of the iso1 promoter. Plant Cell 15:2076–2092

Sung S, Amasino RM (2004) Vernalization in Arabidopsis thaliana is mediated by the PHD finger protein VIN3. Nature 427(6970):159–164

Talanian R, Mcknight C, Kim P (1990) Sequence-specific DNA-binding by a short peptide dimer. Science 249:769–771

Tanaka W, Hirano H (2020) Antagonistic action of TILLERS ABSENT1 and FLORAL ORGAN NUMBER2 regulates stem cell maintenance during axillary meristem development in rice. New Phytol 225:974–984

Tanaka W, Ohmori Y, Ushijima T et al (2015) Axillary meristem formation in rice requires the WUSCHEL ortholog TILLERS ABSENT1. Plant Cell 27 (4):1173–1184

Tsubosa Y, Sato H, Tachimori Y et al (1997) Functional analysis of the Arabidopsis thaliana SBP-box gene SPL3: a novel gene involved in the floral transition. Plant J 12:367–377

Ueda M, Zhang Z, Laux T (2011) Transcriptional activation of Arabidopsis axis patterning genes WOX8/9 links zygote polarity to embryo development. Dev Cell 20(2):264–270

Ulker B, Somssich I (2004) WRKY transcription factors: from DNA binding towards biological function. Curr Opin Plant Biol 2004(7):491–498

Van der Graaff E, Laux T, Rensing SA (2009) The WUS homeobox-containing (WOX) protein family. Genome Biol 10(12):248

Vollbrecht E, Veit B, Sinha N et al (1991) The developmental gene Knotted-1 is a member of a maize homeobox gene family. Nature 350(6315):241–243

Wang Q, Liu J, Wang Y et al (2015) Systematic analysis of the maize PHD-finger gene family reveals a subfamily involved in abiotic stress response. Int J Mol Sci 16(10):23517–23544

Wei W, Zhang Y, Tao J et al (2015) The Alfin-like homeodomain finger protein AL5 suppresses multiple negative factors to confer abiotic stress tolerance in Arabidopsis. Plant J 81(6):871–883

Xi J, Qiu Y, Du L et al (2012) Plant-specific trihelix transcription factor AtGT2L interacts with calcium/calmodulin and responds to cold and salt stresses. Plant Sci, pp 274–280

Xiao C, Zhang J, Xu Y et al (2013) Cloning and functional analysis of *Phyllostachys edulis* MYB transcription factor PeMYB2. Hereditas 35:1217–1225

Xie L (2019) Function analysis of senescence-associated Genes PheORE1s and PheNAPs in *Phyllostachys edulis*. Chinese Academy of Forestry Science

Xie Z, Zou H, Lei G, et al (2012) Soybean Trihelix transcription factors GmGT-2A and GmGT-2B improve plant tolerance to abiotic stresses in transgenic Arabidopsis. PloS one 4(9): e6898

Xie L, Cai M, Gao J et al (2020) Overexpression of NAC transcription factor gene PheNAC3 from *Phyllostachys edulis* promotes leaf senescence and enhances abiotic stress tolerance in transgenic Arabidopsis. Peer J 8:e8716

Yamasaki K, Kigawa T, Inoue M et al (2004) A novel zinc-binding motif revealed by solution structures of DNA-binding domains of Arabidopsis SBP-family transcription factors. J Mol Biol 337:49–63

Yamashino T, Yamawaki S, Hagui E et al (2013) Clock-controlled and *FLOWERING LOCUS T* (*FT*)-dependent photoperiodic pathway in *Lotus japonicus* I: verification of the flowering associated function of an *FT* homolog. Biosci Biotechnol Biochem 77:747–753

Yanagisawa S (2002) The Dof family of plant transcription factors. Trends Plant Sci 7:555–560

Yanagisawa S, Izui K (1993) Molecular cloning of two DNA-binding proteins of maize that are structurally different but interact with the same sequence motif. J Biol Chem 268:16028–16036

Yang K, Li Y, Wang S et al (2019) Genome-wide identification and expression analysis of the MYB transcription factor in moso bamboo (*Phyllostachys edulis*). Peer J 6(9):e6242

Yoshikawa T, Tanaka S, Masumoto Y et al (2016) Barley NARROW LEAFED DWARF1 encoding a WUSCHEL-RELATED HOMEOBOX 3 (WOX3) regulates the marginal development of lateral organs. Breeding Sci, p 16019

Zhang YJ, Wang LJ (2005) The WRKY transcription factor super family: its origin in eukaryotes and expansion in plants. BMC Evol Biol 5:1

Zhang X, Zong J, Liu J et al (2010) Genome-wide analysis of WOX gene family in rice, sorghum, maize, *Arabidopsis* and poplar. J Integr Plant Biol 52 (11):1016–1026

Zhang Y, Jiao Y, Jiao H et al (2016) Two-step functional innovation of the stem-cell factors WUS/WOX5 during plant evolution. Mol Biol Evol 34(3):640–653

Zhang H, Ying Y, Wang J et al (2018) Transcriptome analysis provides insights into xylogenesis formation in moso bamboo (*Phyllostachys edulis*) shoot. Sci Rep 8(1):3951

Zhao H, Gao Z, Wang L et al (2018) Chromosome-level reference genome and alternative splicing atlas of moso bamboo (*Phyllostachys edulis*). GigaScience 7 (10):giy115

Zimmermann I, Heim M, Weisshaar B et al (2004) Comprehensive identification of Arabidopsis thaliana MYB transcription factors interacting with R/B-like BHLH proteins. Plant J 40(1):22–34

MicroRNAs of Moso Bamboo

9

Wei Ge and Zhanchao Cheng

Abstract

Moso bamboo has been featured by its rapidly growing culms, unstable flowering habit, and low sexual reproduction. microRNAs (miRNAs), the small RNAs, have been suggested to participate in regulating numerous processes, such as cell differentiation, growth and development, plant hormone signals, and biotic/abiotic stresses. A variety of miRNAs-targeting TFs are involved in the critical regulation pathways. During recent years, more and more attention has been paid to the roles of miRNAs as well as their target genes in regulating moso bamboo growth. Researchers have found that a number of miRNAs and their target genes exert vital parts in fast development, flower formation, and stress tolerance of moso bamboo. Taken together, the studies have shed new lights on the exact mechanisms related to moso bamboo growth at molecular mechanism.

W. Ge (✉)
Beijing University of Agriculture, No. 7, North Agricultural Road, Zhu Xinzhuang, Changping District, Beijing 102206, China
e-mail: gewei19870615@163.com

Z. Cheng
Key Laboratory of Bamboo and Rattan Science and Technology, International Center for Bamboo and Rattan, State Forestry and Grassland Administration, Beijing 100102, China
e-mail: czc@icbr.ac.cn

9.1 Introduction

MicroRNAs (miRNAs) possibly exert vital parts in the regulation of moso bamboo growth (Gao et al. 2015). miRNAs, the non-coding single-stranded RNAs around 21–22 nucleotides (nt) in length, can be encoded via the internal genes. miRNAs can bind to complementary sites to modulate target gene expression in plants and animal at post-transcriptional level, and many of their targets are transcription factors (TFs) (Mitsuda and Ohme-Takagi 2009; Kamthan et al. 2015). miRNAs are able to suppress the expression of target genes, whereas TFs exert vital parts in the regulation of target genes through promoting or suppressing the specific expression (Istrail et al. 2007). In addition, miRNAs show interactions with TFs to form a complicated regulation network, which is also related to mechanisms underlying plant growth at molecular level (Samad et al. 2016).

9.2 miRNAs Could Regulate Reproductive Growth of Moso Bamboo

The miRNAs regulate the growth and development of reproductive organs such as flowers in *Phyllostachys edulis*. It is not easy to investigate the flowering of moso bamboo due to the unclear flowering interval and rare flowering samples.

J. Gao (ed.), *The Moso Bamboo Genome*, Compendium of Plant Genomes,
https://doi.org/10.1007/978-3-030-80836-5_9

For identifying miRNAs and investigating the specific expression levels in moso bamboo flowering period, the Illumina technology was adopted to sequence miRNAs in non-flowering leaf samples at 4 flowering periods. Altogether, 409 known miRNAs together with 492 new miRNAs with differential expression were detected from moso bamboo. Altogether, 308 targets were predicted for 64 of those known miRNAs with differential expression in flowering samples compared with non-flowering ones. This study carried out a comprehensive analysis to reveal miRNAs with differential expression as well as the respective targets related to different biological pathways, which exerted important parts in regulating the flowering of moso bamboo (Gao et al. 2015).

We chose the important regulatory pathways and then examined the miRNA expression in diverse moso bamboo flowering periods. Moreover, we also screened and analyzed certain potential genes related to the interaction between plant and pathogen, protein processing within the endoplasmic reticulum (ER), and plant hormone signaling (Ge et al. 2017). As a result, miR159a.1, miR160a, miR168-3p, miR390a, miR393, and miR5139 were screened according to DGE analysis on the respective target genes (Fig. 9.1).

The above screened genes were differentially expressed in flowering samples compared with non-flowering leaves. As indicated by clustering analysis, our screened 6 miRNAs were down-regulated. Specifically, as far as miRNAs were concerned, their differential expression, but not their composition, had more impacts on the regulation of moso bamboo flowering. Typically, the miRNA expression profiles in moso bamboo shed important lights on the corresponding biological activities as well as the underlying regulatory mechanism. Therefore, we carried out qRT-PCR to investigate the 6 screened miRNAs for their relative expression. As a result, miR159a.1, miR160a, miR168-3p, miR390a, miR393, and miR5139 exhibited increased expression within non-flowering leaves compared with that within 4 flowering samples, which suggested that these genes exerted certain functions in regulating the flowering of moso bamboo. Besides, miR390a and miR5139 were possibly associated with the interaction between plant and pathogen. miR390a suppressed *ARF2*/*ARF3*/*ARF4* expression through modulating target gene expression while inducing numerous

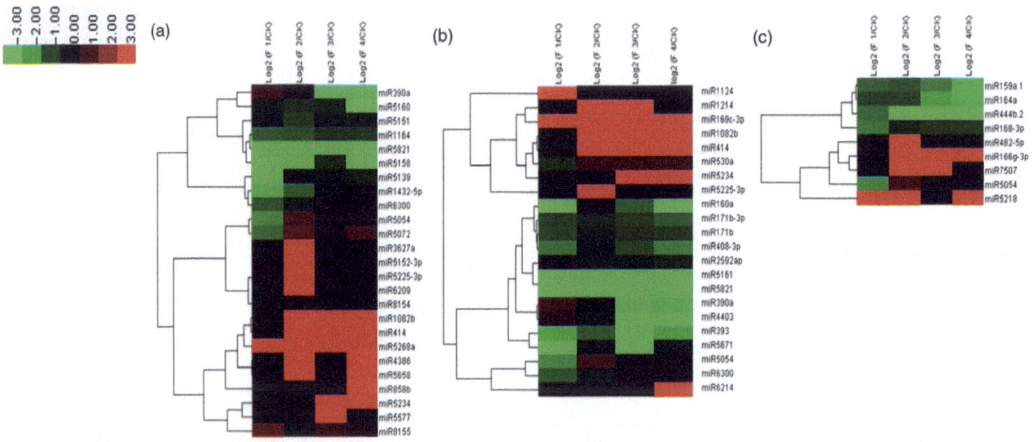

Fig. 9.1 Differentially expressed miRNAs related to the major regulatory pathways. Heatmap showing the clustering analysis on associated miRNAs. Bar scale = target gene expression (log 2). **a** Differentially expressed miRNAs related to the interaction between plant and pathogen. **b** Differentially expressed miRNAs related to plant hormone signaling. **c** Differentially expressed miRNAs related to protein processing within ER. From Ge et al. (2017)

phenotypic alterations. Further, miR390a exerted a vital part in regulating the transition of growth cycle and development of organ. miR390a overexpression had potential impact on the transition to reproductive growth from vegetative growth. In addition, geminivirus Rep-interacting motor protein was the miR5139 target, and miR5139 expression decreased upon stress; therefore, stress might be associated with the flowering of moso bamboo. miR160a and miR393 exert important parts in plant hormone signaling. ARF was predicted as the miR160a targets. miR160a was involved in the auxin signaling through targeting ARF while impacting the plant reproductive and vegetative growth. The miR393 targets were the F-box family members, which modulated the auxin perception in plants and enhanced the sensitivity to auxin. miR159a.1 and miR168-3p were possibly associated with protein processing within ER. miR159a.1 exerted a vital role in regulating MYB expression and directly degrading the *MYB* mRNA, thus resulting in the flowering delay. On the other hand, miR168 played a critical part in plant tissue differentiation and stress responses. It was also the only miRNA participating in cross-border regulation in plants and animals. Therefore, over-expressed levels of the 6 screened miRNAs possibly suppressed the flowering of moso bamboo (Ge et al. 2017).

Cheng and colleagues (2020) applied degradome, transcriptome, and small RNA analyzes for characterizing the critical miRNAs together with the corresponding targets within the moso bamboo floral organs. The authors aligned the selected miRNAs to the known miRNA databases and identified certain conserved and certain new miRNAs. Altogether, 16 conserved miRNAs showed specific and high expression levels within stamens, which included miR159 and miR166. According to in situ hybridization (ISH) analysis, miR159 showed high expression in the moso bamboo flowering period. Besides, no detectable *PheMYB42* or *PheMYB98* expression was observed in the floral organs (Fig. 8.2), indicating the possible role of miR159 in regulating *PheMYB42* and *PheMYB98* expression during flower formation as well as floral organ growth stages. There were 2 MYB targets predicted in moso bamboo, which were under the negative regulation via miR159 among diverse reproductive organs, suggesting the vital part of miR159 in the fertility of plants.

Besides, *Phe*-miR159 shows a certain effect on recovering the miR159ab double-mutant phenotypes. *Phe*-miR159 overexpression induced anther dehisce failure, which prevented the dispersion of mature pollens, thereby reducing the fertility of Arabidopsis. The anthers of WT and *Phe-MIR159* overexpression lines (b2–3) in the flowering period were monitored by semi-thin section. Relative to WT, complete degradation of connective tissue and inner layer was observed in b2–3 line, and an anther was formed, which comprised 2 rooms filled with mature pollens encompassed by one thin skin layer. As revealed by semi-thin section results, secondary thickening was lacking in the endothelial layer in anther of *Phe-MIR159* overexpression plants, giving rise to the restricted force to open anther (Fig. 9.2). *Phe*-miR159 has potential effect on regulating secondary thickening-related gene expression by negatively regulating *AtMYB33* expression, thus impacting the opening of anther (Cheng et al. 2020).

Fan (2012) identified and analyzed moso bamboo's miRNAs through high-throughput sequencing and bioinformatics analysis. The highest expression was miR168. The conservative miRNA families miR156, miR159, and miR172, and the non-conservative miRNA families miR535 and miR528 were highly expressed miRNA families in *P. edulis*. The target genes of 22 new miRNAs were predicted to play a wide role, such as enzymes, membrane proteins, and transcription factors. The expression of 21 known miRNAs and 3 new miRNAs in different tissues of *P. edulis* was tissue-specific. The expression of SPL, AP2, HD-ZIPIII, and PLA in different tissues of miR156, miR172, miR166, and miR528 showed negative correlation with their corresponding miRNA. The results showed that miR168 was differentially expressed in striped bamboo and normal bamboo, which may be related to the formation of striped bamboo. Fluorescence quantitative PCR analysis showed

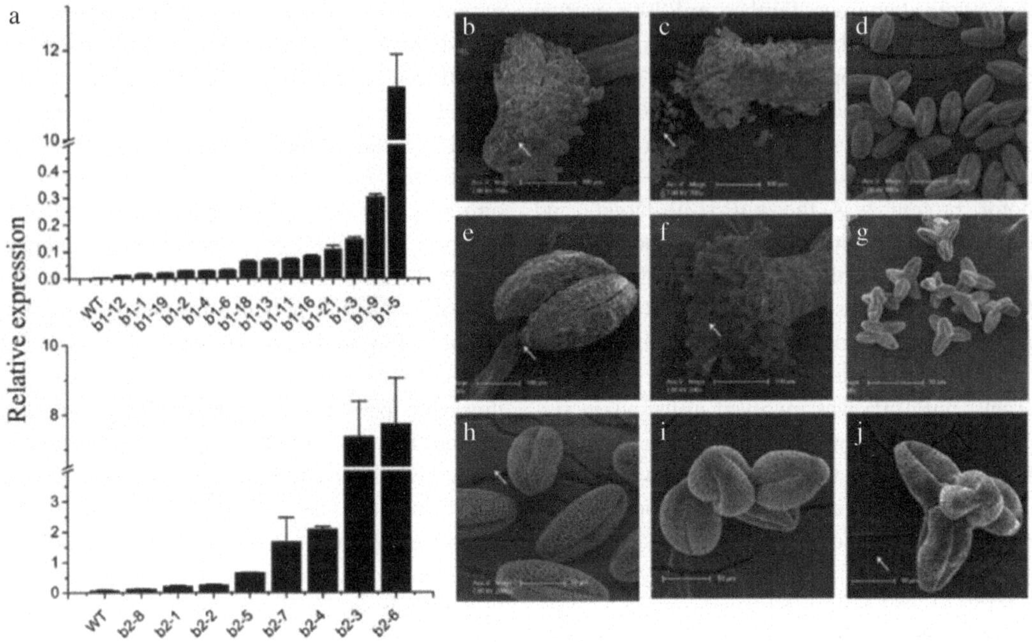

Fig. 9.2 Expression analysis for *Phe-MIRNA159* overexpression Arabidopsis plants and scanning electron microscopy (SEM) for the developing pollens. **a** Expression in *Phe-MIR159* overexpression plants; (**b**, **c**, **d**) anthers, stigma, and pollens in WT, respectively; (**e**, **f**, **g/i-j**) anthers, stigma and pollens in *Phe-MIR159* overexpression plants (b2–3). Gene expression was measured by inflorescence, while pollen and anther patterns were detected by flowers at full-blooming stage. Error bars stand for SD from 3 technical replicates. Data were displayed in the manner of means ± SE from 3 replicates. From Cheng et al. (2020)

15 known miRNAs along with 3 novel miRNAs with differential expression within the flowering, upcoming flowering, and flowering *Ph. edulis*. The expression patterns of miR156 and miR172 in *P. edulis* showed corresponding expression patterns. miR172 was highly expressed in mature stems, while miR156 was poorly expressed. On the contrary, miR172 expression patterns of different members of miR156 family were different. At the same time, miR172 and miR156 have a corresponding way to jointly regulate flowering. Through the analysis of the expression difference of their target genes, miR156 and miR172 have a corresponding relationship with their respective target genes SPL and AP2. Therefore, it is suggested that miR172 and miR156 play an important role in controlling flowering of *P. edulis*.

9.3 miRNAs Could Regulate Vegetative Growth of Moso Bamboo

Xu and colleagues (2014) adopted miRCat to investigate 2 libraries to identify the possible novel miRNAs. Generally, the above algorithm can identify candidate hairpin structures within the genome mapping sequence flanking regions, determine the proportion of miRNA/miRNA* reads of the overall read number mapping to the pre-miRNA hairpin sequence, and examine the sequencing of the desired miRNA* sequence. The authors discovered 7 candidate novel miRNAs based on the former 2 criteria, whose overall normalized read numbers were over 100 in the 2 libraries. For certain conserved miRNAs, they did not have miRNA* sequences in the 2 libraries, including 2 candidate novel miRNAs with no sequenced miRNA* (New-6 and New-7), since the respective miRNA* possibly exist in sRNA libraries in the future. New-1, the candidate novel miRNA, showed great read numbers in the 2 libraries. New-2 was highly expressed in root library, whereas New-3, New-4, and New-5 showed higher abundances in leaf library. The above-mentioned 5 novel miRNAs were verified through Western blotting assays, and they displayed matched expression patterns with those expected from sequencing results. We also adopted the coding sequence database of published bamboo genome to be the input to identify the candidate miRNA targets in bamboo. Numerous mRNAs related to diverse biochemical processes were predicted to be the conserved miRNA targets. Most conserved miRNA targets predicted showed conservation, as verified in additional plant varieties. *Phe*-miR528/397/408/444 and miR1432 accounted for the miRNAs with the highest differential expression degrees between 2 libraries, and the increased abundances were detected in leaves. Of them, *Phe*-miR528/444 and miR1432 have been discovered to be specific to monocot at present. None of the predicted targets was verified in additional monocot varieties. However, for miR528 and miR397, 2 miRNAs with the highest abundances, their candidate targets included some laccase precursor family members. Besides, for miR528 and miR408, their predicted targets included polyphenol oxidase, some plastocyanin-like/multicopper oxidase domain-containing proteins.

Wang and colleagues (2019) built an integrative regulation network of miRNAs, mRNA, and circular RNAs (circRNAs) with differential expression identified from the 84 hub genes, for the sake of generating a candidate regulation network model in the shoot development of moso bamboo. As suggested by the built network, miRNAs targeted several circRNAs or hub genes, including bamboo-miR396-3p and bamboo-miR159a. Meanwhile, various miRNAs might also target circRNAs, including 2 circRNAs from PH01000012:1,644,903–1,645,281 and PH01004819:22,971–32,413. As indicated by the above findings, circRNAs were related to moso bamboo shoot development through the interaction between hub gene and miRNA expression. For better understanding, the relationship of circRNA transcript levels with those of miRNAs, the author conducted transgenic analysis on miR156 overexpression, with the circ-TRF-1 target sites being included (PH01000853:19,599–22,658), and circ-TRF-1 could tolerate digestion by RNase R. As a result, miR156 overexpression markedly reduced the TRF-1 linear transcript and circ-TRF-1 transcript levels, suggesting the possible regulation of circ-TRF-1 via miR156 within moso bamboo. As revealed by the expression patterns, circRNAs might interact with miRNAs for regulating the homologous linear mRNAs, while this was supported by the decreased circ-TRF-1 through miRNA156 overexpression (Fig. 9.3).

miRNAs participate in the fast culm development within moso bamboo. As suggested by anatomical studies, cell division plays an important role in culm development in early periods, while in the middle-late periods, cell elongation plays a dominant role. Those 4 main internal hormones can strongly accelerate culm development. He and colleagues (2013) adopted the RNA-seq technology based on next-generation sequencing (NGS), together with miRNA and mRNA expression profiling to generate a transcriptome as well as a post-transcriptome possessing numerous annotated genes of moso bamboo, which laid the molecular foundation for the sequential elongation of internodes between the base and the top. In addition, some critical pathways, including signal transduction, environmental adaptation, metabolism, translation, and transport, have been suggested to participate in fast bamboo culm elongation. For investigating the miRNA components in small RNAs along with miRNAs dynamic changes in fast moso bamboo culm development, the authors purified culms from G1M-G4M, G3B, and G3T moso bamboo lines and used Solexa high-throughput sequencing to sequence the respective small RNAs. Different

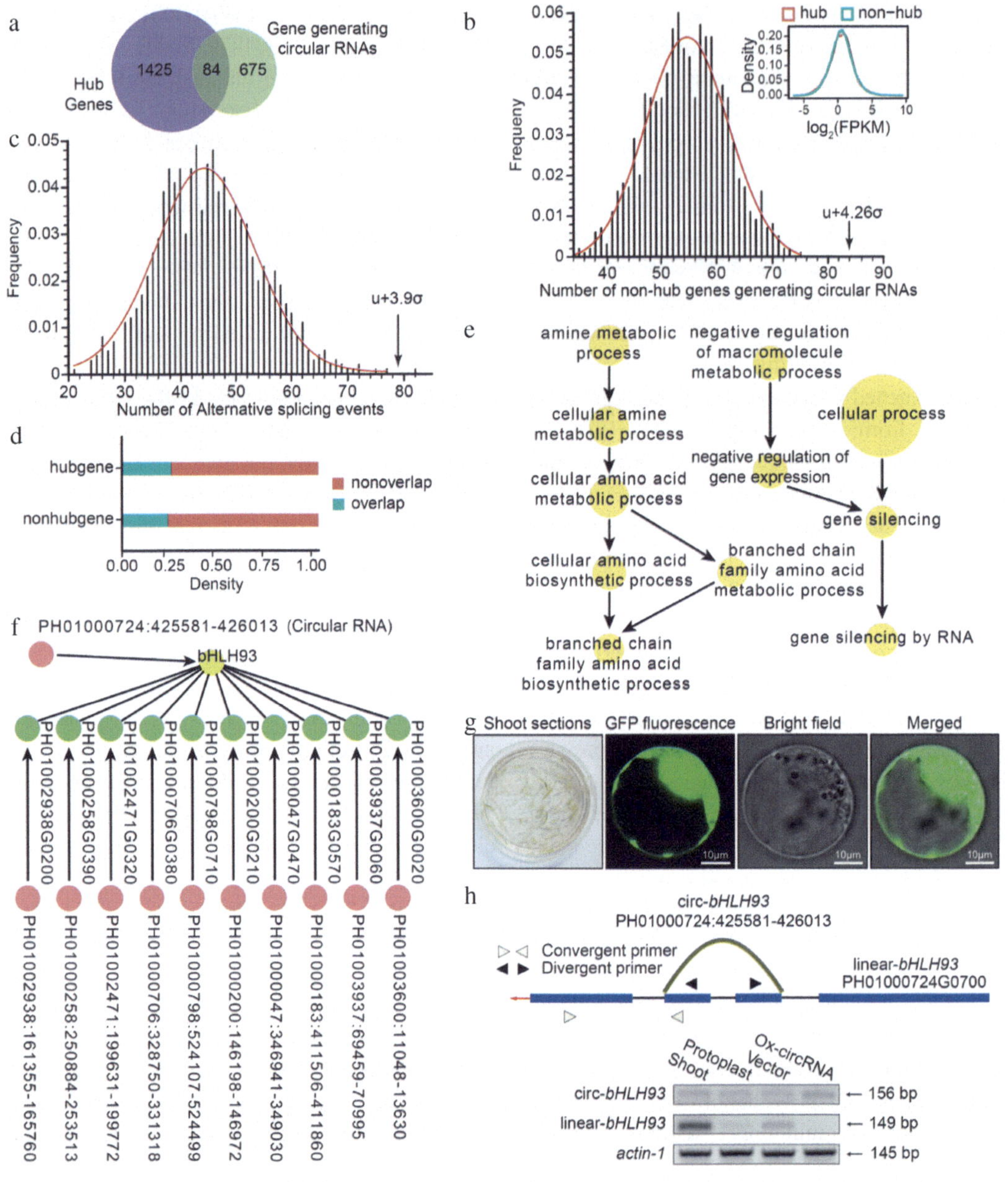

Fig. 9.3 Circular RNAs generated from the hub gene (Wang et al. 2019). **a** Venn diagrams displayed the intersected genes between circRNAs and hub genes. **b** The circRNAs generation frequency by non-hub genes. As observed from the density plot in the top right panel, those 1509 non-hub genes (green) chosen at random with 1000 times showed close expression patterns to hub genes (red). **c** Distribution of AS event frequency in genes chosen at random with no generation of circRNAs. **d** AS origin from non-hub and hub genes that generated circRNAs suggested a decreased frequency of AS events-overlapped circRNAs (green) compared with that of circRNAs beyond the AS events (red). **e** GO annotations of hub genes that generated circRNAs. **f** Visualization for the hub gene network consisting of *bHLH93* (yellow) and the tightly related genes. **g** Protoplast was obtained from the young moso bamboo shoot (left). Confocal laser-scanning microscopy was conducted to check GFP fluorescence in reference construct, which indicated that the circ-*bHLH93*-containing plasmid was successfully transformed. **h** Schematic diagram regarding the parent gene locus of circ-*bHLH93*

from differentially expression genes (DEGs) analysis results, there were more differentially expressed miRNAs in early stages compared with those in the later stages, and more up-regulated miRNAs were detected than the down-regulated counterparts, except for G1M versus G2M. Nonetheless, hierarchical clustering and PCA revealed that miRNAs (n = 165) were consistent with mRNAs (n = 5837); in other words, 6 samples were classified as 2 groups, including G1M/G2M/G3T and G3M/G4M/G3B groups, conforming to results of protein (n = 213) clustering. Moreover, relative to miRNA expression profiles, mRNAs were more consistent with proteins in their expression. After integration/intersection analysis between positive/negative correlations and significant STEM, we acquired 73 pairs, which included 15 proteins, 64 genes, and 55 miRNAs (21 novel and 34 known), and 15 proteins. Among these 73 pairs, 34 (including 27 genes) were significantly correlated ($p \leq 0.05$). GO annotations were performed on 64 genes related to development and reproduction (in particular tissue formation and development of phloem and xylem, leaf morphogenesis), ABA- and auxin-mediated signal transduction, cell processes (like cell division, cell death, and ribosome biogenesis), metabolic processes (such as photorespiration, protein ubiquitination, proteolysis, glycolysis, energy, and glycan metabolism, particularly the metabolism of hemicellulose), bioregulation (including ARF GTPase activity regulation, anthocyanin biosynthesis regulation), along with stimulus responses (cadmium ion and cold). In addition, the pathways involved were shown as follows, metabolism (glycan biosynthesis, lipid/carbohydrate/amino acid/nucleotide/polyketides/terpenoids/glycan/energy metabolism, additional secondary metabolite metabolism), processing of genetic information (translation), processing of environmental information (membrane transport, signal transduction), environmental adaptation (plant circadian rhythm), and cell processes (catabolism, transport).

According to Zhao et al. (2014), miR397, miR1432, and miR7748 were the highly conserved miRNAs specific to moso bamboo. They further determined the numbers of diverse conserved miRNA family members. In moso bamboo, altogether 7 conserved miRNA families had over 3 members. Besides, over 1/2 families had at most 3 members. In the present work, the diverse percentages of lowly conserved miRNAs suggested that a larger number of lowly conserved miRNAs were required for regulating the novel genes, and they facilitate the adaptation of moso bamboo to the dynamic environmental changes. As revealed by qRT-PCR analysis, miR396, miR397, miR1432, and miR7748 showed differential expression in moso bamboo leaves, with miR396 showing the most significant level, whereas miR1432, miR397, and miR7748 ranked the 2nd, 3rd, and 4th places, respectively.

In order to study the role of microRNAs during the process of seed germination and embryo morphogenesis in moso bamboo, the expression pattern of miRNAs during different germination stages was analyzed and verified. The results showed that the *Phe*-miR156k, *Phe*-miR159b, *Phe*-miR160a, and *Phe*-miR396e had a rising expression trend during three stages of the seed germination. Through the association analysis between miRNA and their targets, 4 miRNAs were found that mainly negatively regulate their potential target genes *Phe*-miRNA156-T1, *Phe*-miR156-T2, *Phe*-miR159b-T1, *Phe*-miR160-T1, *Phe*-miR396-T1, and *Phe*-miR396-T2. And these target genes were predicted to be probably participated in the progress of bamboo seed germination, cell differentiation, and the formation of organs and other life activities. In addition, *Phe*-miR396e that with a high expression level in root and its target genes showed simultaneous up-and down-regulation of expression during bamboo seed germination. *Phe*-miR396e maybe achieve control of the process of bamboo germination through inhibit its two target genes which are related to growth factor. These results showed that these four miRNAs possibly exerted vital parts in the germination of bamboo, which laid certain foundation for future research on the bamboo breeding at molecular level (Huang et al. 2017).

9.4 Functions of Moso Bamboo miRNAs in Abiotic Stress Responses

Some miRNAs of *P. edulis* were cloned and expressed, such as miR398, miR408, miR397, and miR1432. The base conservation of miR1432 family was higher than that of miR397 family. The upstream regulatory regions of both bamboo precursors contain the basic acting elements of promoters, such as TATA box and CAAT box. At the same time, there are many stress-related response elements (light, drought, temperature, etc.) and hormone response elements, which mean that *Phe*-miR397 and *Phe*-miR1432 may be regulated by stress and hormones. *Phe*-miR397 and *Phe*-miR1432 expressed the highest abundance in leaf, the lowest abundance *Phe*-miR397 appeared in young stem, and *Phe*-miR1432 in leaf. The expression of *Phe*-miR397 and *Phe*-miR1432 in leaves was down-regulated by the stress of strong light, dark, high temperature, low temperature, and NaCl, while that of *Phe*-miR397 was down-regulated by drought and ABA, while that of *Phe*-miR1432 was upregulated by GA3, while that of *Phe*-miR397 was upregulated by GA3. *Phe*-miR397 and *Phe*-miR1432, as two conserved miRNAs in *P. edulis*, were down-regulated or upregulated in varying degrees under the stress of light, temperature, drought, NaCl, and the treatment of ABA and GA3, suggesting their vital parts in regulating the abiotic stress tolerance in *P. edulis*, which are also associated with internal hormone regulation (Wang et al. 2015).

In order to understand the expression of miR398 and miR408 in *P. edulis*, the precursor sequences of miR398 and miR408 were isolated from the leaves of *P. edulis* and their expression patterns were studied by real-time quantitative PCR. The results showed that the length of miR398 and miR408 precursor sequences in *P. edulis* were 83 bp and 92 bp, respectively. Both of them could form stable stem ring structure, and the mature miRNA sequences (*ped*-miR398 and *ped*-miR408) were located on the 5 'end arm. The expression of *ped*-miR398 and *ped*-miR408 was composed, and the highest expression was found in the leaves of *P. edulis*. The expression of *ped*-miR398 and *ped*-miR408 was up-regulated in leaves treated with strong light, sucrose and GA3, down-regulated in leaves treated with $CuSO_4$ and ABA, upregulated in leaves treated with dark, NaCl and 4 °C, down-regulated in leaves treated with the former and the latter. Therefore, *ped*-miR398 and *ped*-miR408 may play different regulatory roles in the adaptation of *P. edulis* to stress (Li et al. 2017).

This study also analyzed *PheSNAC1*, the stress-associated *NAC1* in *P. edulis*. It is ectopically expressed in Arabidopsis, indicating its role in regulating stress resistance and organ boundaries together with ped-miR164b. NAC (NAM, CUC2, and ATAF1/2) are associated with numerous processes to regulate the growth and stress responses in plants. Altogether, 125 NACs are identified from moso bamboo, yet more studies are needed to examine their functions. More studies should focus on the *ped*-miR164b-targeted *PheSNAC1*. As revealed by tissue-specific expression profiling, *ped*-miR164b was down-regulated gradually from root, sheath, leaf, to stem, while *PheSNAC1* expression exhibited an opposite tread. This study produced transgenic Arabidopsis with *ped*-miR164b (OX-ped-miR164b) or *PheSNAC1* (OX-PheSNAC1) overexpression induced by *CaMV35S* promoter. Of them, OX-ped-miR164b Arabidopsis exhibited close phenotype to *cuc2* mutants, and the development was markedly inhibited. Relative to Col-0, the sense OX-PheSNAC1 Arabidopsis showed rapid growth and early flowering, while the antisense counterparts exhibited slow growth and postponed flowering. The sense OX-PeSNAC1 Arabidopsis had most lateral roots, whereas the antisense OX-ped-miR164b and OX-PeSNAC1 Arabidopsis had decreased lateral roots compared with Col-0. Upon PEG6000 and NaCl stresses, the survival rate increased, whereas the *Fv/Fm* ratio decreased at a slow rate in sense OX-PheSNAC1 Arabidopsis compared with Col-0, whereas opposite results were seen in antisense OX-ped-miR164b and OX-PheSNAC1 Arabidopsis. The above results suggested that *PheSNAC1* targeted by *ped*-miR164b possibly exerted a vital part in plant growth and salt/drought stress resistance (Wang et al. 2016).

References

Cheng Z, Hou D, Ge W et al (2020) Integrated mRNA, microRNA transcriptome and degradome analyses provide insights into stamen development in moso bamboo. Plant Cell Physiol 61(1):76–87

Fan C (2012) Identification of small RNA and its functional analysis by high throughput sequencing. Chinese Academy of Forestry Science

Gao J, Ge W, Zhang Y et al (2015) Identification and characterization of microRNAs at different flowering developmental stages in moso bamboo (*Phyllostachys edulis*) by high-throughput sequencing. Mol Genet Genomics 290(6):2335–2353

Ge W, Zhang Y, Cheng Z et al (2017) Main regulatory pathways, key genes and microRNAs involved in flower formation and development of moso bamboo (*Phyllostachys edulis*). Plant Biotechnol J 15(1):82–96

He CY, Cui K, Zhang JG et al (2013) Next-generation sequencing-based mRNA and microRNA expression profiling analysis revealed pathways involved in the rapid growth of developing culms in moso bamboo. BMC Plant Biol 13:119

Huang Z, Li C, Chen L et al (2017) Research of four miRNAs response during embryo morphogenesis progress in *Phyllostachys edulis*. Chin J Trop Crops 11:19

Istrail S, De-Leon BT, Davidson EH (2007) The regulatory genome and the computer. Dev Biol 310:187–195

Kamthan A, Chaudhuri A, Kamthan M et al (2015) Small RNAs in plants: recent development and application for crop improvement. Front Plant Sci 06:208

Li L, Sun H, Yang Y et al (2017) Cloning and expression analysis of miR398 and miR408 of moso bamboo. J Trop Subtrop Bot 25(03):241–249

Mitsuda N, Ohme-Takagi M (2009) Functional analysis of transcription factors in Arabidopsis. Plant Cell Physiol 50:1232–1248

Samad AFA, Ali NM, Ismail I, et al (2016) Analysis of miRNAs targeting transcription factors in *Persicaria minor* induced by *Fusarium oxysporum*. Faculty of Science and Technology

Wang L, Zhao H, Chen D et al (2016) Characterization and primary functional analysis of a bamboo NAC gene targeted by miR164b. Plant Cell Rep 35 (6):1371–1383

Wang L, Zhao H, Sun H et al (2015) Cloning of miR397 and miR1432 and expression analysis of their responses to stress. For Sci 51(06):63–70

Wang Y, Gao Y, Zhang H et al (2019) Genome-wide profiling of circular RNAs in the rapidly growing shoots of moso bamboo (*Phyllostachys edulis*). Plant Cell Physiol 60(6):1354–1373

Xu P, Mohorianu I, Yang L, et al (2014) Small RNA profile in moso bamboo root and leaf obtained by high definition adapters. PLoS One 9(7):e103590

Zhao H, Wang L, Dong L, et al (2014) Discovery and comparative profiling of microRNAs in representative monopodial bamboo (*Phyllostachys edulis*) and sympodial bamboo (*Dendrocalamus latiflorus*). PLoS One 9(7):e102375

Mitochondrial Genome of Bambusicolous Fungus- *Shiraia bambusicola*

10

Cheng-Lin Hou and Xiao-Ye Shen

Abstract

Shiraia bambusicola represents one of the pathogenic fungi in bamboo, whose fruiting bodies have been used in folk medicine. It could cause bamboo diseases and even destroy bamboo forest severely. The phylogenetic position and host range of *S. bamabusicola* have been disputed for a long time, and the complete mitogenomic DNA would assist to resolve the outstanding issue and improve the understanding of the relevant host, habitat, bioactive substance, and utility. By comparative genomics, *S. bambusicola* shows particular characteristics unlike the rest Pleosporales species, in particular within those coding regions surrounding *rns* (*trnR-trnY*). In addition, the mitogenome gene orders are compared in Pezizomycotina, which discovers that each encoded gene was located in the identical strand in mtDNAs of many Pezizomycotina species. To be specific, Dothideomycetes genes show various directions, gene locations, and colocalization (including *cox1-cox2*, *cox3*, *nad2–nad3*). Such differences have been deemed to be the class-specific characteristics. It is interesting that, there are 2 partial *atp6* gene copies in the diverse mtDNA strands, which is not found in additional fungi analyzed before fungal species. This part first thoroughly examined the Dothideomycetes mitogenomes, where numerous non-studied species were included.

C.-L. Hou (✉) · X.-Y. Shen
College of Life Science, Capital Normal University, Beijing 100048, China
e-mail: chenglin-hou@cnu.edu.cn

X.-Y. Shen
e-mail: shenxiaoye2009@cnu.edu.cn

10.1 Introduction

Shiraia bambusicola represents a vital pathogenic fungus in bamboo, which can cause bambusicolous disease and negatively affect plant development. *S. bambusicola* shows high specificity, and it is generally confined to infection of *Brachystachyum densiflorum* as well as the associated species across China or the *Bambusa* species across Japan (Li et al. 2000). Noteworthily, its fruiting bodies have been extensively utilized across southern China to mitigate inflammation, sciatica, and apoplexy. There are abundant bioactive components in its mycelia and sporophores, like hypocrellins, and they can serve as the photosensitizers with the effects of antibiotics, anticancer, anti-inflammation, and antivirus.

10.2 Mitogenome Summary

Mitochondrion is a vital cellular organelle, which exerts an important part in energy production. Mitogenome is originally derived from

J. Gao (ed.), *The Moso Bamboo Genome*, Compendium of Plant Genomes,
https://doi.org/10.1007/978-3-030-80836-5_10

the α-proteobacteria-originating DNA transposable elements within the eukaryotic host, even though many genes that encoded mtDNAs are transported to nuclear chromosomes in the evolutionary process. Mitogenomes have great copy number, fact evolutionary rate, and absent recombination; as a result, they have been identified to be the efficient markers to study evolution of fungi. Mitogenomics in fungi substantially improve recently after the novel sequencing technology is applied, making mitogenomes available, which contributes to solving different problems in terms of the evolutionary history. To take an example, the greatest fungal mitogenome from *Agaricus bisporus* was reported in February 2013, which was 135,005 bp long (Ferandon et al. 2013). After few months, an mitogenome from *Rhizoctonia solani* was reported, which was 235,849 bp long (Losada et al. 2014). Trans-splicing was originally detected in the organelles of *Gigaspora margarita* through analyzing the entire mitogenome sequence (Pelin et al. 2012). Similarly, the mitochondrial group I introns of *Gigaspora rosea* exhibited an analog phenomenon, which presented the special characteristic, namely the impact of the third helper RNA fragment on trans. Research on *Pneumocystis jirovecii*, the vital opportunistic pathogen related to AIDS or immunodeficiency disorders, suggests that mitogenomic genes across the tightly associated species are arranged in a unique manner. Therefore, in-depth study on mtDNA sequences may make it possible to identify novel therapeutic targets. Individual analysis on gene arrangement in orders Sordariales (van de Sande 2012) and Helotiales (Youssar et al. 2013) indicates significant difference in conservation patterns; in addition, mitogenomes in fungi show significant variations in and between the main fungal phyla with regard to the gene order, which is found through comparing 38 full-length mtDNA sequences previously reported (Aguileta et al. 2014).

10.2.1 General Description of Dothideomycetes

There are over 1900 species in class Dothideomycetes, and this class has been recognized to be the greatest one with the greatest diversity of Ascomycete fungal class (Schoch et al. 2009). In taxonomy, those species can be divided as 11 orders, including Dothideales, Capnodiales, Hysteriales, Myriangiales, Mytilinidiales, Jahnulales, Botryosphaeriales, Trypetheliales, Pleosporales, Patellariales, and Microthyriales. Among them, Pleosporales represents the order with the greatest fungal order diversity in Dothideomycetes, which takes up 1/4 of the dothideomycetous species. These species appear across a variety of habitats, and 105 generic types have been identified through multigene phylogenetic analysis.

There are over 165 fungal mitogenomes imported to the NCBI database; most sequences are obtained from Ascomycete mitochondria, and only 2 from class Dothideomycetes have been officially published, including one from Pleosporales (Hane et al. 2007) and the other from Capnodiales (*Mycosphaerella graminicola*) (Torriani et al. 2008).

10.2.1.1 Fast Development of Mitogenomes

More and more fungal mitogenomes are completely or incompletely sequenced. Such published sequences can possibly accelerate the development of breeding, genetics evolution, and classification engineering for respective mycetes. The present work analyzed the new mitogenomic sequence in *S. bambusicola*, and 4 other known sequences not reported previously. Each Dothideomycete species for reference was the known plant pathogen obtained from a variety of hosts, like *Phaeosphaeria nodorunm* from wheat, *Leptosphaeria maculans* from oilseed rape, or *Bipolaris maydis* from southern corn. These sequences attracted wide attention because they affected the forestry and agriculture industries. Besides, it was discovered that hypocrellins from *S. bambusicola* (Dothideales, Dothideomycetes) possessed a unique perylenequinonoid compound structure, which was also detected from *Cladosporium phlei* (Capnodiales, Dothideomycetes) as phleichrome, *Elsinoë* spp. (Myriangiales, Dothideomycetidae) as elsinochromes, as well as *Cercospora* spp. (Capnodiales, Dothideomycetes) as cercosporin. Comprehensive analyses on mitogenomes may help to understand the evolutionary biology of fungi and shed morel lights on Dothideomycete pathogens-induced infection and toxicity.

10.3 Definition of *S. bambusicola* and Its Distribution

10.3.1 Variational Definition of *S. bambusicola*

During the more than 100-year taxonomic study, the *S. bambusicola* position is reclassified repeatedly. Genus *Shiraia* was originally identified to belong to Nectriaceae, Hypocreales, and Pyrenomycetes in 1900. After 2 years, *Shiraia* was identified to belong to Hypocreaceae family according to the larger fleshy stroma base. Such opinion was then accepted for few decades, till it was discovered that the ascus was not unitunicate but bitunicate; since then, *Shiraia* was reclassified into class Loculoascomycetes, order Pleosporales, and family Pleosporaceae. It was found from the fungal dictionary (9th edition) that, *Shiraia* belonged to Dothideales species, but its family affiliation was not determined yet. Using DNA sequencing combined with morphological evidence, recent articles have been conducted to phylogenetically analyze *Shiraia* for its taxonomic position. To be specific, the ITS-5.8S rDNA and 18S rDNA regions were sequenced, which suggested the affiliation of genus *Shiraia* to Phaeospheriaceae, Pleosporales. In addition, Liu and colleagues (2013) constructed a novel family Shiraiceae in Pleosporales, which accommodated *Shiraia* in line with its incomplete 28S nrDNA nucleotide (nt) sequence. Noteworthily, just one typical species is available in this group, no matter whether it is classified into a family or a genus; besides, differences across fungal isolates obtained from diverse bamboo hosts were not significant.

10.3.2 Expansive Distribution of *S. bambusicola*

As a well-known bambusicolous fungus in East and Southeast Asia, *S. bambusicola* usually appears in the southern provinces of China, such as Jiangsu, Zhejiang, Anhui, Jiangxi, Hubie, Hunan, Henan, and Sichuan. In the later reports, Yunnan and Guizhou also display the corresponding fruit bodies from bamboos (Zhao and Liang 2005; Xu et al. 2005). As to Japan, the neighbor and close country of China, there is also the similar distribution of *S. bambusicola.* It is noticeable that the *Shiraia* species were discovered in the Guangxi Zhuang Autonomous Region, where they only emerged as endophytes, not the forementioned sporophores or pathogens (Shen et al. 2014).

10.4 Fascinating Properties of *S. bambusicola*

10.4.1 Exceptional Bambusicolous Fungus and Its Special Host

From the older materials, there was tentatively a broad range of the hosts associated with *S. bambusicola*, including *Phyllostachys* Sieb. et Zucc., *Pleioblastus* Nakai, *Bambusa* Schreber, *Fargesia* Franch, and *Indosasa shibataeoides* Mcclure. However, Lai et al. (Lai and Fu 2000) had taken the all reports of strains from different southern parts of China into consideration and found that *S. bambusicola* always limited to some members of *Brachystachyum* Keng group, which covered *B. ensiflorum* (Pendle) Keng, *B. ensiflorum* var. *villosum* S. L. Chen et C. Y. Yao, *B. albostriatum* G. H. Lai, and four other species without official names.

Furthermore, *Shiraia* sp. strains are separated from moso bamboo (*Phyllostachys edulis*) and cultured. Moso bamboo belongs to Bambusoideae

(Poaceae), and it represents a special monopodial vegetative bamboo species that is originally grown in subtropic areas across China (Shen et al. 2014). *Shiraia* sp. can generate high levels of hypocrellin A, unlike additional reported strains. It is interesting that, its fruiting bodies are usually observed in bamboos across China, which suggest corresponding symptoms. However, the fruiting bodies are not seen in *P. edulis* or additional Phyllostachys spp.. *Shiraia* sp. is suggested to live on a variety of bamboos as a kind of asymptomatic endophyte with no production of fruiting bodies, because of the restricted conditions like host structure and nutrition.

The incomplete taxonomic system has inhibited the identification of the species from Bambusoideae and, especially, blocked the development of the hosts around *S. bambusicola.* In addition, the misunderstandings of *Engleromycetes goetzii* P. Henn. and *Hypocrella bambusae* (Berk. et Broome) Sacc. would also make trouble to the judgment of these host plants. Thereof the evaluation of the bamboo hosts for other genus should receive further supports in the future study.

10.4.2 Amazing Function of Natural Products

For few centuries, the *S. bambusicola* fruiting bodies are used in folk medicine for the treatment of pertussis, rheumatic arthritis (RA), skin lesions, and tracheitis. Hypocrellins, as the main secondary metabolites of this strain, played an important role in the traditional therapy and were well-known as the new generation of photosensitizers. They were at first explored from *Hypocrella bambusae* and now mainly extracted from *S. bambusicola.* As so far four different components, hypocrellin A, B, C, D have appeared in this group, and they displayed various bioactivities in the medicinal, food, and agricultural industries. Hypocrellin A (HA) is the most striking constituent of them and has been revealed to possess excellent light-induced antimicrobial, antiviral, and anticancer activities. It can also significantly kill the HIV virus. These activities of HA belong to typical properties of photodynamic therapy (PDT). Some studies utilized Poly (lactic-co-glycolic) acid nanoparticles or synthesized a targeting drug delivery system (TDDS) to solve the difficulty of poor water-solubility of HA, enhanced the ability of PDT of HA. In other word, although hypocrellin B (HB) has a similar structure with HA, the difference of one hydroxyl group predicted to play the special role. By inducing apoptosis of cancer cells and inhibiting cell viability, HB or HB derivatives were proved to take special part in antitumor activity. For instance, HB-LED PDT has been found to significantly induce KFB apoptosis while reducing their viability.

It is noteworthy that one group of perlenequinonoid compounds are usually limited to one speices, such as Cercosporins from *Cercospora.* However, *S. bambusicola* strangely could cover diverged groups (Tong et al. 2017). In some subgroups, their crude extracts are not only rich in hypocrellins, but also contained the certain content of elsinochromes. Hypocrellin A, B, C, D, and elsinochrome A, B, C, D have separate structure and demonstrated various bioactivities. Once several agents co-occurred in the final products, it would be no longer suitable of the previous assay to only focus on hypocrellin A or crude perlenequinonoid compounds.

10.5 Mitogenome Structure of *S. bambusicola*

10.5.1 Small Mitogenome Size

The *S. bambusicola* mitogenome was sequenced by adopting the Illumina Hiseq 2000 system. Then, PCR amplification was performed to assemble 8 scaffolds in the representative circular DNA for the successful spanning of each gap. As a result, the *S. bambusicola* mitogenome showed a circular DNA molecule that was 39,030 bp long, the smallest one among all the tightly associated species. In addition, those sequenced Dothideomycetes mitogenomes had significantly different sizes, between 39,030 bp

(*S. bambusicola*) and more than 154,863 bp (*L. maculans*). For *S. Bambusicola*, it has the smallest mitogenome size across all the mtDNAs of Dothideomycetes analyzed, which included orders Botryosphaeriales, Capnodiales, and Pleosporales. Such tremendously altered lengths were mostly associated with the existence of hypothetical proteins, along with changes in intergenic regions and introns; for instance, there were 3 ORFs and 1 intron within the *S. bambusicola* mitogenome intergenic region.

10.5.2 Explicit Sequence Structure

AT was abundant in the sequence, and the total G + C content was just 25.19% (Table 10.1). The G + C content in protein-encoding gene regions was 27.1%, while that in RNA genes was 35.4%. Generally, the *S. bambusicola* mitogenome was compact, and coding regions were contained in 68.96% of its genome.

The protein-encoding gene regions took up 47.33% of the entire mitogenome, which included 17 genes that encoded proteins. Such genes were responsible for encoding the complex F0 subunit of ATP synthase (*atp6*), 7 complex I subunits for electron transport (*nad1-6*, *nad4L*), 1 complex III subunit (*cob*), 3 complex IV subunits (*cox1-3*), 4 putative proteins (*orf250*, *orf262*, *orf322* and *orf352*), and 1 ribosomal protein (*rps3*) (Fig. 10.1) (Table 10.2). On the other hand, such genes existed in the 2 strands unbiasedly. There were 13 typical mitochondrial genes related to the respiratory chain complexes (OXPHOS), which were highly conserved in their sequence relative to additional filamentous fungal species. To be specific, *cox1* was located near *cox2* with no intergenic region, while the ATG initiation codon in *nad5* was located immediately behind the termination codon in *nad4L,* and one base was intersected.

Those tracks stand for: (1) Forward tRNA and CDS; (2) Reverse tRNA and CDS; (3) %GC plot (Yellow indicates >50%, Violet indicates <50%); (4) GC skew [(G − C)/(G + C)] (Yellow indicates + , Violet indicates -). Genes were colored based on the respective functional groups, with color schemes being presented on the left side of the circle. There was an intron (IB) in cox1. tRNA genes were labeled by brackets; at the same time, anticodons were appended to gene names. Table 10.2 lists the exact locations of introns and genes.

There were 3 ORFs (*orf250*, *orf262*, *orf322*) identified from intergenic regions, whereas one (*orf352*) from intron. Many ORFs were detected from the negative strand, except for *orf352* that

Table 10.1 General features in the mitogenome of *Shiraia bambusicola*

Genomes features	Value
Genomes size (bp)	39,030
G + C content (%)	25.19
No. of protein-coding genes	17
G + C content of protein-coding genes (%)	27.1
Structural proteins coding exons (%)	47.33
No. of rRNAs/tRNAs	2/32
G + C content of RNA genes (%)	35.4
rRNAs + tRNAs (%)	18.92
Coding regions (%)	68.96
Intergenic regions (%)	30.51
No. of introns	1
No. of intronic ORFs	1
Introns (%)	3.24

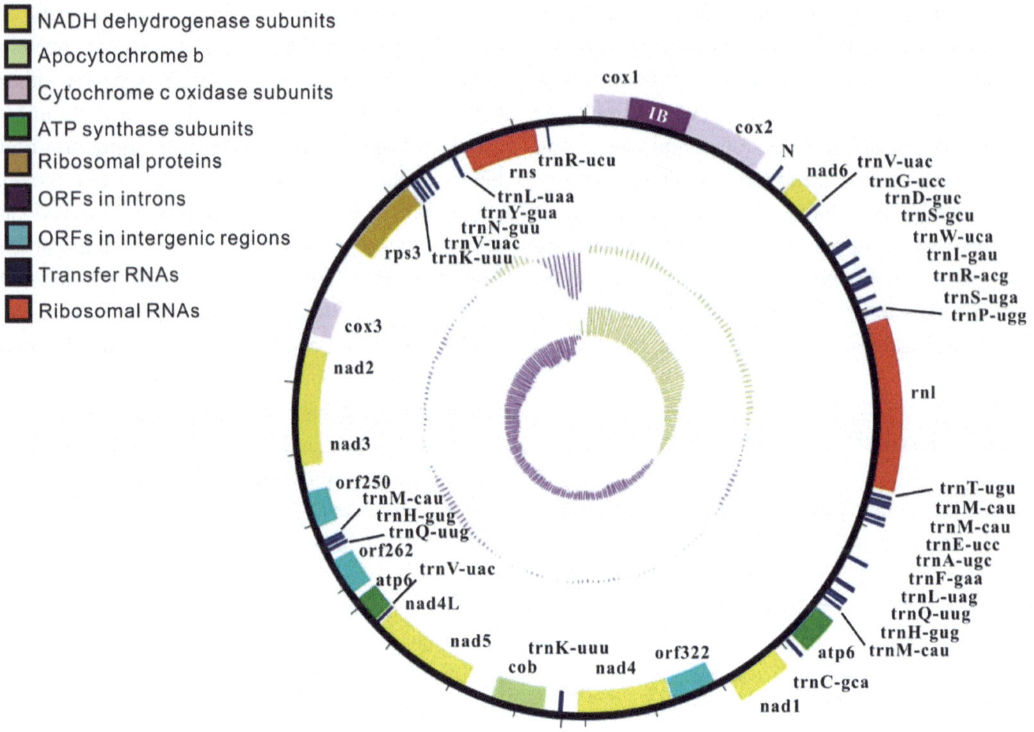

Fig. 10.1 Circular mapping for the complete *S. bambusicola* mitogenome

was detected from the *cox1* intron. Remarkably, just 1 group I intron was found throughout the whole mitogenome, and just 1 intronic ORF that encoded a protein belonging to the LAGLI-DADG endonuclease family was found, which was highly similar to additional filamentous fungal species, such as *Candida oxycetoniae* and *Talaromyces marneffei*. As for the *orf322*-encoded putative protein, it was partially homologous to YP 001427397, the domain-containing ribosomal S5-like protein belonging to the SNF7 superfamily from *P. nodorum* (alignment range, 31–314 aa; length, 323 aa; e-value, 0.24; identity, 67/291 (23%)). Notably, *orf262* protein showed slight similarity at protein aa level to an undefined protein product (accession No., XM 003024124). It was also discovered from the nuclear genome in *Trichophyton verrucosum*, the zoophilic dermatophytes (alignment range, 81–141 aa; length, 263 aa; e-value, 1.0; identity, 27/75 (36%)). On the other hand, *orf250* protein comprised 251 aa, and it was not similar to the rest fungal proteins, but was mildly analogous to the putative protein in *Thiohalocapsa* sp., as revealed by BLASTx (alignment range, 29–110 aa; e-value, 2.4; identity, 45/83 (54%)).

In the study on codon use frequency, the candidate mitochondrial genes (*atp6*, *cox1-3*, *nad1*-6, *cob*, *nad4L*, *rpS3*, *orf250*, and *orf262*) were used to encode the functional proteins. We utilized the *orf352* sequence for evaluating the codon use in intronic genes. As a result, the "AUG" initiation codon showed the highest use frequency (Table 10.3), except for *atp6* with "AUU" and *cox2* with "AAU." *orf352* was begun with the unique "UUG" codon, and it was not detected from additional genes. Besides, the UGA codon terminated the *cox1* coding region, while UCC codon terminated that of *cob*, the UAG codon terminated those of *cox2* and *nad4*, with UAA being the termination codon utilized in the remaining genes, and the intronic *orf352* was included.

Table 10.2 Gene organization of the mitogenome

Gene	Start position	Stop position	Length (nt)	Length (aa)	Start Codon	Stop codon
cox1	162	3046	2885	526	ATG	TGA
cox2	3047	3793	747	249	AAT	TAG
trnN	4169	4239	71			
nad6	4563	5129	567	189	ATG	TAA
trnV	5201	5273	73			
trnG	6116	6188	73			
trnD	6191	6262	72			
trnS	6502	6581	80			
trnW	6761	6832	72			
trnI	6934	7005	72			
trnR	7010	7081	72			
trnS	7288	7372	85			
trnP	7564	7636	73			
rnl	7808	11,191	3384			
trnT	11,299	11,369	71			
trnM	11,392	11,462	71			
trnM	11,468	11,540	73			
trnE	11,719	11,791	73			
trnA	11,824	11,895	72			
trnF	12,675	12,747	73			
trnL	13,201	13,283	83			
trnQ	13,459	13,530	72			
trnH	13,534	13,607	74			
trnM	13,655	13,726	72			
atp6	13,946	14,719	774	258	ATG	TAA
trnC	14,824	14,895	72			
nad1	15,187	16,302	1116	372	ATG	TAA
rps5	17,580	16,612	969	323	ATG	TAA
nad4	19,667	17,622	2046	682	ATG	TAG
trnK	20,096	20,002	95			
cob	21,573	20,416	1158	386	ATG	TCC
nad5	24,190	22,205	1986	662	ATG	TAA
nad4L	24,459	24,190	270	90	ATG	TAA
trnV	24,563	24,491	73			
atp6	25,228	24,599	630	210	ATT	TAA
orf262	26,126	25,338	789	263	ATG	TAA
trnM	26,415	26,344	72			
trnH	26,538	26,465	74			
trnQ	26,613	26,542	72			

(continued)

Table 10.2 (continued)

Gene	Start position	Stop position	Length (nt)	Length (aa)	Start Codon	Stop codon
orf250	27,677	26,925	753	251	ATG	TAA
nad3	29,026	28,259	768	256	ATG	TAA
nad2	30,778	29,027	1752	584	ATG	TAA
cox3	31,883	31,074	810	270	ATG	TAA
orf564	34,882	33,188	1695	565	ATG	TAG
trnK	35,050	34,979	72			
trnV	35,152	35,080	73			
trnN	35,288	35,218	71			
trnY	35,468	35,384	85			
trnL	36,120	36,038	83			
rns	37,944	36,333	1612			
trnR	38,242	38,172	71			

As observed from Table 10.3, isoleucine and serine were the top 2 aa with the highest use frequency among those 17 protein-encoding genes. It was observed from Table 10.1 that, mitochondrial genes showed strong bias to codons with AT (72.9%) and preferred to A and U residues, conforming to the total codon use. To be specific, "A" and "U" were exclusively included in codons with the highest use frequency, as shown below, UAA (7.03%), AAA (5.29%), UAU (4.92%), UUU (4.90%), AUU (4.84%), AUA (4.43%), AAU (4.37%), and UUA (2.89%). On the other hand, Gs and Cs mostly comprised codons with the lowest use frequency as follows, (Table 10.3): CCG (0.03%), UCG (0.05%), CGG (0.06%), CGA (0.08%), GCG (0.10%), and GCC (0.11%).

Altogether 32 tRNAs and 2 rRNAs were discovered from the genome through comparing ARWIN, tRNAscan-SE, AGAGORN, BLAST, and RNAWEASEL with additional fungal mitogenomes, which corresponded to the entire 20 aa (Table 10.4). Figure 10.1 presents the ideogram describing gene classification and genomic organization. The genes were found in both strands.

There were 32 tRNAs (length, 70–94 bp) discovered from *S. bambusicola* mitogenome, which were broadly classified as 3 groups (Fig. 10.1). These genes harbored codons for the entire 20 aa, among which, some were presented in the form of multiple tRNAs (Table 10.4). Altogether 4 *trnM-CAU* tRNA gene copies were obtained for methionine, while 3 for valine with

Table 10.3 Codon usage of protein-coding genes in the mitogenome of *Shiraia bambusicola*

AA		Codon	%
Stop	End	UAA	7.03
K	Lys	AAA	5.29
Y	Tyr	UAU	4.92
F	Phe	UUU	4.9
I	Ile	AUU	4.84
I	Ile	AUA	4.43
N	Asn	AAU	4.37

(continued)

Table 10.3 (continued)

AA		Codon	%
Stop	End	UAA	7.03
L	Leu	UUA	2.89
R	Arg	AGA	2.79
F	Phe	UUC	2.74
S	Ser	AGU	2.55
I	Ile	AUC	2.47
Y	Tyr	UAC	2.42
T	Thr	ACU	2.37
C	Cys	UGU	2.29
W	Trp	UGA	2.29
S	Ser	UCU	2.18
T	Thr	ACC	2.14
H	His	CAU	2.03
N	Asn	AAC	2.01
S	Ser	AGC	1.95
E	Glu	GAA	1.77
L	Leu	UUG	1.69
V	Val	GUU	1.66
M	Met	AUG	1.62
S	Ser	UCC	1.38
V	Val	GUA	1.38
T	Thr	ACA	1.35
R	Arg	AGG	1.33
A	Ala	GCU	1.3
C	Cys	UGC	1.28
K	Lys	AAG	1.28
G	Gly	GGU	1.19
S	Ser	UCA	1.15
L	Leu	CUU	1.12
D	Asp	GAU	1.09
Stop	End	UAG	1.02
Q	Gln	CAA	0.97
P	Pro	CCU	0.78
G	Gly	GGA	0.71
A	Ala	GCA	0.7
H	His	CAC	0.55
L	Leu	CUC	0.55
W	Trp	UGG	0.55
T	Thr	ACG	0.45
P	Pro	CCC	0.42
P	Pro	CCA	0.41
V	Val	GUG	0.41

(continued)

Table 10.3 (continued)

AA		Codon	%
Stop	End	UAA	7.03
G	Gly	GGG	0.37
G	Gly	GGC	0.37
V	Val	GUC	0.36
L	Leu	CUA	0.29
L	Leu	CUG	0.23
R	Arg	CGU	0.23
D	Asp	GAC	0.19
Q	Gln	CAG	0.16
R	Arg	CGC	0.16
E	Glu	GAG	0.13
A	Ala	GCC	0.11
A	Ala	GCG	0.1
R	Arg	CGA	0.08
R	Arg	CGG	0.06
S	Ser	UCG	0.05
P	Pro	CCG	0.03

identical anticodon (*trnV-UAC*). Besides, there were 2 distinct tRNA genes for serine (*trnR-UGA* and *trnR-GCU*), arginine (*trnR-ACG* and *trnR-UCU*), and leucine (*trnL-UAA* and *trnL-UAG*). Moreover, 2 copies of *trnQ-UUG*, *trnN-GUU*, *trnK-UUU*, and *trnH-GUG* were detected within diverse regions; meanwhile, just 1 copy was discovered in the rest 11 tRNA genes. Noteworthily, 2 copies of 3 consecutive genes (*trnH-GUG* and *trnM-CAU* and *trnQ-UUG*) were isolated by the great gaps in diverse DNA strands. According to tRNAscan, each tRNA showed the typical cloverleaf structure.

Table 10.4 tRNAs in the mitogenome of *Shiraia bambusicola*

AA	Anticodon	Numbers
Ala	UGC	1
Arg	ACG	1
Arg	UCU	1
Asn	GUU	2
Asp	GUC	1
Cys	GCA	1
Gln	UUG	2
Glu	UUC	1
Gly	UCC	1
His	GUG	2
Ile	GAU	1
Leu	UAG	1
Leu	UAA	1
Lys	UUU	2
Met	CAU	4
Phe	GAA	1
Pro	UGG	1
Ser	GCU	1
Ser	UGA	1
Thr	UGU	1
Trp	UCA	1
Tyr	GUA	1
Val	UAC	3

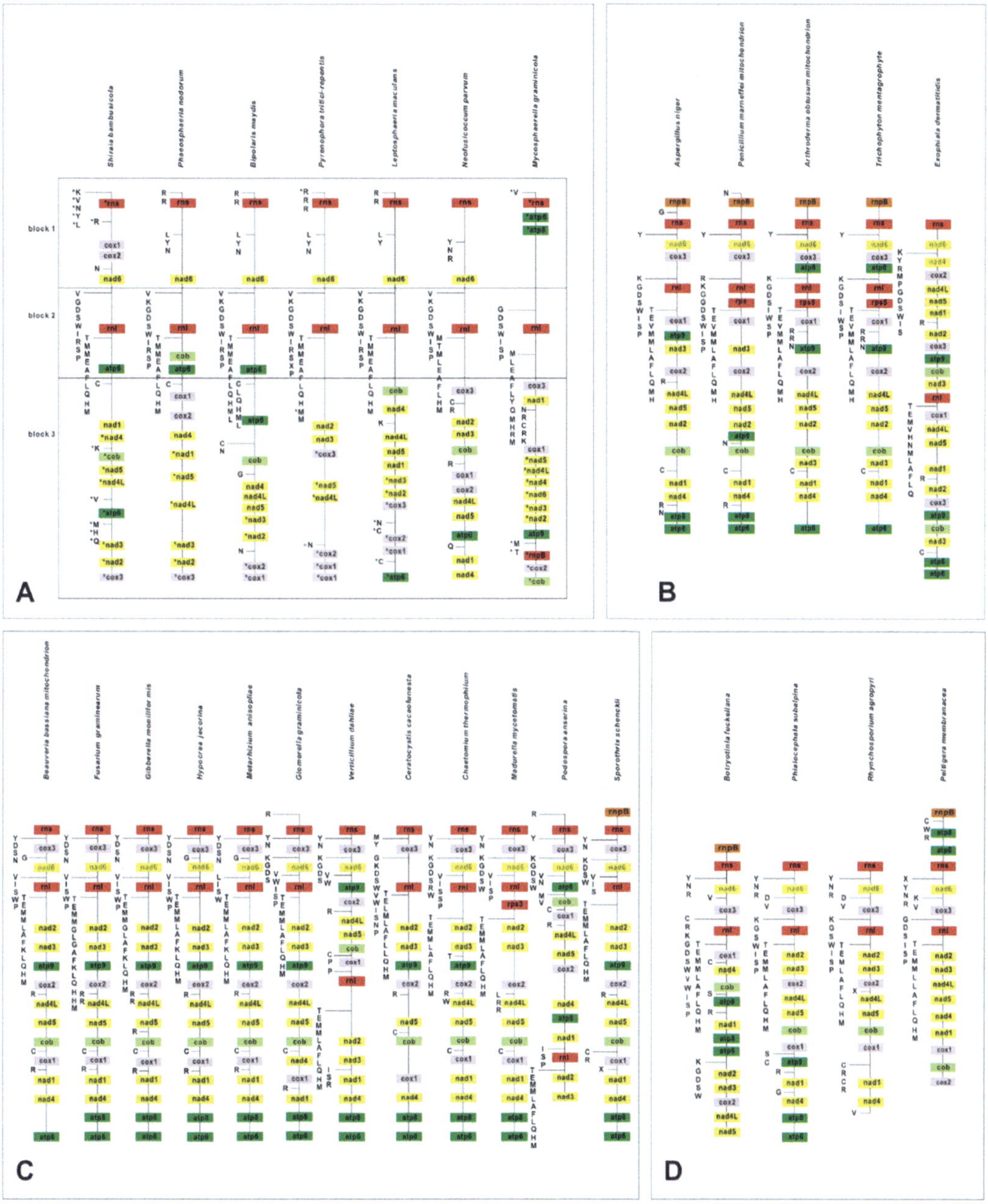

Fig. 10.2 Mitochondrial gene orders of 29 fungal species from 4 classes

10.6 Interesting Genes from Mitogenome

10.6.1 Special Gene Order

Mitogenomes can serve as the efficient approach for driving the commonly seen evolutionary route for fungi, and they experience complex rearrangements. This study compared the gene order between *S. bambusicola* and Dothideomycetes species with thoroughly characterized mitogenomes. Altogether 7 mitogenomes were chosen in the typical species, including 5 from Pleosporlaes (*B. maydis*, *L. maculans*, *P. tritici-repentis*, *P. nodorum*, *S. bambusicola*), 1 from Capnodiales (*M. graminicola*), while 1 from Botryosphaeriales (*N. parvum*). According to Fig. 10.2, protein-coding genes were sequenced, which indicated the presence of significantly conserved regions; meanwhile, Dothideomycetes species displayed remarkable synteny in gene order, in particular between the Pleosporlaes species.

Those 4 classes included Dothideomycetes (A), Eurotiomycetes (B), Sordariomycetes (C), and Leotiomycetes (D). All genes were colored according to the respective functional groups observed from Fig. 10.1, whereas tRNA gene positions were exhibited by the corresponding 1-letter amino acid (aa) codes. Noncoding regions (NCRs) were not shown. Asterisk (*) stands for the reverse direction of gene transcription.

For *S. bambusicola*, its gene order within mitogenome was detected to identify 3 typical regions, including blocks 1–3 (*nad6-atp6*, *nad1-cox3* and genes surrounding *rns*, respectively) (Fig. 10.1). With regard to block 1 genes, they had less diverse gene rearrangements relative to the remaining Dothideomycetes species. On the contrary, block 2 genes experienced complicated rearrangements across diverse species, along with different directions and patterns of relative gene positions (*nad4-nad1-nad5-nad4L*). Noteworthily, block 3 genes were arranged in a unique manner different from those 4 surrounding species belonging to the identical order. Additionally, certain genes clustered together, which indicated that they were strongly related to each other. For instance, the *nad2–nad3* gene pair was always related among those 7 examined mitogenomes; at the same time, among the 6 species, the *cox1-cox2* gene pair was not modified (Fig. 10.2a).

For investigating the presence of an analogous pattern within additional tightly associated classes, we chose the Eurotiomycetes, Sordariomycetes, and Leotiomycetes species to analyze the genomic synteny. Mitogenomes of Pezizomycotina have been thoroughly or almost thoroughly characterized, which are restricted to 3 classes, consistent with our results from Dothideomycetes. In addition, the gene order showed close synteny among Chaetothyriales, Onygenales, and Eurotiales (Fig. 10.2b). Similar phenomenon was also seen when Sordariomycetes species were also enrolled for analysis (Fig. 10.2c). As for gene arrangement, there was no significant difference between Sordariomycetidae and Hypocreomycetidae subclasses, except for the *Podospora anserine* and *Verticillium dahliae* with special positions of *nad1*, *nad4*, *nad4L*, *nad5*, *cob*, *cox1*, *cox2*, *atp6,* and *atp8.* For 4 Leotiomycetes species, the arrangements of their mitogenome genes were highly conserved within block 1 (*cox3- trnM*) as well as block 3 (genes surrounding *rns*) genes, but block 2 genes displayed an unstable order in comparison with diverse species (Fig. 10.2d). Interestingly, genes in many known mtDNAs of Pezizomycotina were encoded into one strand, except for Dothideomycetes species, while the different *cox3* locations might serve as the class-specific characteristic. Of those 29 mitogenomes in 4 classes, Sordariomycetes and Eurotiomycetes exhibited the highest gene arrangement conservation level. The 4 mitogenomes in Leotiomycetes revealed 3 gene order patterns in block 2 genes. In the Dothideomycetes mtDNAs, block 2 genes were not regularly arranged; in the meantime, certain species exhibited specific positions of genes surrounding *rns*. Moreover, mitogenomic genes were distributed among diverse strands, which added to the mtDNA sequence complexity. A larger number of

Dothideomycetes mitogenomes are necessary to carry out in-depth study, but the current public data reveal the diverse mitochondrial gene orders within Dothideomycetes class compared with Pezizomycotina species.

10.6.2 Introns and the ORFs in Introns

As far as Dothideomycetes mitogenomes are concerned, their intronic elements are variable in their sequences, with irregular frequency of intron insertions within coding genes. The intron number and length, as well as intronic region content, were significantly different, especially in the ORFs-encoding regions.

Many mitogenomes in fungi are sequenced at present, which possess one or more group I introns along with a small number of group II introns. According to our results, *P. tritici-repentis* had the most mitochondrial introns (n = 39) in Pezizomycotina subphylum (where each of the published mitogenomes was included), whereas *M. graminicola* is at present the sole filamentous fungal species with complete lack of mitochondrial introns.

For *S. bambusicola*, its sequence of *cox1* gene contains one intron only; meanwhile, insertions of mobile element homologs have been detected at an analogous location (*cox1*) in *B. maydis*, *L. maculans*, *N. parvum,* and *P. tritici-repentis* UCRNP2. Such intronic ORFs (*orf318*, *orf321*, *orf324-1*, *orf324-2*, and *orf352*) exhibit great sequence identity. Noteworthily, introns from the Dothideomycetes *cox1* genes have been constantly discovered to possess such hypothetical proteins. The complete sequence appeared repeatedly. In the *L. maculans* and *P. tritici-repentis* mitogenomes, the full-length *orf324* sequence occurs repeatedly. In additional species, such as *B. maydis* that was in the identical family (Pleosporaceae) to *P. tritici-repentis*, the sequence was not identical. In addition, in *rns* gene, *L. maculans* intron II (domain V) occurred in the identical position in *N. parvum*, with low identify of intronic sequences. Interestingly, the gene-encoding intronic ORFs occurred within the *P. nodorum* and *S. bambusicola* mitogenomes at lower frequencies, and close mobile elements existed frequently in additional Dothideomycetes species. There are numerous gene-encoding intronic ORFs not identified within *B. maydis*, *L. maculans*, *P. tritici-repentis*, and *N. parvum*. These were partially associated with additional fungal species, since invasive ORFs showed greater comparative identity with those unknown proteins in the distant filamentous fungal relative species or even mushrooms.

10.6.3 Unknown ORFs and Conserved ORFs Within Intergenic Regions

The *S. bambusicola* mitogenomes contain functional genes commonly detected in additional species; but the specific ORFs can be found within the intergenic regions in unknown proteins. With regard to the ORFs number discovered, there were 3 in strain *S. bambusicola*, 64 in *B. maydis*, 5 in *P. nodorum*, 28 in *L. maculans*, 36 in *P. tritici-repentis*, 15 in *M. graminicola,* and 8 in *N. parvum*. Such strains possessed different ORFs numbers estimated, which were between 3 (*S. bambusicola*) and 64 (*B. maydis*). Such differences were partially related to the variable genome sizes. After the percent identity for every genome was calculated, it suggested that such differences were mainly detected within intergenic regions.

In *L. maculans* mitogenome, there was one ORF (*orf221*) in an intergenic region. The candidate homolog for the putative protein could be detected from the mitogenome of the tightly associated *P. tritici-repentis* (*orf493*). Similarly, close ORFs could also be found from the mitogenomes of *P. tritici-repentis* (*orf205*, *orf243*) and *L. maculans* (*orf158*, *orf207*, *orf535*). Notably, *orf221* (*orf493*) possessed a conserved region to encode the LAGLIDADG endonuclease detected in additional mycelial species (like *Annulohypoxylon stygium*, *Ceratocystis cacaofunesta*), the sequences of *orf158*, *orf205*, *orf207*, *orf243*, and *orf535* also possessed the special sequence structural characteristics, and there was no apparently matched region in additional NCBI-derived species.

10.6.4 Additional Obvious Characteristics

Numerous repeats were detected within the coding sequence (CDS) and intergenic spacer (IGS) regions in *S. bambusicola*, and they were divided into reverse ($n = 6$), forward (direct, $n = 17$), tandem ($n = 25$), and inverted (palindromic n = 22) repeats. Typically, there were 3 especially long repeats (P1, P2, and P3), among which P1 had the greatest length (485 bp) and was detected within CDS region in *atp6*. It was interesting that, just 2 partial *atp6* gene copies were found from the mitogenome, but they did not encode the intact *atp6* protein. In addition, P2 was detected within IGS region in *trnL-trnQ* as well as IGS in *trnM- trnH*, whereas P3 was detected within IGS sequences in *nad3*, *orf250*, *cox1*, and *trnR*. Repeats have been well-recognized to induce genetic recombination, where inverted and direct repeats in loop-out resulting in submolecules, whereas the flip-flop mechanism causing inversion.

In the present work, *atp8* and *atp9* merely co-existed within the *M. graminicola* mitogenome, and *atp9* might also be detected in *N. parvum*. For the 5 Pleosporales species, *atp8* and atp9 were not detected from their mtDNA; therefore, just 12 protein-encoding genes, which were associated with oxidative phosphorylation and did not contain 2 ATP synthase subunits, were selected to carry out phylogenetic analysis. Noteworthily, one *atp6* gene pair was discovered from *S. bambusicola* mitogenome; CDS was unable to encode the entire ATP synthase F0 subunit 6 by its own, while combining these 2 fragments contributed to assembling the intact gene. The crossed region between these two fragments was 543 bp long, and 2 incomplete genes occurred in diverse strands. PCR and traditional sequencing approaches confirmed such phenomenon.

In the ATP synthase in fungi, 2 genetic origins are reported, namely mitochondrial and nuclear origins. Generally, the mtDNA-derived *atp6* gene can encode a necessary subunit of proton translocating ATP synthase domain. This study detected 2 incomplete fragments of duplications in diverse strands in *S. bambusicola* mitogenome. With regard to organelle genes, trans-splicing is discovered in the higher fungi and plants; however, as far as we know, there is no report regarding trans-splicing within fungal ATP synthase genes. Therefore, our next study will analyze the expression and activity of this gene, so as to examine the underlying mechanism.

10.7 Evolutionary Status of *S. bambusicola*

S. bambusicola represents a greatest fungal class with the highest ecological diversity, and its integrative phylogenic data are obtained through combining 5 genes (nucLSU rDNA, nucSSU, RPB1, RPB2, and TEF1) for 356 isolates from 41 Dothideomycetes families (Schoch et al. 2009). In a previous study, the genome characteristics of 18 species are compared and analyzed. There was an increased number of genes in order Pleosporales relative to Capnodiales, which might suggested that diverse pathogenic patterns were used (Ohm et al. 2012). Nonetheless, little research is available on mitogenomic analysis. In general, mitogenomic analysis serves as an efficient way to analyze evolution, but it is just used in Dothideomycete and Pleosporales species at present. As far as we know, our study first compared Dothideomycetes mitogenomes within species. Altogether 7 whole or almost whole mitogenomic sequences from Dothideomycetes species were utilized, including 4 (*C. heterostrophus L. maculans*, *P. tritici-repentis*, and *N. parvum*) obtained from online contigs, 2 (*M. graminicola* and *P. nodorum*) obtained from published studies (Hane et al. 2007; Torriani et al. 2008), and 1 (*S. bambusicola*) sequenced by our team. Additionally, the NCBI-derived sequence data were used to analyze mitogenomes in *C. heterostrophus L.*

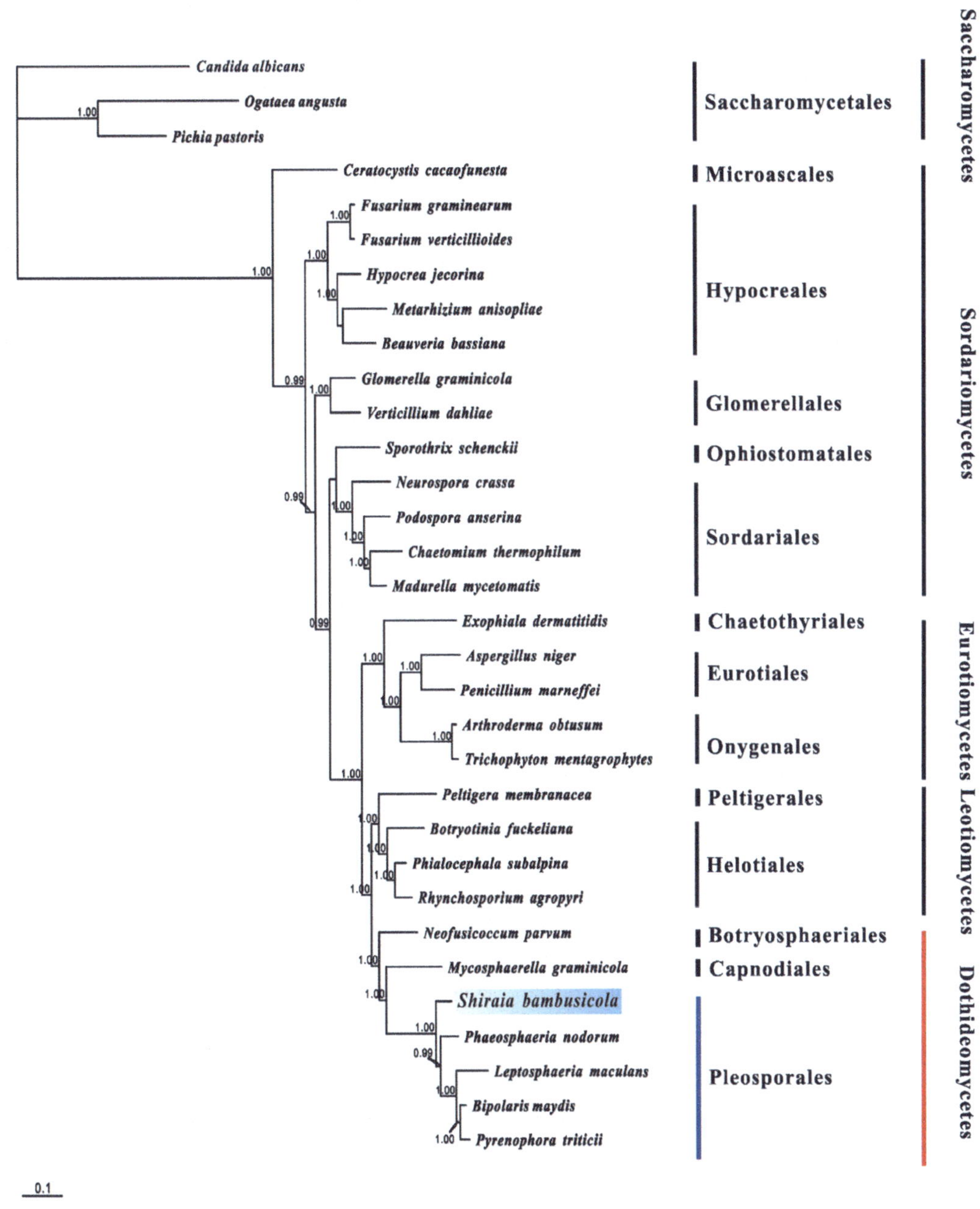

Fig. 10.3 Bayesian-based phylogenetic analysis on aa sequences of Ascomycota mitogenomes

maculans, *P. tritici-repentis*, and *N. parvum*. Therefore, the present work (*S. bambusicola*) and prior articles (*M. graminicola* and *P. nodorum*) did not verify these sequences. Our future work will conduct PCR assays to screen and correct the potential errors across the sequence data.

For obtaining more evidence for Dothideomycetes species classification and understanding the evolution of its mitogenome, we adopted the intact concatenated aa sequences from 12 standard mitochondrial genes (*atp6*, *cox1-3*, *cob*, *nad1-6*, *nad4L*) to construct a phylogenetic tree through using maximum parsimony (Fig. 10.3).

This phylogenetic tree was constructed on the basis of 12 OXPHOS proteins, including atp6, cob, cox1-3, nad1-6, and nad4L. We acquired sequences from the GenBank database, as shown below, *Leptosphaeria maculans* (FP929115), *Bipolaris maydis* (AIDY01000043, AIDY01000067), Beauveria bassiana (NC010652), *Phaeosphaeria nodorunm* (NC009746), *Pyrenophora tritici-repentis* (NW002475730), *Mycosphaerella graminicola* (NC010222), *Neofusicoccum parvum* (AORE01000551), *Fusarium fujikuroi* (NC016687), *Fusarium graminearum* (NC009493), *Metarhizium anisopliae* (NC008068), *Hypocrea jecorina* (NC003388), *Chaetomium thermophilum* (NC015893), *Cochliobolus heterostrophus* (JX185564), *Madurella mycetomatis* (JQ015302), *Podospora anserine* (NC001329), *Neourospora crassa* (KC683708), *Glomerella graminicola* (CM001021), *Sporothrix schenckii* (NC015923), *Annulohypoxylon stygium* (NC023117), *Verticillium dahliae* (NC008248), *Trichophyton mentagrophyte* (NC012826), *Penifillium marneffei* (NC005256), *Exophiala dermatitidis* (CM001238), *Arthroderma obtusum* (NC012830), *Aspergillus niger* (NC007445), *Phialocephala subalpina* (NC015789), *Botryotinia fuckeliana* (KC832409), *Ogataea angusta* (NC014805), *Peltigera membranacea* (NC016957), *Rhynchosporium agropyri* (NC023125), *Pichia pastoris* (NC015384), and *Candida albicans* (NC002653). Among them, *Candida albicans*, *Ogataea angusta,* and *Pichia pastoris* served as the outgroups. Bayesian posterior probabilities were predicted and labeled on top of branches ($\geq$99%).

With three Saccharomycetales species under the Saccharomycetes class (*Candida albicans*, *Ogataea angusta* as well as *Pichia pastoris*) being outgroups, four classes belonging to Pezizomycotina species had been recognized (Dothideomycetes, Eurotiomycetes, Leotiomycetes as well as Sordariomycetes). Under the Dothideomycetes class, the clade of five Pleosporales species could be separated from Botryosphaeriales and Capnodiales species clustering in the same clade, showing a posterior probability support of 95%. *S. bambusicola* ranked among species in the Pleosporales order, and its bootstrap support reached up to 100%. It belonged to the sister sequence of the other four species of Pleosporales.

The phylogeny of *Shiraia* and associated genera remain controversial, since group relative position demonstrated particular characteristics in accordance with marker gene sequences (LSU rDNA, ITS regions as well as tub2). In recent years, *Shiraia* species were considered as the new family belonging to the Pleosporales order (Liu et al. 2013). The mtDNA encoded protein-based phylogenetic tree of this research defined four species as sister clades of *S. bambusicola* from the Pleosporales cluster. The novel regulation for *S. bambusicola* relative position supports the assumption of classifying *Shiraia* species into the Pleosporales order.

Comparison on gene arrangements is usually conducted to infer evolutionary route. Though *S. bambusicola* comes under the Pleosporales order, genomic organization in *S. bambusicola* was greatly different from that of other species (Fig. 10.3). Moreover, the gene and tRNA order of Dothideomycete species was seemingly less conserved compared to that of other Pezizomycotina species (Eurotiomycetes, Leotiomycetes as well as Sordariomycetes). Protein-coding genes and tRNAs rose on various strands, which deduced more complicated relative position organization. For investigating mitochondrial gene order's high variability in Dothideomycetes species, more research should be made to develop extra mitogenomic sequences. tRNA genes clustered as groups had been considered unique characteristics in fungal mitochondrial genomes, in which associated contents and positions usually demonstrated alike characteristics among closely linked species (Fig. 10.3). For mitogenomes in Dothideomycetes species, tRNA genes contained three groups. Great tRNA gene clusters surrounded *rnl* gene, and such conserved pattern could be seen from Eurotiomycetes, Leotiomycetes as well as Sordariomycetes species.

tRNA genes around *rns* included consensus RRLV in Pleosporales species. Except for special characteristics in *S. bambusicola*, other Ascomycete species had no such similar order. The analysis on mitogenomic sequences in Eurotiomycetes, Leotiomycetes as well as Sordariomycetes showed that genes were on positive strands, and protein-coding and tRNA gene arrangement showed high sequence conservation, either individually or mutually. Note that Dothideomycete species mostly include multiple genes (*nad2*, *nad3*, *cox3,* etc.) on various strands, particularly *M. graminicola* (Capnodiales) and *S. bambusicola* (Pleosporales), in which genes had even distribution on two strands (particularly *rns* and *rnl*). No similar pattern exists in other Pezizomycotina mtDNA sequences. Kouvelis et al. (2004) thought gene pairs *nad2-nad3*, *nad1-nad4*, *nad4L-nad5*, *atp6-atp8*, and *cob-cox1* are often joined in Ascomycetes, Eurotiomycetes, Leotiomycetes as well as Sordariomycetes. However, for the majority of mitogenomes of Dothideomycetes species, *atp8-9* genes did not exist, and *cytb-cox1* and *nad1-nad4* genes were not coupled. Merely two gene pairs were coupled (*nad2-nad3* and *nad4L-nad5*) on sections, while *cox1-cox2* gene pair was considered the typical feature of Dothideomycetes species, excluding *M. graminicola* from Capnodiales.

10.8 Conclusions and Future Prospects

As the largest order of Dothideomycetes, Pleosporales contained multiple species, including epiphytes, endophytes, and parasites from living leaves or stems, hyperparasites from fungi or insects and lichens, saprobes on dead plant stems, leaves or barks. Fungi under the class Dothideomycetes usually inhabit in soil, wood, or dung. Aforementioned seven pathogens are from living plant tissues. *Shiraia* is kind of parasite on bamboos which infects *Brachystachyum densiflorum*, *Bambusa*, as well as *Phyllostachys edulis species*. By comparing ORFs in intergenic regions with intronic ORFs of *S. bambusicola* and other species, both plant and animal fungi included, we did not find highly similar mobile elements. Seven Dothideomycete species involved in the research are usually plant pathogens, and complexity in gene arrangement of relevant mitogenomes speculated the possible impression of the host for the parasite. Further mitogenomic analysis helps understand plant-Dothideomycete pathogen interactions, as well as interactions between bamboo and bambusicolous fungus.

References

Aguileta G, de Vienne DM, Ross ON et al (2014) High variability of mitochondrial gene order among fungi. Genome Biol Evol 6:451–465

Ferandon C, Xu JP, Barroso G (2013) The 135 kbp mitochondrial genome of *Agaricus bisporus* is the largest known eukaryotic reservoir of group I introns and plasmid-related sequences. Fungal Genet Biol 55:85–91

Hane JK, Lowe RG, Solomon PS et al (2007) Dothideomycete plant interactions illuminated by genome sequencing and EST analysis of the wheat pathogen *Stagonospora nodorum*. Plant Cell 19:3347–3368

Kouvelis VN, Ghikas DV, Typas MA (2004) The analysis of the complete mitochondrial genome of *Lecanicillium muscarium* (synonym *Verticillium lecanii*) suggests a minimum common gene organization in mtDNAs of Sordariomycetes: phylogenetic implications. Fungal Genet Biol 41:930–940

Lai GH, Fu LY (2000) Study on the main host plants of *Shiraia bambusicola*. Chin Wild Plant Resour 19:8–11

Li X, Gao J, Yue Y, Hou C (2000) Studies on systematics, biology and bioactive substance of *Shiraia bambusicola*. Forest Res 22:279–284

Liu YX, Hyde KD, Ariyawansa HA et al (2013) Shiraiaceae, new family of Pleosporales (Dothideomycetes, Ascomycota). Phytotaxa 103:51–60

Losada L, Pakala SB, Fedorova ND et al (2014) Mobile elements and mitochondrial genome expansion in the soil fungus and potato pathogen *Rhizoctonia solani* AG-3. Fems Microbiol Lett 352:165–173

Ohm RA, Feau N, Henrissat B et al (2012) Diverse lifestyles and strategies of plant pathogenesis encoded in the genomes of eighteen Dothideomycetes fungi. PLoS Pathog 8:e1003037

Pelin A, Pombert JF, Salvioli A et al (2012) The mitochondrial genome of the arbuscular mycorrhizal fungus *Gigaspora margarita* reveals two unsuspected trans-splicing events of group I introns. New Phytol 194:836–845

Schoch CL, Crous PW, Groenewald JZ et al (2009) A class-wide phylogenetic assessment of Dothideomycetes. Stud Mycol 64:1–15

Shen XY, Cheng YL, Cai CJ et al (2014) Diversity and antimicrobial activity of culturable endophytic fungi isolated from moso bamboo seeds. PLoS One 9: e95838

Tong Z, Mao L, Liang H, Zhang Z et al (2017) Simultaneous determination of six perylenequinones in *Shiraia* sp. Slf14 by HPLC. J Liq Chromatogr R T 40:536–540

Torriani SF, Goodwin SB, Kema GH et al (2008) Intraspecific comparison and annotation of two complete mitochondrial genome sequences from the plant pathogenic fungus *Mycosphaerella graminicola*. Fungal Genet Biol 45:628–637

van de Sande WW (2012) Phylogenetic analysis of the complete mitochondrial genome of *Madurella mycetomatis* confirms its taxonomic position within the order Sordariales. PLoS One 7:e38654

Youssar L, Gruning BA, Gunther S et al (2013) Characterization and phylogenetic analysis of the mitochondrial genome of *Glarea lozoyensis* indicates high diversity within the order Helotiales. PLoS One 8:e74792

Zhao D, Liang ZQ (2005) Reviews of studies on isolation and culture of *Shiraia bambusicola* Henn. J Fungal Res 3:53–57

11 Breeding Status and Strategies of Moso Bamboo

Jian Gao

Abstract

Moso bamboo (*Phyllostachys edulis*) is not only the largest dual-purpose bamboo species for timber and shoot in Gramineae, but also used as raw material for feed and traditional Chinese medicine. The main distribution area is concentrated in the subtropical region of the northern hemisphere, mainly in Asia, and China has the largest planting area in the world. However, the flowering of moso bamboo is rare, and the flowering date is uncertain. It may take 60–120 years or more from seedling to flowering, so the traditional cross breeding is very difficult. With genome information, it is possible to create new moso bamboo Germplasm by using modern molecular breeding or gene-editing technology. The breeding methods, research techniques, and breeding objectives of moso bamboo were addressed, and the prospect of the research on moso bamboo breeding was put forward.

J. Gao (✉)
Key Laboratory of Bamboo and Rattan Science and Technology, International Centre for Bamboo and Rattan, State Forestry and Grassland Administration, Beijing 10012, China
e-mail: gaojian@icbr.ac.cn

11.1 Main Propagation Methods and Techniques of Moso Bamboo

11.1.1 Transplanting of Mother Bamboo

Transplanting of mother bamboo (*Phyllostachys edulis*) is the most important method of moso bamboo forest cultivation, and it is also the fastest method of forest formation. It is the most widely used method in China.

Selection of mother bamboo: 1–2 years old, 3–6 cm in diameter, strong growth, low branches, luxuriant branches and leaves, normal bamboo nodes, no diseases and pests (Fig. 11.1) (Zhou et al. 1998).

11.1.2 Seedling of Moso Bamboo

Traditional moso bamboo forest afforestation is carried out by transplanting mother bamboo. However, under the new normal economy, the shortage of labor force and the aging of labor force are becoming more and more serious. However, the development of moso bamboo resources is faced with difficulties in obtaining and transporting mother bamboo, and the survival rate of mother bamboo transplanting is low, especially in the steep slope afforestation in the middle and upper reaches of the Yangtze River in

J. Gao (ed.), *The Moso Bamboo Genome*, Compendium of Plant Genomes,
https://doi.org/10.1007/978-3-030-80836-5_11

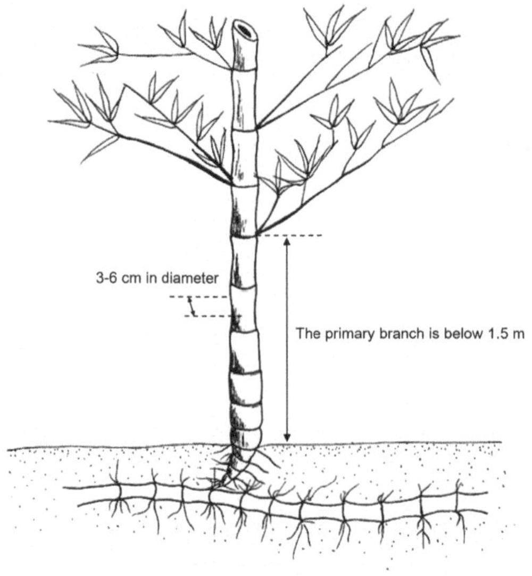

Fig. 11.1 Specifications of mother bamboo for moso bamboo plantation. Excavation of mother bamboo: dig 30–50 cm away from the mother bamboo, find the bamboo whip, and then cut along the two sides of the mother bamboo, with the length of 30–40 cm and 40–50 cm on both sides of the mother bamboo (Fig. 11.2) (Zhou et al. 1998)

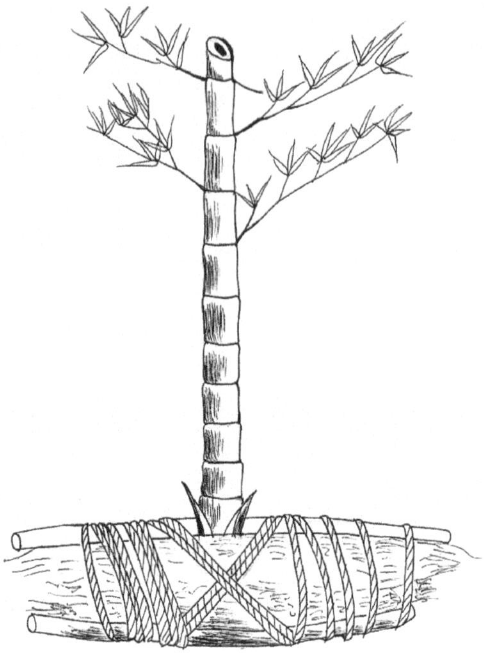

Fig. 11.2 Packing of mother bamboo for transportation. Transportation of mother bamboo: prevent damage at screw. Long distance binding, loading and unloading, reduce evaporation on the way to spray water. Planting of mother bamboo: in the hole that has been prepared, first use topsoil to pad the bottom, generally 10–15 cm thick, carefully put the mother bamboo into the hole, spread the rhizome root, connect the lower part with the soil tightly, and make the root and soil closely (Fig. 11.3) (Zhou et al. 1998)

southwest mountainous areas and the barren mountain afforestation in poor site conditions. The container seedlings of moso bamboo are small in volume and light in weight, which can relieve physical deficiency and reduce the labor cost of afforestation. The efficiency of seedling raising was improved by using bamboo container seedlings. Through the application of white net bag and light substrate, the seedling lifting and transportation are convenient, the demand for labor is reduced, and the rotten root rate and burn rate of seedlings are reduced.

Compared with digging and planting mother bamboo, this technology can save labor and time, save the economic cost of more than 90%, and is superior to mother bamboo transplanting in afforestation, transportation, survival and forest formation.

Technical points:

(1) Garden selection. The selection of breeding nursery should be based on leeward and sunny, convenient irrigation and drainage, and the flat land close to the water source. The soil acidity and alkalinity should be moderate, and there should be fertile and loose soil. Generally speaking, in the process of mountain nursery, the new nursery should dig deeply and turn over the soil within 2–3 months before sowing, and clean it up. After about a month of weathering, the soil was dug 1–2 times before sowing, and the broken soil was used as the seedbed for seedling cultivation. The height of the bed is generally 12 cm, the width is 1 m, and the width of the sidewalk should be kept within 40 cm. Flatbeds may be used in well-drained areas. The surface soil of the seedbed should be fine, the seedbed should be compact, the sidewalk should be straight,

Fig. 11.3 Erecting support after planting mother bamboo (Zhou et al. 1998)

and the drainage ditch should be connected; potash fertilizer, compost, burnt soil and human and animal manure should be applied, and calcium magnesium phosphate, or superphosphate, etc. should be mixed with soil compost and miscellaneous fertilizer, which can be used after composting.

(2) Seed disinfection. Soak in water for 0.5 h, disinfect with 0.3% potassium permanganate solution for 2–4 h, and then wash. Finally, the seeds can be sown after soaking in clean water for 24 h, and then being removed from the water and dried for 1–2 h.

(3) Sowing method. Seed dressing with wet sand to promote germination. When the seed is white, it can be sown on the seedbed, in order to prevent the phenomenon of back bud rotten seed, it should be often appropriate watering. In actual sowing operation, sowing can be divided into hole sowing and drill sowing. Among them, the sum of hole sowing is exceptionally little, and the distribution of bamboo seedlings is moderately uniform and flawless, which is helpful for future administration; the spacing between plants and columns ought to be controlled at 30 cm × 40 cm, the hole distance across ought to be almost 5–6 cm, and the depth ought to be controlled at 2–3 cm; approximately 8–10 moso bamboo seeds ought to be equally sown in each gap; the seedling container should be sown on demand, and the nutrition bowl ought to be 10–15 cm, and each bowl ought to be equally sown 6–4 seeds. Then, cover it with fire soil or straw drenched with water, the thickness of covering seed is 1.5 cm, and no seed is suitable.

(4) Sowing time. The seeds of moso bamboo ought to not be stored for more than half a year at room temperature. As long as the temperature is suitable, they ought to be sown in autumn or spring.

(5) Container selection. White degradable light matrix nutrition net bag or nutrition bowl ought to be chosen for holder seedlings, and the size and diameter of bowl ought to be 10–15 cm. Compared with the dark nourishment bowl of the same size, the spoiled root rate and burn rate of seedlings were reduced, the seedling lifting and transportation were helpful, and the request for labor was reduced.

(6) Garden management. The soil moisture should be maintained from sowing to emergence, but the amount of watering should not be too much to avoid hardening of seedbed. At the same time, covering some chaff and sawdust around the bamboo seedling can not only play the role of drought resistance, but also effectively reduce the harm of bamboo seedling mud. When the weather is overcast and rainy, the dense bamboo sticks with soil should be transplanted into the hole of lacking seedlings to ensure the uniform distribution and robust growth of bamboo seedlings, so as to improve the yield of seedlings. After the seedlings unearthed, attention should be

paid to weeding and loosening the soil. When the soil is cultivated, it should not be too thick and the roots should not be exposed, and proper amount of watering and topdressing should be carried out. When winter comes, wind barriers should be set around the nursery, and the bamboo clumps should be covered with 5–10-cm thick soil and scattered with straw.

(7) Secondary transplanting of seedlings. The experiment was carried out on rainy days with 3–5 leaves, 10–12 cm seedling height and before tillering. It is required to plant as soon as possible, with more perennial soil. When planting, the roots should be extended, the depth should be appropriate (about 0.5 cm deeper than the original soil), the spacing between plants and rows should be 30 cm × 30 cm, keep 1–2 plants in each hole, drench enough fixed root water, and then cover a layer of loose soil to the seedling stem position.

(8) Nursery management. After the bamboo seedlings unearthed, weeding should be started in time. When it is dry, water should be drenched. It is better to apply fertilizer with relatively thin human manure water, once or twice a month. In order to make the bamboo seedlings evenly distributed and increase the yield of bamboo seedlings, we can take advantage of overcast and rainy weather to transplant the over dense bamboo seedlings into the ground hole with 1–2 seedlings per hole.

(9) Take off the grass and remove the film. After the emergence of most of the sowing holes, 1/2 of them were removed, and then all of them were removed after 7–10 days. Remove the film during the day, cover it at night to achieve the purpose of refining seedlings, and finally completely remove it.

(10) Seeding transplantation. First, leave 2 seedlings with 3–4 seedlings in each hole, and transplant the surplus ones into the hole without seedlings with roots and soil with bamboo sticks to make them even. This work should be carried out before the beginning of summer, and at the same time, it should be often drenched to ensure survival.

(11) Reasonable farming. After two years of container planting, the disease and insect pests of moso bamboo seedlings were reduced.

11.1.3 Propagation by Burying Rhizomes

The main method of propagation of moso bamboo is transplanting mother bamboo. In order to speed up the propagation of moso bamboo, it is better to burying rhizomes of moso bamboo.

(1) Burying Time: The best time to raise seedlings is from late winter to early spring, from late December to early February. When the next spring is warm, the seedlings can germinate and be unearthed. It is beneficial to cultivate one-year-old strong seedlings for transplanting.

(2) Digging Bamboo Whip: A 2–3-year-old strong bamboo rhizome is selected from the Phyllostachys praecox forest, and a section of 40–50 cm long is cut. Generally, a bamboo seedling can grow on each rhizome node. 40–50 cm long can produce more than 10 continuous bamboo seedlings.

(3) Burying Whip in Seedbed: For burying the whip, the acid soil with flat terrain, good drainage, close to water source and deep and fertile soil layer should be selected. Seedbed soil should be finely broken, bed width 1.3 m, bed height 17 cm, dredge drainage ditch. After mixing with the soil, the bamboo rhizome is dipped with yifuyuan germinating and rooting solution, and then spread into the nursery ditch to cover the soil. The soil layer is about 10-cm thick, and then cover the grass and drench the soil.

(4) Management After Excavation: After the bamboo seedling was born. It is necessary to carry out intertillage and weeding to make

the seedbed free of weeds. To keep the seedbed moist and moderate, drainage should be done when there is water, and watering should be done when there is drought. After one year, the height of the seedlings is about 1 m, and the seedlings can be lifted and transplanted.

(5) Cutting and Transplanting: If you want to cut-off the female rhizome one-by-one, you can plant a live one for each plant by keeping its own root system. If you plant a live one in spring, you can send out new shoots from the new rhizome and grow into a new bamboo in the next year.

11.2 Main Breeding Methods and Techniques of Moso Bamboo

Since the beginning of the twenty-first century, the field of moso bamboo breeding in China has entered a new stage of development. Based on the traditional selective breeding, the continuous development of biotechnology has incredibly influenced the development trend of moso bamboo breeding. Transgenic and molecular marker technology have been applied in the research of *Phyllostachys edulis* breeding, and various high-tech means have been applied in the breeding work, so that the breeding of *P. edulis* has developed into a new era.

11.2.1 Hybrid Breeding Technology

Hybrid breeding technology refers to the process of population selection and individual selection for fine characters of offspring and selection of fine lines in line with breeding objectives after artificial hybridization.

At present, due to the special biological characteristics of moso bamboo, the flowering cycle is long, which may be from 60 to 120 years, and the natural seed setting rate is low. It's just a long-term clonal propagation, and there are many variations, forming the diversity of moso bamboo, as described in the second chapter of this book.

(1) **Stamen Development**: Under normal circumstances, the setting rate of moso bamboo is low, and the setting rate is only 9%. In order to explore the reasons for the low setting rate, the stamen development was studied. With the help of modern molecular biology methods, Prof. Dr. Gao Jian and her research group from ICBR deeply studied the problem of stamen abortion in moso bamboo, and revealed the molecular mechanism of regulating stamen development (Cheng et al. 2020). Characterization of critical miRNAs was achieved by small RNA, degradomic and transcriptomic analyses from the multi-omics perspective. The miRNA targets in floral organs of moso bamboo were elucidated as well. Apart from a total of 13,051 differentially expressed genes, they also confirmed 109 recognized miRNAs among 26 families. By aligning these miRNAs to the established databases, a few conserved miRNAs were found, as well as novel miRNAs. High and specific expressions of 16 conserved miRNAs, such as miRNA159 and miRNA166, were noticed in stamens. The crucial function of miRNA159 in the stamen growth adjustment was suggested by the in situ hybridization, and *PheMYB98* and *PheMYB42*, the targets of miRNA159, were lowly expressed. Besides, *Phe-MIRNA159* enabled partial restoration of mir159ab double-mutant phenotypes, and its over-expression led to possible anther dehisce defects. As a result, mature pollen dispersion was impossible, thus causing Arabidopsis fecundity decline. Insufficient secondary thickening of anther endothelial layer was noted by semithin section observation for the *Phe-MIRNA159* overexpressing lines, so that the force for dehiscing anther was deficient. Probably, by negatively adjusting *AtMYB33*, the expression of secondary

thickening-associated genes was mediated by Phe-miR159, consequently impacting anther opening.

(2) **Hybrid Combinations and Their Seedlings**: Prof. Tang Dingqin, from Zhejiang Agroforestry University, explored the cross breeding of monopodial bamboo species of the same genus and hybrid combinations were obtained (Lu et al. 2009).

Crosss breeding has great significance for the formation of new species. Due to the long flowering interval of most bamboo species, the crossbreeding process of bamboo is underway. The hybrids from the sympodial bamboo species with excellent characters were selected, and some new varieties with advantages have been cultivated. The hybrids from uniaxial bamboo species are infrequent. The hybrid female parent moso bamboo has the advantages of fast growth, strong adaptability, thick stem and low-planting cost. It is an excellent bamboo species for timber, shoot and garden Paternal parent *Phyllostachys violascens* is a unique and excellent economic bamboo species in China. It shoots early, with long period of shoots harvest, shoots delicious, low cost, high benefit, high yield and market demand.

Interspecific hybridization between *P. edulis* and *Phyllostachys violascens* is expected to produce hybrid progenies which are superior to their parents in terms of bamboo shoot taste, shoot material yield, plant growth vigor, growth activity, stress resistance and adaptability, creating great social, economic and ecological value. Using moso bamboo and *Phyllostachys violascens*, which are newly blooming and have excellent growth characteristics and shoot material, to carry out interspecific hybridization which is expected to provide a reference for the future hybridization between monopodial bamboo plants and breakthrough of new varieties in the bamboo industry.

Field cross breeding was carried out with *P. violascens* as the male parent and moso bamboo as the female parent. The half-flowering *P. violascens* was dug in campus of Zhejiang Agroforestry University and the small-scale flowering *P. violascens* was cultivated by the method of buried scourge propagation. It was transported to Dajing Yao township, Linchuan County, Guilin city, Guangxi Zhuang Autonomous Region. The flowering moso bamboo, in full bloom stage was selected as the female are not, and field hybridization was carried out through the processes of artificial emasculation and pollination. In order to increase the yield of the hybrid seeds, the water and fertilizer management method of spraying branch and leaf and bamboo was adopted. During the three-year period (2017–2019), a total of 2650 florets were pollinated and 473 seeds were obtained. The average pregnancy rate was 17.18%. And a total of 48 suspected hybrid seedlings were obtained in 2017–2018.

The average length, diameter, 1000-grain weight, germination rate, and emergence rate of the suspected hybrids obtained from field crossed were significantly less than those of the self-cultivated seeds of moso bamboo. Compared with the control group, the average leaf length, leaf width, and plant height of the suspected hybrid seedlings obtained in 2017 were slightly lower, but the difference was not significant ($P > 0.05$), in comparison to the control, markedly higher ground diameter, overall number and bamboo number were found ($P > 0.01$), as well as prominently higher overall shoot number ($P < 0.05$), but there was no significant difference between the shoot height and the growth trend of bamboo shoots ($P > 0.05$). At the same time, the suspected hybrids had earlier shoots, and some of the shoot sheaths appeared similar wrinkles of the male parent. The average leaf length, leaf width, tiller and plant height of the suspected hybrid lines obtained in 2018 compared with control were slightly lower, and the difference was not significant ($P > 0.05$). In summary, compared with the control, there was no significant difference in the growth phenotypic indicators of the one-year suspected hybrids. This phenomenon may be due to the long growth period of the loose bamboo plants and the shorter culture period. The growth of suspected hybrids began to manifest itself. After preliminary screening, the cultivars with the potential of exploiting the superior germplasm resources were screened out, which were screened out, numbered as a2, a3, a6, a7, a9, a11, a12. The advantages of rapid propagation of the three suspected hybrids numbered a2, a6, a9 and all are extremely significant. The total number of tillers and shoots of the series of plants is 40%-200% higher than that of the control. The hybrids numbered a7 has a greater advantage in terms of growth. The hybrids of this series have great potential for growth. The total number of shoots and the number of bamboo shoots in the series are compared with those of young shoots. In the control, the self-crossing gain advantage is as high as 40%-120%. The three suspected hybrids, number a3, a12 are highly advantageous in most aspects of eight phenotype characteristics such as leaf length, leaf width and plant height. The gain values of all aspects range from 1 to 181%. Both growth and rapid propagation have enormous potential for germplasm mining.

11.2.2 Radiation Mutation Breeding

Radiation mutation breeding refers to the use of chemical and physical factors (chemical inducer, artificial radiation, space mutation) to induce the genetic characteristics of animals and plants, and then through artificial selection, screening excellent mutants or individuals, and then breeding new varieties or germplasm. Prof. Dr. Gao Jian and her research group and others have made great achievements in the research of $^{60}Co\gamma$ radiation mutation breeding of moso bamboo, and analyzed its breeding process and mutation mechanism. It is considered that the combination of radiation mutation technology and biotechnology will become an important direction of its development. Zhejiang Agriculture and Forestry University and Fujian Agriforestry University also carried out radiation breeding research. The materials, dosage and rate of mutagenesis and the breeding of seedlings after mutagenesis were introduced.

Moso bamboo dry seeds were irradiated by different doses of $^{60}Co\gamma$ rays, the effects of the radiation on seed vigor and seedling growth characters were investigated by four testing methods, i. e., germination testing indoor, electrical conductivity, TTC vigor testing, and growth classification of seedlings. The results showed that $^{60}Co\gamma$ ray (100 Gy) could accelerate the germination process and improve the germination rate of seeds, and higher dose of $^{60}Co\gamma$ ray (<100 Gy) could significantly inhibit the process of seed germination, reduce seed vigor, induce seed embryo rupture, cell division, growth was inhibited, seedling height and root length were reduced. With the increase in radiation dose, the inhibition effect was obviously enhanced. The optimal range of radiation dose for radiation breeding of moso bamboo dry seeds was 100 Gy to 175 Gy. Linear relationships were existed in electrical conductivity after dipping in water for 24 h in germination rate (G), germination index (GI), vigor index (VI), the height of seedlings and length of roots. EC after 24 h and height of seedlings were chosen to test the change of seeds vigor and the effect of the radiation of $^{60}Co\gamma$ rays on moso bamboo seeds.

In order to explain the formation of the above phenomenon, the effect of mutagenesis was studied at the cytological level. Moso bamboo seeds were irradiated by different doses of $^{60}Co\gamma$rays and the effects on chromosomal aberration, nuclear aberration and mitosis of the root tip cells were studied. The results indicated that γ-rays irradiation with lower doses could promote and stimulate mitosis of root tip cells, and with high doses could inhibit mitosis of root tip cells. Furthermore, γ-rays irradiation could induce many types of chromosomal and nuclear aberration, such as chromosome bridges, chromosome fragment, laggard chromosome (clump), free chromosome (clump), chromosome conglutination, micronucleus, minornucleus, double nucleus, nuclear budding, nuclear bleb, nuclear fragmentation and so on. The relationship between dosages and frequencies of chromosomal and nuclear aberration of root tip cells was fitted in the model of $Y = 010390X - 013{,}756$, $Y = 010530X - 010{,}783$, respectively (Wang at al. 2008). The correlation coefficient was 018,934 and 0,199,821. The relationship between frequency of chromosomal aberration and frequency of moso bamboo micronucleus of root tip cell was fitted in with the model $Y = 018836X$ - 0,145,251. The correlation coefficient was 018,345, $F = 25.22 > F_{0.01}$.

The main growth characters of moso bamboo seedlings whose seeds were irradiated by 5 different doses of $^{60}Co\gamma$ rays were investigated and explored by using the principal component analysis method, for the sake of finding new excellent gene resources and widening the hereditary basis of moso bamboo. The results indicated that near 90% information of seven growth characters (average single—plant ground diameter, average single—plant seedling height, maximum ground diameter, maximum seedling height, number of tillers, canopy diameter, and root breadth) could be condensed into three principal components. The first principle component mainly consisted of plant height and ground diameter factors. 16 superior seedlings were finally screened out by calculating and comparing the values of key principal components of various varieties, and the results were similar to the practical phenotype of bamboo seedlings in the nursery.

From 2007 to 2010 years, the seedling stage test was carried out in Feixi county, and 16 superior seedlings were screened out. Then, the afforestation experiment was carried out from 2011 to 2017. The 6 growth traits of 16 superior seedlings were analyzed for 6 consecutive years by using unirradiated bamboo seedlings as the control. Difference analysis, correlation analysis, principal component analysis and subordinate function values analysis were used to evaluate the quality and stability of different superior plants. The results showed that the differences of 6 growth indexes were significantly different each other in different periods, and different superior plants with two-factor ANOVA, eight single indexes were grouped into two principal components by principal component analysis, and the cumulative contribution rate was 94.9% The growth quality of superior plants was evaluated by subordinate function value D. The growth characteristics of CK-1 and superior 10 Gy were the best, and these were similar to the growth phenotypes in afforestation experiments. This study provideed a feasible method for the selection of excellent radiative plants of moso bamboo, and laid a foundation for further screening of excellent gene types induced by radiation from moso bamboo.

After 15 years of excellent plant selection, the ICBR applied for three new varieties of moso Bamboo in 2020 according to the international DUS test. It is hoped that new varieties of moso bamboo can be obtained after approval.

Furthermore, in 2020 moso bamboo seeds were carried out on Chang'e-5 lunar probe.

11.2.3 Molecular Marker Assisted Breeding

With the development of modern biotechnology, RAPD, AFLP, SSR, SNP, and other molecular markers have been widely used in bamboo breeding. Many scientists believe that the application of molecular marker technology and genomics in crop breeding has significantly changed the selection methods and steps of breeding progeny, making the selection of hybrid parents more efficient, the selection of breeding progeny more accurate, and the convergence of beneficial gene loci in breeding population faster. Ding Yulong and others think that the biological breeding technology produced by molecular marker technology has been widely used in many fields, among which SNP, SSR and RFLP technology are used more frequently. Analysis of marker technology can effectively divide the heterosis group of moso bamboo population, reveal the genetic basis of heterosis, and improve the efficiency of female parent selection.

China has a long history of and vast land for moso bamboo cultivation (Zhou 1998). In the long course of moso bamboo forest production and management, there have been a series of varieties. Records say there are a total of 15 varieties or cultivars (Ding 1998). Recently, there are new varieties or cultivars were found. In the second chapter of this book, 25 varieties or cultivars of moso bamboo are recorded and illustrated. Most of these morphological varieties show differences in culm type, culm color and leaf color. From these types of morphological variation, we can see the diversity of morphological variation within the moso bamboo species. Methods of plant molecular biology can not only obtain molecular-level data, but can also explore the genetic structure of plant populations to help study their genetic diversity and systematic evolution pattern. Molecular markers are extensively applied in the studies on bamboo germplasm resource identification, classification, genetic relationship, genetic diversity as well as origin and evolution of bamboo plants (Xing 2004; Zhang 2007).

Genetic Diversity in Moso Bamboo

Studies on bamboo genetic diversity first began in 1991: Friar and Kochert (1991) compared the genetic diversity of 51 samples of 24 bamboo species under 5 genera by using the RFLP technology that was considered advanced at that time. Lai and Hsiao (1997) used RAPD technology to analyze the genetic structures of the moso bamboo population and *Yushania niitakayamensis* population in Taiwan and found that although it has just been less than 200 years since moso bamboo was introduced to Taiwan, great genetic diversity were detected among different populations, which was presumably due to the geographical differentiation caused by habitat difference. Shi (2002) also used RAPD technology to distinguish successfully seven varieties and two cultivars of moso bamboo and proved that *P. edulis* was most genetically distant from other moso bamboo species. Ruan (2008) used AFLP and ISSR molecular marker technology to explore the genetic diversity of moso bamboo and proved the feasibility of these two molecular marker technologies in studying the genetic diversity of moso bamboo. Jiang (2013) used 20 pairs of SSRs to conduct an in-depth study of the genetic diversity and population structure of 34 representative populations of moso bamboo in China's existing distribution areas before moving on to explore the current distribution patterns and possible evolutionary characteristics of moso bamboo.

Twenty pairs of microsatellite fluorescent markers (SSRs) were used to study the genetic diversity and population structure of 34 representative *P. edulis* populations in China.

Microsatellite Polymorphism

20 pairs of polymorphic microsatellite (SSR) loci suitable for population genetic research of *P. edulis* were developed by using the cDNA sequences obtained from NCBI. The percentage of polymorphism (PP) of all SSR loci at the species level was 100%. A total of 169 alleles were detected in 804 individuals. The number of alleles at each locus ranged from 4 to 15, with an average of 8.5; the pic value ranged from 0.034 to 0.663, with an average of 0.318; the length of allele fragments at SSR loci ranged from 12 to 51 BP, and 75% of the loci followed any allele mutation pattern In addition, 20% and 5% of the loci may be inserted or deleted according to the step-by-step mutation model (SSM) and SSR flanking region, respectively. (2) Genetic diversity and differentiation. The average allele richness and Nei's gene diversity of 34 populations were 2.31 and 0.320, respectively. There was no significant correlation between the degree of genetic variation within the population and geographical location, so it was difficult to find the phylogenetic relationship between different populations. The genetic diversity and gene differentiation coefficient of *P. edulis* were 0 376 and 0. 162, which was consistent with the conclusion that 15. 45% of genetic variation existed among populations. Although most of the genetic variation (84.55%) existed in the population, compared with most forest populations (FST 0.15), the degree of genetic differentiation among the populations was still high; most of the loci in the population deviated significantly from hardy Weinberg equilibrium ($P < 0.05$) The fixed index of total population level and individual

population level were −0.192 and −0.440, respectively, which indicated that heterozygotes were excessive both at the overall level and among individuals within the population, which reflected that heterozygotes had stronger vitality and could adapt to selection pressure. Meanwhile, heterozygotes were more dominant in the long generation cycle species of *P. edulis*. Mantel test showed that there was a significant positive correlation between genetic distance and geographical distance ($r = 0.247/0.215$, $P < 0.05$). According to FST, the gene flow of moso bamboo was estimated to be 1.203, which indicated that there was less gene exchange among populations. Although artificial transplanting might increase the gene flow among populations, it only affected the genetic variation pattern of moso bamboo within a certain distance (transplanting nearby).

Clonal Diversity

The clonal diversity of *P. edulis* is lower than that of the species with sexual and vegetative reproduction, which may be related to the unique biological characteristics of *P. edulis,* that is, strong vegetative reproduction, long flowering cycle and low-natural seed setting rate. A total of 261 multi locus genotypes were detected in 804 individuals of 34 *P. edulis* populations, and the most widely distributed clones were found in 14 populations with 107 clonal plants. Based on the results of Bayesian clustering, 34 populations were divided into two groups according to their genetic composition. The large group (C1) included the majority of Moso Bamboo populations, and the small group (C2) mainly included Zhouzhi population in Shaanxi, Huaihua population in Hunan, Renhua population in Guangdong and Liping population in Guizhou. The Central South China region was the mixed region of the two gene pools, which was consistent with the results of cluster analysis based on genetic distance. According to the results of component analysis (PCoA), adjacency analysis (NJ) and structure clustering at $k = 3$, the *P. edulis* population in the large group can be divided into two subgroups, one of which mainly includes 17 populations from Yunnan, Guizhou, Sichuan border, Hubei, Anhui Huoshan and Fujian, the other mainly includes 13 populations from Guangxi, Jiangxi, Zhejiang, Jiangsu and Anhui Guangde. Although there is no geographical division among the subgroups, most of the populations from the same geographical origin belong to the same subgroup. *P. edulis* has a very long history of cultivation. SSR analysis revealed that the source of each cultivated population was different, and some populations with low-genetic diversity had a single source. Zhouzhi of Shaanxi Province, and Laoshan of Shandong Province were introduced and cultivated populations during the period of bamboo migration from south to north. The genetic diversity of Zhouzhi of Shaanxi Province was very high. According to historical data, the introduction sources of Zhouzhi of Shaanxi Province were Hunan, Hubei, etc., which were confirmed by fluorescence SSR Analysis. According to the results of cluster analysis, Laoshan population of Shandong Province was also confirmed to be from Zhejiang and Jiangxi. According to the records, *P. edulis* in Yunnan, Guizhou were introduced and cultivated from Fujian in 1769. This study further confirmed that *P. edulis* at the junction of Yunnan, Guizhou and Sichuan might come from Wuyishan, Fujian, and after a long period of development and differentiation and adaptation to the environment, it was found that *P. edulis* in Xingwen, Sichuan and Chishui city, Guizhou province were in the same population, which had some specific alleles.

Protection Strategy

The study of Xia (2014) showed that as a clonal plant, moso bamboo manifested rich phenotypic diversity and relatively high-genotypic diversity and that there was significant genetic differentiation among various populations and sources of moso bamboo. For populations with rich genetic variation, such as Huaihua in Hunan Province, Liping in Guizhou Province and Renhua in Guangdong Province, priority should be given to protect the species from deep damage. For populations with high endemic gene richness, such as Xingwen in Sichuan Province, as well as special germplasm distributed in each population, efforts should be made to conserve and utilize them in other places.

Biological Characteristics by Molecular Markers

The studies on the biological characteristics of moso bamboo began by using AFLP marker technology to study the flowering characteristics of moso bamboo and eventually determine that the flowering cycle of moso bamboo is 67 years (Isagi et al. 2004).

Molecular Identification of Hybrids

Three hybrid bamboo candidates (*Pleioblastus simonii* × *Phyllostachys violascens*, *Sasa tokugawana* × *S. borealis* and *Sinobambusa tootsik* × *Pl. distichus*) were identified using 10 SSR markers developed from GSS of moso bamboo in GenBank which were used for amplifying the cross-species. DNA fingerprinting of the 3 hybrid bamboo candidates showed complementary pattern from ♀and ♂parents at PBM014, PBM025 and (or) PBM018 locus. The sequencing and alignment of fragment DNA bands at the mentioned-above loci approved homology between the hybrid bamboo candidate and their parents, indicating the validity of hybrid bamboo results. This is the first report for identification of bamboo hybrid using DNA markers.

Using TP-M13-SSR technology into bamboo plants, 15 pairs of SSR primers were screened out from 98 pairs of SSR markers that have been developed, it can be used for TP-M13-SSR technology to identify hybrid cultivar, 13 pairs of SSR primers were screened in the remaining primers for routine capillary electrophoresis Using TP-M13-SSR fluorescent labeling capillary electrophoresis technology and conventional capillary electrophoresis technology, the suspected hybrids numbered 8 lines have parental bands, which are *P. violascens* and *P. edulis*, hybrid progeny provides molecular evidence. Due to the different inserted fragments of the gene locus of the male parent, it is still necessary to select suitable primers for the identification of the suspected hybrids that have not yet been identified.

In the process of developing molecular marker, many molecular markers in moso bamboo were established. It has strong generality in *Phyllostachys* and other bamboo species. These SSR markers were used to identify 20 suspected hybrid bamboo species. Molecular identification was carried out. The hybrid species of *Phyllostachys* can be successfully identified by SSR markers. Among other genera seven hybrid combinations could be identified by 5 SSR markers. At the same time, sequencing confirmed that the 5 SSR loci of moso bamboo was also common in hybrid bamboo species nature in short.

11.3 Molecular Breeding

Bamboo research largely lagged behind due to the lack of efficient genetic manipulation tools. In 2004, Zhuo Renying et al. transformed the callus containing GUS gene into *Dendrocalamus latiflorus* by *Agrobacterium* tumefaciens, and successfully obtained 6 regenerated plants. This research has initially established the transgenic technology system of *D. latiflorus*. In 2012, Zhang Ling et al. used calli induced by anther of *D. latiflorus* as acceptors, and used *Agrobacterium* mediated method to study different infection time, pre culture time, hyg concentration, AS concentration and other conditions, and successfully obtained regenerated plants with codA gene. RT-PCR detection showed that codA gene had been expressed in *D. latiflorus* genome. In 2017, Shanwen et al. optimized the CMM composition and established *Agrobacterium* mediated genetic transformation system of *D. latiflorus*, which was verified by PCR and GUS staining.

In general, due to the lack of mature gene transformation system of bamboo plants in recent years, the progress of transgenic breeding of bamboo plants is slow, especially moso bamboo. The main technical bottleneck is that it is difficult to establish an efficient regeneration system after callus dedifferentiation, and the research on transgenic breeding of bamboo plants needs to be developed.

11.3.1 Tissue Culture of Sympodial Bamboo Species

Clonal propagation with classical techniques are available for bamboo species, such as rhizome and culm cuttings, clump division and seed propagation, but the success has been limited for mass scale propagation due to various factors, such as large nursery and infrastructure required for mass production, limited availability of seeds and high-investment costs. Micropropagation through tissue culture can be a good alternative to conventional systems. The first successful report on micropropagation of bamboo (*Bambusa* sp.) using zygotic embryo explants was described by Alexander and Rao (1968). Efficient in vitro regeneration of plants, such as somatic embryogenesis and indirect organogenesis has been a major bottleneck in micropropagation of bamboo (Ye et al. 2017). Notwithstanding, successful plant regeneration using micropropagation and somatic embryogenesis has been established for moso bamboo which is important since genetic transformation for trait management requires an efficient tissue based regeneration system for successful transgene incorporation and plant establishment.

11.3.2 Moso Bamboo Tissue Culture

Although the tissue culture of sympodial bamboos has been successful all over the world, the tissue culture of large-scale monopodial bamboos is very difficult. It has been difficult to obtain callus and browning seriously after obtaining. After years of research, it was finally successful in 2013. The callus regeneration system of *P. edulis* was established by Yuan Jin-ling and her research group in 2013.

In moso bamboo, nodal segments and inflorescences are generally used as explant sources for organogenesis. Plant growth regulators (PGRs) such as thidiazuron (TDZ), 2,4-dichlorophenoxyacetic acid (2,4-D), 1-aminocyclopropane-1-carboxylic acid (ACC) and 1-naphthaleneacetic acid (NAA) are used for shoot regeneration, in vitro flowering and rooting. In somatic embryogenesis, in vitro spikelets, nodal and internodal segments and zygotic seed embryos are used for induction of somatic embryo. TDZ, 2,4-D, Kinetin, Zeatin and coconut milk are used for somatic embryo formation. TDZ is optimal hormone for somatic embryo regeneration and in vitro flowering, and NAA is the best for rooting (Yuan et al. 2013).

11.3.3 Genetic Transformation

Only a few reports are available on bamboo genetic transformation. The *Agrobacterium* and particle bombardment have successfully been used in bamboo species, but the transformation frequency was limited. An efficient plant regeneration and transformation system of ma bamboo (*D. latiflorus* Munro) started from young shoot was explanted. Recently, Prof. Zhu and his research group (Ye et al. 2017) developed the successful *Agrobacterium* transformation protocol for ma bamboo. It took 8 months, however, to get the healthy calluses, and 7–8 months for the transformation procedure. The low efficiency and slow regeneration are the major limitations for successful transformation in the bamboo species. To the best of our knowledge, no report is available on moso bamboo genetic transformation. Clustered regularly interspaced short palindromic repeats (CRISPR)-associated endonuclease Cas9 system (CRISPR/Cas9) technology has not been established for Bamboo. It requires directed effort to study the molecular mechanism of somatic embryogenesis and regeneration to shorten the time for successful transformation.

11.4 Gene Editing

It is difficult to improve the moso bamboo by conventional plant breeding. Modern biotechnology tools such as molecular biology, genetics, bio-informatics, etc. will be helpful for moso bamboo species productivity as part of sustainable development. The now available chromosome level reference genome for moso bamboo can effectively be used for bamboo trait development.

Although *Agrobacterium* mediated transformation and particle bombardment have been used in bamboo, the transformation frequency was reported limited. These studies have set the foundation for future research wherein actual crop improvement through genetic transformation in bamboos has been envisaged. Recently, Ye et al. (2017) developed a successful *Agrobacterium* transformation protocol for ma bamboo, a hexaploid species, by taking 8 months to get healthy calluses, and 7–8 months for the transformation. As far as the current information goes, the reports of improvement of agronomic traits in bamboo through genetic engineering remain limited only to one species, ma bamboo. Low efficiency and slow regeneration are major limitations for successful transformation in bamboo.

Latest advancements in gene editing technologies have found increasing use of clustered regularly interspaced short palindromic repeat (CRISPR)–CRISPR-associated endonuclease (Cas9) system in several crop species (Doudna and Charpentier 2014). Very recently, Ye et al. (2020) established a CRISPR/Cas9 system in ma bamboo by targeting all homo-alleles or specifically one allele of a gene. Using this technology, they could demonstrate the plant height changes by knocking out one gibberellin-responsive gene in ma bamboo.

To the best of our knowledge, no report is available either on the successful genetic transformation or on gene editing in moso bamboo, although plantlet regeneration via embryoid germination through callus induction in moso bamboo has been demonstrated (Yuan et al. 2013).

Zhu et al. (2019) established genetic transformation protocol and optimized the CRISPR/Cas9 system in bamboo protoplast. They first improved the protoplast preparation methods and could isolate 3.09106 protoplasts/g fresh leaves. Next, they improved the PEG-mediated transformation method and reached efficiencies of 53.3% for a single plasmid and 29.8% for two cotransformed plasmids, which is sufficient for optimizing the CRISPR/Cas9 system. The maize UBI promoter was used to drive Cas9 expression (Ye et al. 2017). Three polymerase III-dependent promoters from rice (OsU6a/OsU6b/OsU6c) were selected to express the sgRNA cassettes (Ma et al. 2015), as bamboo exhibits high-genomic similarity with rice (Peng et al. 2013). To check the effectiveness of CRISPR/Cas9 constructs, a frame shift mutated GFP (mGFP) containing an additional 'guanine' thereby produces no fluorescence signal was simultaneously cotransformed with CRISPR/Cas9 plasmids. Around 1.8% of the proto-plasts transformed with theUBI-Cas9/OsU6b-sgRNAconstructshowed strong signals within 72 h, indicating that the mGFP function was restored by the CRISPR/Cas9 system through deleting the additional 'guanine'. TheOsU6aandOsU6cpromoters work as well, however, with lower efficiency than theOsU6bpromoter, as positive signals were only occasionally observed with more than 10 replicates. Taking together, theUBI-Cas9/OsU6b-sgRNA construct effectively works in bamboo protoplast and was used for the following endogenous gene editing in Ma bamboo.

Three bamboo PSY1 alleles (DlmPSY1-A, DlmPSY1-B and DlmPSY1-C) were identified and cloned by a homology cloning strategy. To mutate all copies of DlmPSY1, sgRNA1 targeting a conserved site among all DlmPSY1 loci were designed. In addition, the sgRNA2 target site containing 2–3single-nucleotide polymorphisms (SNPs) in the spacer region among three DlmPSY1 homoeo alleles was selected to test the tolerance of sgRNA mismatches. A total of 1600 bamboo calluses induced from stem were transformed as described previously (Ye et al. 2017). In total, 34 independent transgenic lines were confirmed positive (2.1%) by PCR. Based on Sanger sequencing results, 22 (100%) and 10 (83.3%) independent T0 lines were edited in the sgRNA1 and sgRNA2 regions, respectively, indicating that both constructs effectively induce endogenous gene editing. The editing profiles were further analyzed by sequencing. Eighteen lines (81.8%) contained putative homozygote/biallelicmutations in all subgenomes at the sgRNA1 target site. In some lines, putative homozygote/biallelic mutations exist in one subgenome, while heterozygote or chimeric mutations appear in other subgenomes (T0-10 and T0-26). Eight mutation types were identified from 590 independent clones. The most frequent mutation type was deletion (75%), of which 59.1% are small deletions (<2 bp). The ratios of large fragment deletions ($\geq$ 14 bp), insertions and combined indels were 15.9%, 2.21%, and 7.82%, respectively. Since bamboo propagates through asexual budding, those homozygote/biallelic mutations will remain in the genome of their offspring clones during breeding. sgRNA2 that perfectly targets DlmPSY1-A1, but not DlmPSY1-B1 or DlmPSY1-C1 was designed to study the recognition specificity. The ratios of deletions, insertions and combined mutations were 86%, 9%, and 5%, respectively. The mutations were predominantly short nucleotide changes (1–26 bp), and 22.7% were 1 bp nucleotide deletions. Those data demonstrated the successful application of the CRISPR/Cas9 system in mutating a specific DlmPSY1 allele. Eighteen lines (81.8%) with homozygote/biallelic mutations in all subgenomes at the sgRNA1 site exhibited albinophenotypes, which appeared at an early stage during tissue culture and persisted at the plantlets stage. Those results suggest that genome editing takes place at an early stage in embryonic cells and led to the loss of function of all DlmPSY1 alleles. In the case of sgRNA2, although DlmPSY1-A was mutated, no visible phenotypic change was observed due to the existence of the wild-type DlmPSY1-B and DlmPSY1-C alleles. Next, Zu et al. applied this technology in bamboo molecular research. Bamboo is the tallest grass in the world, while the underlying mechanism is unknown. Previously, they identified several Gibberellin-responsive genes including GRG1 (GA-responsive gene 1,PH01004823G0070) that potentially acts in controlling bamboo height (Zhang et al. 2018). Here, two homozygotegrg1 mutants (efficiency 40%) in Ma bamboo were produced using their optimized CRISPR/Cas9 technology. Mutation in GRG1 increased plant height, mostly due to elongated internodes. Sequencing results confirmed that the *grg1* mutant has the putative homozygous mutation in A1 subgenome, biallelic mutation in B1 subgenome and homozygous mutation in C1 subgenome, indicating the loss of function of GRG1 in transgenic bamboo.

This is the first example of controlling bamboo height through gene manipulation, which will contribute to subsequent studies on the molecular mechanisms behind the fast growth of bamboo. In summary, for the first time Mankind engineered the hexaploid ma bamboo through CRISPR/Cas9 technology. The homozygote mutations were obtained in the first generation of transgenic lines, which are extremely important for bamboo species due to its long vegetative growth periods. Zu et al. also confirmed the albino phenotype of *dlmpsy1* mutant in bamboo and generated a bamboo mutant with altered plant height. This demonstrates the applicability of CRISPR/Cas9 in bamboo and thereby boosts future bamboo research and breeding.

11.5 Prospect of Future of Moso Bamboo Breeding

11.5.1 Main Problems in Moso Bamboo Breeding

At present, the main problems of moso bamboo breeding are as follows. (1) The research of germplasm resources lags behind due to the narrow Germplasm Base. (2) There is a lack of long-term and stable scientific research teams. (3) Insufficient attention is paid to breeding research and less research on special-purpose moso bamboo. (4) There is insufficient investment made in breeding research and the scientific research force is relatively scattered. (5) Scientific breeding technologies, methods and equipment need to be improved. (6) There is no breakthrough in the regeneration system and transformation system.

11.5.2 Proposal of Moso Bamboo Breeding

Strengthen Germplasm Improvement and Material Innovation, Broaden the Base of Germplasm Resources: Through the selection of breeding, cross breeding and other conventional breeding techniques, the breakthrough new germplasm of moso bamboo was created, the existing excellent resources were fully exploited and utilized, and the outstanding characteristics were further improved and reused. Key factors for the sustainable development of the bamboo industry are the characteristics of bamboo resources and their geographical distribution (Zhao and Wei 2018). Therefore, conservation of bamboo germplasm and genotype improvement are considered as key targets in the development programs.

Stress Resistant Breeding Should Be Strengthened: In order to meet the needs of current bamboo production, stress resistance breeding was carried out by artificially strengthening stress and selecting stress.

Special Bamboo Varieties Should Be Cultivated: Increase the cultivation of special bamboo varieties such as large diameter clums, long internodes and bamboo wall thickness.

In addition, applications such as transcriptome, co-expression networks, identification of TFs, miRNA, alternative splicing, proteomics, etc. have tremendously advanced the basic understanding of the moso bamboo genomes including molecular pathways and process. Focused research in several areas of moso bamboo such as flower development, rapid growth, in vitro regeneration, and genetic transformation constitutes a great opportunity for the advancement of ecological bioeconomy.

11.5.3 The Main Goal and Direction of Breeding Research

A special concerted effort is required to study the molecular mechanism of somatic embryogenesis and regeneration for achieving a successful transformation system to reap the benefits of molecular breeding and gene editing.

The flowering mechanism of bamboo plants is not clear yet, so in future, the genetic improvement of bamboo should still be based on conventional selection breeding and cross-breeding. In scientific research, the main research content is to analyze the flowering mechanism of moso bamboo, and the goal is to regulate the flowering cycle of moso bamboo.

In future research of bamboo hybrid breeding, researchers should have a clear breeding goal, according to the market demand and the biological characteristics of bamboo species, choose flowering bamboo with excellent characters for hybridization, in order to cultivate new high-quality planting resources with great economic, social and ecological value. For the practical development of bamboo products at an industrial scale, an inclusive development model should consider product innovation, product structure optimization, capacity building, product demonstration, policies, and multi take holder participation, besides ensuring the relentless flow of raw materials species specificity of vascular bundle strength (Ahvenainen et al. 2017), lignin accumulation, and pith cavity formation (Guo et al. 2019; Wang et al. 2008) can be genetically manipulated at the cellular level with the critical understanding of the genomics of such traits. Therefore, genetic manipulation can bring in advantages for industrial applications resulting in various categories of products such as food, energy resources, biochemicals, construction materials and household products.

Strengthen the introduction of new breeding technology, master and apply breeding technology, such as transgenic, molecular markers, space breeding technology and gene editing research and application, improve the level of bamboo breeding. In addition, the government should also increase investment, establish a stable research team and improve the promotion system.

References

Cai C, Gao J, Mu S (2007) Effects of ^{60}Coγ rays radiation on seed vigor and young seedling growth of Phyllostachys edulis. J Nucl Agric Sci 21(5):436–440, 455

Cao Z, Wu Z, Wang F et al (2019) Comprehensive evaluation of growth characteristics. of superior *Phyllostachys pubescens* seedlings with radiation-treated mutagenesis. J Central South Univ Forest Technol 39(2):34–40, 58

Cheng Z, Hou D, Ge W et al (2020) Integrated mRNA, MicroRNA transcriptome and degradome analyses provide insights into stamen development in moso bamboo. Plant Cell Physiol 1:1–12

Doudna JA, Charpentier E (2014) The new frontier of genome engineering with CRISPR-Cas9. Sci 346, Article ID: 1258096. https://doi.org/10.1126/science.1258096

Guo Z, Ma P, Yang G et al (2019) Genome sequences provide insights into the reticulate origin and unique traits of woody bamboos. Mol Plant 12:1353–1365

Isagi Y, Shimada K, Kushima H et al (2004) Clonal structure and flowering traits of a bamboo [*Phyllostachys pubescens* (Mazel) *Ohwi*] stand grown from a simultaneous flowering as revealed by AFLP analysis. Mol Ecol 13:2017–2021

Lai C, Hsiao J (1997) Genetic variation of *Phyllostachys pubescens* (*Bambusoideae*, Poaceae) in Taiwan based on DNA polymorphisms. Bot. Bull. Acad. Sin. 38:145–152

Lu J, Yoshinaga K, Fang W, Tang et al (2009) Identification of the Hybrid Bamboo F1 by SSR Markers. Scientia Silvae Sinicae 45(3):29–34

Ma X, Zhang Q, Zhu Q et al (2015) Arobust CRISPR/Cas9 system for convenient, high-efficiency multiplex genomeediting in monocot and dicot plants. Mol Plant 8:1274–1284

Peng Z, Lu Y, Li L et al (2013) The draft genome of the fast-growing non-timber forest species moso bamboo (*Phyllostachys heterocycla*. Nat Genet 45:456–461

Wang X, Gao J, Yang P, Zhou L (2008) Cytological effects induced by $^{60}Co\gamma$ rays-irradiation on moso seed. J Nucl Agric Sci 22(6):770–775

Ye S, Cai C, Ren H et al (2017) Anefficient plant regeneration and transformation system of ma bamboo (*Dendrocalamus latiflorus Munro*) started from young shoot as explant. Front Plant Sci 8:1298

Ye S, Chen G, Kohnen MV, Wang W, Cai C, Ding W, Wu C, Gu L, Zheng Y, Ma X, Lin C, Zhu Q (2020) Robust CRISPR/Cas9 mediated genome editing and its application in manipulating plant height in the first generation of hexaploid Ma bamboo (*Dendrocalamus latiflorus Munro*). Plant Biotechnol J 18:1501–1503

Yuan JL, Yue JJ, Wu XL, et al (2013) Protocol for callus induction and somatic embryogenesis in Moso bamboo. PLoS ONE 8(12):e81954. https://doi.org/10.1371/journal.pone.0081954

Zhou F (1998) Bamboo forest cultivation. China Forestry Press, Beijing

The manufacturer's authorised representative in the EU is Springer Nature Customer Service Centre GmbH, Europaplatz 3, 69115 Heidelberg, Germany. If you have any concerns regarding our products, please contact ProductSafety@springernature.com

Printed and bound by CPI Group (UK) Ltd, Croydon, CR0 4YY
30/09/2026
02225462-0020